Determination of
Organic
Structures by
Physical Methods

VOLUME 5

Contributors

Larry K. Blair
John I. Brauman
N. W. G. Debye
J. W. Faller
V. P. Feshin
J. Karle
Ronald G. Lawler
Shin-Ichi Sasaki
M. G. Voronkov
Harold R. Ward
J. J. Zuckerman

Determination of Organic Structures by Physical Methods

VOLUME 5

Edited by

F. C. NACHOD
Sterling-Winthrop Research Institute
Rensselaer, New York

and

J. J. ZUCKERMAN
Department of Chemistry
State University of New York
  at Albany
Albany, New York

1973
ACADEMIC PRESS
New York and London

COPYRIGHT © 1973, BY ACADEMIC PRESS, INC.
ALL RIGHTS RESERVED.
NO PART OF THIS PUBLICATION MAY BE REPRODUCED OR
TRANSMITTED IN ANY FORM OR BY ANY MEANS, ELECTRONIC
OR MECHANICAL, INCLUDING PHOTOCOPY, RECORDING, OR ANY
INFORMATION STORAGE AND RETRIEVAL SYSTEM, WITHOUT
PERMISSION IN WRITING FROM THE PUBLISHER.

ACADEMIC PRESS, INC.
111 Fifth Avenue, New York, New York 10003

United Kingdom Edition published by
ACADEMIC PRESS, INC. (LONDON) LTD.
24/28 Oval Road, London NW1

LIBRARY OF CONGRESS CATALOG CARD NUMBER: 54-11057

PRINTED IN THE UNITED STATES OF AMERICA

Contents

LIST OF CONTRIBUTORS	ix
PREFACE	xi
CONTENTS OF OTHER VOLUMES	xiii

1. ELECTRON DIFFRACTION
J. KARLE

I. Introduction	1
II. Instrumentation	3
III. Theory and Analysis	6
IV. Reading an Electron Diffraction Paper	14
V. Regularity of Bond Lengths	20
VI. Single Bonds	23
VII. Single and Double Bonds	28
VIII. Conjugated Systems	39
IX. Triple Bonds	49
X. Conformers	51
XI. A Variety of Topics	60
XII. Conclusions	68
References	68

2. SPIN SATURATION LABELING
J. W. FALLER

I. Introduction	75
II. Theory	76
III. Applications and Examples	81
IV. Measurement Techniques	90
V. Sources of Error	93
VI. Practical Aspects	95
References	96

3. CHEMICALLY AND ELECTROMAGNETICALLY INDUCED DYNAMIC NUCLEAR POLARIZATION
RONALD G. LAWLER AND HAROLD R. WARD

I. Introduction	99
II. Overhauser and Related Effects	103
III. Chemically Induced Dynamic Nuclear Polarization (CIDNP)	109
References	144

4. ION CYCLOTRON RESONANCE SPECTROSCOPY
JOHN I. BRAUMAN AND LARRY K. BLAIR

I. Basic Principles of Ion Detection	152
II. The Ion Cyclotron Double Resonance Technique for Studying Ion–Molecule Reactions	153

III.	Interpretation of the Double Resonance Experiment	154
IV.	Applications	155
V.	Thermochemical Quantities	156
VI.	Proton Transfer Reactions. Double Resonance Spectra	157
VII.	Acidities from Proton Transfer	161
VIII.	Reaction Pathways	162
IX.	Structures of Ions	163
X.	Concluding Remarks	165
	References	166

5. NUCLEAR QUADRUPOLE RESONANCE IN ORGANIC AND METALLOORGANIC CHEMISTRY
M. G. VORONKOV and V. P. FESHIN

I.	Introduction	169
II.	Distribution and Assignment of Frequencies in the Spectral Bandwidth	174
III.	The Nature of the Chemical Bond	176
IV.	The Correlation of NQR Frequencies of the Type RX, RZX, and RR'R"MX with Substituent Inductive Constants and with Chemical and Physical Properties	177
V.	Transmission of Electronic Effects Across Saturated Carbon and Heteroatomic Systems	184
VI.	Electronic Effects of Alkyl Groups	191
VII.	Alicyclic Compounds	195
VIII.	Aromatic Systems	199
IX.	Araliphatic Compounds	204
X.	Heterocyclic Compounds	205
XI.	Phosphorus Compounds	209
XII.	Sulfur Compounds	210
XIII.	Intra- and Intermolecular Coordination in Metalloorganic Compounds	211
XIV.	Molecular Complexes	216
XV.	Polymeric Compounds	222
XVI.	Some Other Areas of Application of the NQR Method	223
XVII.	Conclusion	224
	References	224

6. MÖSSBAUER SPECTRA OF ORGANOMETALLICS
N. W. G. DEBYE and J. J. ZUCKERMAN

I.	Theory	235
II.	Applications	250
	References	278

7. AUTOMATED CHEMICAL STRUCTURE ANALYSIS SYSTEMS
SHIN-ICHI SASAKI

I.	Introduction	285
II.	Structure Determination of Alkanes with Monomethyl Side Chain	286
III.	Structure Determination of Oligopeptides	289

IV. Structure Determination of Aliphatic Ketones	293
V. Structure Determination of Aliphatic Ethers	299
VI. Construction of Molecular Structure from Suitable Fragments	304
References	320

Appendix to Chapter 5. NEWEST REPORTS ON NQR APPLICATIONS IN ORGANIC AND METALLOORGANIC CHEMISTRY

M. G. VORONKOV and V. P. FESHIN

Text	323
References	337
AUTHOR INDEX	341
SUBJECT INDEX	361

List of Contributors

Numbers in parentheses indicate the pages on which the authors' contributions begin.

LARRY K. BLAIR (151), Department of Chemistry, Stanford University, Stanford, California

JOHN I. BRAUMAN (151), Department of Chemistry, Stanford University, Stanford, California

N. W. G. DEBYE (235),* Department of Chemistry, State University of New York at Albany, Albany, New York

J. W. FALLER (75), Sterling Chemistry Laboratory, Department of Chemistry, Yale University, New Haven, Connecticut

V. P. FESHIN (169, 323), Institute of Organic Chemistry, Siberian Division of the Academy of Sciences of the USSR, Irkutsk-33, USSR

J. KARLE (1), Laboratory for the Structure of Matter, U.S. Naval Research Laboratory, Washington, D.C.

RONALD G. LAWLER (99), Department of Chemistry, Brown University, Providence, Rhode Island

SHIN-ICHI SASAKI (285), Department of Chemistry, Miyagi University of Education, Aoba, Sendai, Japan

M. G. VORONKOV (169, 323),† Institute of Organic Synthesis, Latvian Academy of Sciences, Aizkraukles Lela, Riga, USSR

HAROLD R. WARD (99), Department of Chemistry, Brown University, Providence, Rhode Island

J. J. ZUCKERMAN (235), Department of Chemistry, State University of New York at Albany, Albany, New York

* Present address: Chemistry Department, Temple Buell College, Denver, Colorado
†Present address: Institute of Organic Chemistry, Siberian Division of the Academy of Sciences of the USSR, Irkutsk-33, USSR

Preface

When Professors Rice and Teller introduced their treatise on the structure of matter over 20 years ago [F. O. Rice and E. Teller (1949). "The Structure of Matter." Wiley (Interscience), New York.], they started the introductory chapter as follows:

> At present, atomic physical theory in principle enables us to calculate all of the chemical and most of the physical properties of matter and thus makes the science of experimental chemistry superfluous. If, however, we consider things from an economic point of view, the human labor that must go into such calculations is far greater than that which would be required to make the experimental study.

Nowadays, all chemists are interested in molecular structure some of the time, while some chemists are interested in molecular structure all of the time (*pace* Lincoln). Each group attempts to work the problem out experimentally, the first group of chemists utilizing a combination of physical techniques which they themselves routinely practise. Here data are collected and the interpretation generally supplied by the synthetic chemist himself, with a little help from his friends, the expert authors of the chapters in our volumes which are concerned with the so-called "sporting techniques." The second group of chemists have available techniques which are capable of yielding precise values of internuclear distances and angles. Here the practicing synthetic chemist uses the results of the detailed determination of the shape and structure of the necessarily limited number of molecules as a key to the interpretation of results of routinely used physical techniques for a much larger number of other molecules.

In recent years we have seen the enormous growth of information concerning molecular structure in the important solution phase, aided by precise results from the gas and crystal phases. New physical parameters which vary with chemical state and molecular structure have come to light from work in experimental physics, and information combined from a variety of techniques has opened the possibility of automating the process of structure determination.

As stated in earlier prefaces, credit for the surveys of the status of these arts must go to the individual authors; for sins of omission or commission, none but the editors will share the blame.

F. C. Nachod
J. J. Zuckerman

Contents of Other Volumes

VOLUME 1

Part I: The Determination of Molecular Size
Phase Properties of Small Molecules
    H. F. HERBRANDSON and F. C. NACHOD
Equilibrium and Dynamic Properties of Large Molecules
    P. JOHNSON

Part II: The Determination of Molecular Pattern
Optical Rotation
    W. KLYNE
Ultraviolet and Visible Light Absorption
    E. A. BRAUDE
Infrared Light Absorption
    R. C. GORE
Raman Spectra
    FORREST L. CLEVELAND
Magnetic Susceptibilities
    CLYDE A. HUTCHINSON, JR.

Part III: The Determination of Molecular Fine-Structure
Surface Films
    E. STENHAGEN
Dipole Moments
    L. E. SUTTON
Electron Diffraction
    J. KARLE and I. L. KARLE
X-Ray Diffraction
    J. M. ROBERTSON
Microwave Spectroscopy
    E. BRIGHT WILSON, JR., and DAVID R. LIDE, JR.
Thermodynamic Properties
    J. G. ASTON
Dissociation Constants
    H. C. BROWN, D. H. McDANIEL, and O. HÄFLIGER
Reaction Kinetics
    E. A. BRAUDE and L. M. JACKMAN
Wave-Mechanical Theory
    C. A. COULSON
Author Index–Subject Index

VOLUME 2

Optical Rotatory Dispersion
    GLORIA G. LYLE and ROBERT E. LYLE

Mass Spectrometry
    F. W. McLAFFERTY
Infrared and Raman Spectroscopy
    M. KENT WILSON
Electronic Spectra of Polyatomic Molecules and the Configurations of Molecules in Excited Electronic States
    D. A. RAMSAY
Far and Vacuum Ultraviolet Spectroscopy
    D. W. TURNER
High Resolution H^1 and F^{19} Magnetic Resonance Spectra of Organic Molecules
    W. D. PHILLIPS
Nuclear Magnetic Resonance Spectra of Elements Other than Hydrogen and Fluorine
    PAUL C. LAUTERBUR
Nuclear Magnetic Resonance Spectra of Organic Solids
    R. E. RICHARDS
Electron Paramagnetic Resonance of Organic Molecules
    RICHARD BERSOHN
Electron Paramagnetic Resonance of the Organometallics
    RICHARD E. ROBERTSON
Nuclear Quadrupole Resonance Spectroscopy
    CHESTER T. O'KONSKI
Author Index–Subject Index

VOLUME 3

Photoelectron Spectroscopy
    C. R. BRUNDLE and M. B. ROBIN
X-Ray Diffraction
    ROBERT F. STEWART and SYDNEY R. HALL
Optical Rotatory Dispersion and Circular Dichroism in Organic Chemistry
    PIERRE CRABBÉ
Thermochemistry
    KENNETH B. WIBERG
Mass Spectrometry
    DUDLEY WILLIAMS
Electron Spin Resonance Spectroscopy
    GLEN A. RUSSELL
Configuration and Conformation by NMR
    F. A. L. ANET and RAGINI ANET
Author Index–Subject Index

VOLUME 4

Applications of High-Field NMR Spectroscopy
    W. NAEGELE
Pulsed NMR Methods
    N. BODEN
Nuclear Magnetic Double Resonance Spectroscopy
    W. McFARLANE

^{15}N Nuclear Magnetic Resonance
    ROBERT L. LICHTER
NMR Spectra of the Heavier Elements
    PETER R. WELLS
^{13}C Nuclear Magnetic Resonance
    P. S. PREGOSIN and E. W. RANDALL
^{31}P Nuclear Magnetic Resonance
    JOHN R. VAN WAZER
Author Index–Subject Index

Electron Diffraction   1

J. KARLE

I. Introduction	1
II. Instrumentation	3
III. Theory and Analysis	6
IV. Reading an Electron Diffraction Paper	14
A. Distance Parameters	15
B. Shrinkage	17
C. Errors	18
V. Regularity of Bond Lengths	20
VI. Single Bonds	23
VII. Single and Double Bonds	28
VIII. Conjugated Systems	39
IX. Triple Bonds	49
X. Conformers	51
XI. A Variety of Topics	60
A. Steric Effects	60
B. Free Radicals	61
C. Other Structural Types	64
XII. Conclusions	68
References	68

I. INTRODUCTION

The scattering of electrons by gases has been used for the past 40 years as an effective tool for the investigation of molecular structure. A great deal of information, useful to the organic chemist, has been collected concerning molecular configuration, bond distances, gross internal motion and internal rotation, preferred orientation in conformers, and conjugation and aromaticity. Some systems of mixtures in equilibrium have been studied, and more recently investigations of some free radicals have been made. In the attempt to extend the usefulness and accuracy of structure investigations, more effective use has recently been made of combining structural parameters available from spectroscopic investigations with those from electron diffraction studies.

This review is a sequel to the earlier one on electron diffraction[1] published in 1955. At that time it was possible to report that considerable strides had already taken place in the theoretical, analytical, and experimental aspects of electron diffraction research, resulting in the improved accuracy of interatomic distances and the opportunity to measure average amplitudes of vibration between pairs of atoms. On the theoretical side, the radial distribution function representing the probability of finding interatomic distances in a molecule had been rigorously described by Debye.[2] This was followed by several analytical developments concerning, for example, the manner for handling the total and background scattering intensities in order to obtain a molecular intensity curve which corresponded essentially to that from point scatterers.[3-5] Several criteria were introduced in order to determine the detailed nature of the background intensity to be employed with the total measured intensity. It should be smooth so as not to introduce features into the molecular intensity function,[4,6-8] and it should be shaped so that the resulting radial distribution curve would be nonnegative everywhere[5] since it represents a probability function. As an additional criterion the areas under the peaks in the radial distribution curve should correspond to theoretical values, as initially illustrated in a study of the structures of 1,1-difluoroethylene and tetrafluoroethylene.[9] To facilitate the experimental measurements, a rotating sector had been introduced into the diffraction apparatus by Finbak[10] and P. P. Debye,[11] affording improved accuracy in the microphotometry of the photographic plates by leveling the steeply falling background intensity. Finbak, Hassel and their collaborators at the University of Oslo pioneered in the use of the rotating sector and microphotometry.[6-8] An apparatus for rotating the photographic plates while they were being traced on the microdensitometer was also developed,[12] minimizing the graininess of the photographic emulsions and improving considerably the quality of the densitometer traces. Improvements had also been introduced into the design of the electron optics of the diffraction camera, and the experimental conditions of the gas–electron collision process were analyzed in detail.[5,9,13,14] Additional theoretical progress was made in the study of vibrational motion[2,15] and internal rotation[2,16-19] by electron diffraction and in the computation of root mean squared amplitudes of vibration from spectroscopic data by Morino and collaborators[20,21] for comparison with the results of electron diffraction analyses.

The electron diffraction method which arose from these developments is known as the sector–microphotometer method. The name is evidently based upon some of the special apparatus employed but implies, in addition, the many improvements incorporated into molecular structure determination which have been facilitated by the deeper theoretical insight, the improved analytical procedures, and the greater accuracy of the data. The earlier method for analyzing electron diffraction patterns was known as the visual method. It was based upon the visual measurement of the positions of the maxima and minima of the diffraction pattern and the visual estimate of the molecular intensity. The pioneering work in this field of research was initiated by H. Mark and R. Wierl and further developed by L. O. Brockway and L. Pauling. A review article by Brockway[22] details the theoretical and experimental aspects of this early work. Although it was not generally possible to make highly accurate visual estimates of the diffracted intensities, the positions of the maxima and minima could be measured with a high degree of reliability. Since interatomic distances and molecular configuration are often largely determined by knowledge of the positions of the maxima and minima, good structure research could be carried out by the visual method. Some workers developed a fine facility in the techniques of the visual method, producing many high quality structure determinations. This is exemplified by the investigations of V. Schomaker and his collaborators.

Several developments have occurred since the previous review article[1] was written. These concern improvements in theory, microphotometry, automatic data processing, and computing facilities. Theoretical studies of the limitations of the Born approximation by Schomaker and Glauber[23,24] have led to improved values for the tabulated atomic scattering factors for elastic scattering.[25] More accurate values for the inelastic atomic scattering factors have also been computed.[26] The details of internal molecular motion have also been subject to considerable theoretical investigation. As a consequence of these various developments, the accuracy, detail, and facility with which molecular structure can be determined by electron diffraction has continuously improved, and the incorporation of information from spectroscopic investigations into the analyses has been facilitated.

Several review articles have been prepared on electron diffraction by Bastiansen and Skancke,[27] Bartell,[28] and Bauer.[29] This article will include a brief discussion of the main characteristics of the apparatus and analytical procedure for electron diffraction investigations, and the nature of the results and their evaluation. A general discussion of structural features and regularities in organic molecules will then be presented, illustrated by several examples.

II. INSTRUMENTATION

The design of electron diffraction apparatus has been quite standard since the first experiments were carried out. A source of a fine collimated beam of electrons accelerated to about 50 kV is made to intersect with a fine jet of gas and the intensity of the scattered electrons is then recorded. For high-precision work, the electron beam having a current of a few tenths of a microamp is collimated to about 0.1 mm diameter and intersects a narrow gas jet as closely as possible to the orifice of the nozzle. A rotating sector is used to modify the steeply falling background intensity in order to facilitate and increase the accuracy of the microphotometry. Special apertures, cold traps, and design features are introduced into the apparatus to minimize the extraneous scatter of electrons and the spread of gas sample. It is also important to measure accurately the accelerating voltage and the distance between the nozzle orifice and the photographic plate.

The apparatus shown in Fig. 1 was designed to meet the criteria for precision electron diffraction analysis and was constructed at the Naval Research Laboratory in 1947–1948. The apparatus was briefly described previously.[4] The filament and grid design of the electron gun is similar to that of Brockway.[22] The first magnetic lens is used to provide a crossover in the electron beam, and only a 0.1-mm cross section of the central portion of the ray enters the second magnetic lens to be focused on the photographic plate. The small opening in the second lens isolates the upper portion of the apparatus from the specimen chamber, permitting independent pumping systems and isolation from any stray gas in the specimen chamber. A pressure of 10^{-5} torr can be obtained in the specimen chamber and 10^{-6} torr can be obtained in the upper section including the electron gun. A nozzle orifice of about 0.2 mm maintains a minimal scattering volume. The nozzle is furnished with a Teflon tipped (originally rubber) plunger, closing off the orifice just behind the nozzle tip. The plunger is activated by a solenoid. Through an automatic timing device the electrostatic deflection plates which act as a shutter for the electron beam are grounded for a brief time interval while the solenoid is simultaneously activated to permit a jet of gas to collide with the electron beam. The backing pressure of the gas is often 20–40 mm. Photographs can be taken in less than 1 second but are often taken for longer periods. At lower backing pressures the time of exposure is correspondingly increased. Not shown in Fig. 1 are devices associated with the nozzle for confining and trapping the gas jet on surfaces cooled by liquid nitrogen. Also not shown are strategically placed orifices for confining the beam and shielding against extraneously scattered electrons. The sector was placed as close to the photographic plates as possible, with its center used as a beam trap, a further step in minimizing stray scattering. The accelerating voltage is regulated to vary no more than 1 part/10^4 over time periods which are long compared to the exposure time.

Since commercially produced apparatus suitable for electron diffraction by gases has not been available until recently, individual laboratories have designed and built their own. The newer apparatus at the University of Oslo[30] features a large specimen chamber to facilitate introduction of auxiliary equipment and elaborate cold traps to efficiently condense the gas jet. The apparatus at the University of Michigan[31] features readily cleaned platinum apertures, a platinum tipped nozzle, a well-regulated self-biasing electron gun,

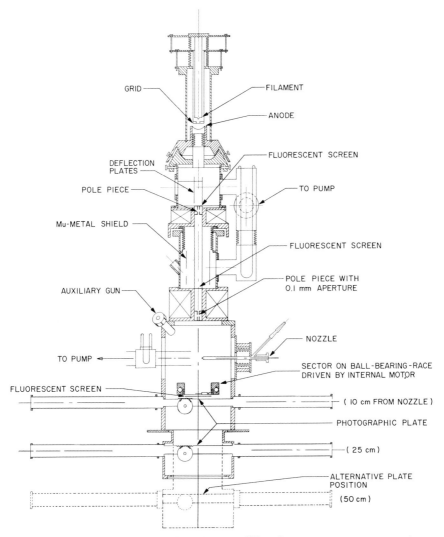

FIG. 1. Schematic diagram of an electron diffraction apparatus constructed at the Naval Research Laboratory.

and a centrifugally controlled beam trap associated with the sector device. An apparatus constructed at the Shell Laboratory in Amsterdam[32] permits the continual adjustment of the film cassette and the rotating sector within the chamber so that the specimen to film distance can be varied to any value between 92 and 302 mm. At the University of Moscow the electron diffraction apparatus[33] features a very high-temperature evaporating furnace for the study of materials of low volatility. A commercial apparatus is manufactured by Balzers A. G., Furstentum Lichtenstein, and the details of its design have been described.[34] A new electron diffraction apparatus at the University of Tokyo[35] features a table in the specimen chamber to which the nozzle and photographic plate can be firmly fixed in order to establish the distance between the two accurately. It also has a special device attached to the sector race to minimize extraneous scattering. The apparatus currently employed at Cornell University[36] is similar to the previously described equipment except for a unique lens system. The sample is irradiated with a parallel beam of electrons, and the diffraction pattern is then projected onto a photographic plate by two large-aperture magnetic lenses. The focus condition is independent of sample position, and therefore the apparatus is particularly suitable for low-density samples which are difficult to confine in a jet. The second magnetic lens is used to magnify the pattern.

III. THEORY AND ANALYSIS

When a beam of fast electrons strikes a jet of free molecules, a portion of the scattering, the molecular scattering, arises from the distribution of distances within the individual molecules. The molecular scattering $I_m(s)$ can be represented by

$$I_m(s) = \sum_{i=1}^{n} \sum_{j=1}^{n} c_{ij} Q_{ij} \quad (i \neq j) \tag{1}$$

where n is the number of atoms in the molecule, the coefficients c_{ij} are characteristic of the ith and jth atoms, and the Q_{ij} are interference functions. If a molecule were absolutely rigid, then the interference function Q_{ij} would be equal to $\sin sr_{ij}/sr_{ij}$, where r_{ij} is the distance between the centers of the ith and jth atoms and

$$s = [4\pi \sin(\theta/2)]/\lambda \tag{2}$$

The scattering angle θ is the angle between the incident and scattered beam, and λ is the wavelength of the electron beam. The atoms in a molecule are, in fact, constantly in relative motion, and therefore the interference function has the form

$$Q_{ij} = \int_0^\infty P_{ij}(\rho)(\sin s\rho/s\rho) \, d\rho \tag{3}$$

where $P_{ij}(\rho) \, d\rho$ is the probability that the distance between the ith and jth atoms has a value in the interval ρ and $\rho + d\rho$. In addition to the molecular scattering intensity, there is a scattered intensity forming a steeply falling background as s increases, $I_b(s)$, upon which the molecular scattering, Eq. (1), is superimposed. The total intensity of scattering, $I_t(s)$, is composed of the sum of the background and molecular intensities.

When Eq. (3) is substituted into Eq. (1), the molecular scattering intensity becomes

$$I_m(s) = \sum_{i=1}^{n} \sum_{j=1}^{n} c_{ij} \int_0^{\infty} P_{ij}(\rho)(\sin s\rho/s\rho) \, d\rho \qquad (i \neq j) \qquad (4)$$

It was shown by Debye[2] that the application of Fourier transform theory to Eq. (4) could be readily facilitated, yielding a simple and significant result, if the coefficients c_{ij} are independent of the scattering variable s. The Fourier transform function defined in terms of $I_m(s)$ is

$$D(r) = (2/\pi)^{1/2} \int_0^{\infty} sI_m(s) \sin sr \, ds, \qquad (5)$$

where $D(r)$ is a function of the distance variable r. For c_{ij} independent of s, the integration of Eq. (5) gives

$$D(r) = (\pi/2)^{1/2} \sum_{i=1}^{n} \sum_{j=1}^{n} c_{ij} P_{ij}(r)/r \qquad (i \neq j) \qquad (6)$$

The calculation implied by Eq. (5) leads to a function $rD(r)$, formed from Eq. (6), which represents a sum whose terms are proportional to the probability distributions for the interatomic distances in a molecule. This is a simple representation of the molecular structure. In the implementation of this theory, several problems arise:

(i) The total intensity of scattering has to be treated in such a way that the experimental molecular scattering intensity function has coefficients c_{ij} which are essentially constant with respect to s.

(ii) The calculation indicated by Eq. (5) requires the molecular scattering to be known over an effectively infinite range for the scattering variable. A physical infinity is attained when the upper limit of s is sufficiently large that the omitted data do not have a significant effect on the accuracy of the calculated function. Experimental data rarely reach such a limit, and so the calculation in Eq. (5) has to be altered to overcome the effect of this limitation.

(iii) The theoretical values for the steeply falling background intensity upon which the molecular scattering is superimposed are not completely applicable for defining the experimental background because of the effects of chemical binding which are not taken into account and the effects of experimental error. It is therefore important to have physical criteria for improving the background intensity by altering the theoretical values somewhat.

(iv) Microphotometry designed to obtain a weak, oscillating signal superimposed on a steeply falling background is difficult to carry out accurately. It is helpful to employ physical devices which facilitate the measurement.

The handling of items (i)–(iv) affords a good insight into the data reduction and the method of analysis for electron diffraction from gases. Although the precise procedure varies among the laboratories carrying out this work, the variations on the main theme are quite minor and the general characteristics can be readily outlined.

Owing to the fact that the molecules in a gas occupy all orientations in space, the electron scattering pattern is radially symmetric (Fig. 2). The total intensity of scattering is

$$I_t(s) = K \left\{ \sum_{i=1}^{n} \frac{S_i(s)}{s^4} + \sum_{i=1}^{n} \frac{|f_i(s)|^2}{s^4} + 2 \sum_{i<j}^{n} \frac{|f_i(s)f_j(s)|}{s^4} \cos[\eta_i(s) - \eta_j(s)] Q_{ij} \right\}$$

(7)

FIG. 2. A radially symmetric electron diffraction pattern obtained from the vapor of hexafluoropropene with an accelerating voltage of ≈ 40 kV. Detail which can be readily seen on the original negative at large angles is lost in the reproduction.

where the constant K is definable in terms of the intensity of the incident beam and a variety of physical constants, $S_i(s)$ is the inelastic scattering function for the ith atom which is tabulated,[26] and f_i/s^2 is the atomic scattering factor for electrons for the ith atom which is also tabulated.[25] The quantities f_i are complex numbers and may be written $f_i = |f_i| \exp[i\eta_i(s)]$. The background intensity $I_b(s)$ may be defined by

$$I_b(s) = K \sum_{i=1}^{n} (S_i(s) + |f_i(s)|^2)/s^4 \tag{8}$$

It was pointed out that by dividing the total intensity by an appropriate function, a molecular intensity curve with essentially constant coefficients over the experimental range of s could be obtained.[3,4] This was carried out[4] by dividing Eq. (7) by Eq. (8) and subtracting unity, giving

$$I_m(s) = [I_t(s)/I_b(s)] - 1 \tag{9}$$

or

$$I_m(s) = 2 \sum_{\substack{i<j \\ 1}}^{n} \sum_{i=1}^{n} \{|f_i(s)f_j(s)|/\sum [S_i(s) + |f_i(s)|^2]\} \cos[\eta_i(s) - \eta_j(s)]Q_{ij} \tag{10}$$

By comparing Eq. (10) with (1), the c_{ij} are seen to be defined by

$$c_{ij} = \{|f_i(s)f_j(s)|/\sum_{i=1}^{n} [S_i(s) + |f_i(s)|^2]\} \cos[\eta_i(s) - \eta_j(s)] \tag{11}$$

Calculations from tabulated values show that the c_{ij} are quite constant and $\eta_i \approx \eta_j$ if the ith and jth atoms do not differ greatly in atomic number, except for the first few values of s. In this inner region of scattering, for which accurate measurements are often not obtained, a theoretical molecular intensity function computed with constant coefficients in the range $0 \leq s \leq s_{\min}$ may be attached to the experimental molecular intensity[4] obtained from Eq. (10) in the range $s_{\min} \leq s \leq s_{\max}$. As the discrepancy in atomic number between atoms i and j increases, the deviation of c_{ij} from constancy increases. This is readily handled by regarding c_{ij} as the coefficient of the cosine term in Eq. (11) and incorporating $\cos[\eta_i(s) - \eta_j(s)]$ into the Q_{ij} function of Eq. (1). The interpretation of the Fourier transform function with this alteration of Q_{ij} has been analyzed by Bonham and Ukaji[37] and by Kimura and Iijima.[38] The small variation of $|f_i(s)f_j(s)|/\sum [S_i(s) + |f_i(s)|^2]$ with s can be corrected. A simple calculation to accomplish this has been described by Bartell et al.[39]

In order to assure that the integrand in the Fourier integral transform converges at the upper limit of the measured data, s_{max}, the function in Eq. (5) is replaced by

$$f(r) = (2/\pi)^{1/2} \int_0^{s_{max}} sI_m(s) \exp(-as^2) \sin sr \, ds \qquad (12)$$

where a in $\exp(-as^2)$ is some suitably chosen value. The latter function was introduced by Degard[40] and by Schomaker.[41] When the molecules undergo harmonic vibrations, the function $rD(r)$ in Eq. (6) can be represented by

$$rD(r) = (\pi/2)^{1/2} \sum_{i=1}^{n} \sum_{j=1}^{n} c_{ij}(h_{ij}/\pi)^{1/2} \exp[-h_{ij}(r_{ij} - r)^2] \qquad (i \neq j) \quad (13)$$

where $(2h_{ij})^{-1/2} = \langle l_{ij}^2 \rangle^{1/2}$ is the root mean squared amplitude of vibration for the pair of atoms denoted by i and j. It has been shown[4,15] that if $rD(r)$ is defined by Eq. (13), $rf(r)$ defined by Eq. (12) has the form

$$rf(r) = (\pi/2)^{1/2} \sum_{i=1}^{n} \sum_{j=1}^{n} c_{ij}(h_{ij}/\pi)^{1/2}$$
$$\times \exp[-h_{ij}(r_{ij} - r)^2/(4ah_{ij} + 1)]/(4ah_{ij} + 1)^{1/2} \qquad (i \neq j) \quad (14)$$

It is seen from Eq. (14) that the effect of introducing $\exp(-as^2)$ is to broaden the peaks representing the interatomic distances in a known way, while preserving the positions of the maxima.

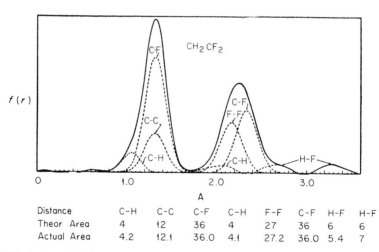

Distance	C-H	C-C	C-F	C-H	F-F	C-F	H-F	H-F
Theor Area	4	12	36	4	27	36	6	6
Actual Area	4.2	12.1	36.0	4.1	27.2	36.0	5.4	7

FIG. 3. A radial distribution function for 1,1-difluoroethylene computed from Eq. (12). Note the essential positivity of the function, its resolution into component peaks, and the good agreement for the areas, even for distances involving hydrogen atoms. (Karle and Karle.[9])

Typical radial distribution functions $f(r)$ computed from experimental data employing Eq. (12) are shown in Fig. 3 for 1,1-difluoroethylene[9] and in Fig. 4 for 1,2-dichloroethane.[42] The functional representation of these curves is given by Eq. (14) as a sum over Gaussian shaped peaks to a good approximation, except for the largest Cl—Cl distance in Fig. 4 which is affected by internal rotation causing it to be asymmetric as described by Eq. (26) of Karle and Hauptman.[18] Composite peaks are seen to be decomposable into their individual components. This is a unique process since both the shape and the area (related to atomic numbers) of the peaks from individual distances are known. However some correlation exists between vibrational amplitudes (related to peak widths) and the positions of the maxima of the components. It is important to take such correlations into account in error estimates. It is seen that distances involving hydrogen atoms make a significant contribution to the radial distribution functions in Figs. 3 and 4. The areas listed for Fig. 3 apply to the function $rf(r)$.

Analysis of the peaks in a radial distribution curve in terms of Eq. (14) can yield Eq. (13) since the value of a inserted into the calculation is known. It is of interest that a calculation involving finite data [Eq. (12)] can ultimately be interpreted in terms of a calculation involving infinite data [Eq. (4)]. This occurs because the functional form of $rD(r)$ is assumed to be accurately represented by Eq. (13). Actually an even more general uniqueness principle holds[4] based on the fact that the $D(r)$ function is interpretable in terms of probability functions, Eq. (6), and must therefore be nonnegative. The requirement of nonnegativity for the radial distribution function is often referred to as the "positivity" principle. The radial distribution curve provides information concerning interatomic distances, r_{ij}, and root mean square

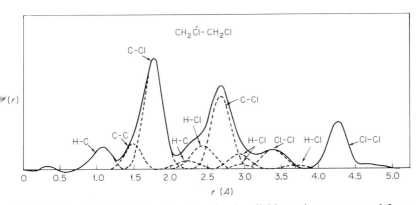

FIG. 4. A radial distribution function for 1,2-dichloroethane computed from Eq. (12). The curve has contributions from both *trans* and *gauche* forms, $\approx 3:1$, respectively. (Ainsworth and Karle.[42])

amplitudes of vibration, $\langle l_{ij}^2\rangle^{1/2}$. In special cases such as with 1,2-dichloroethane, analysis of the radial distribution curve shows that the gas sample is a mixture of conformers (*trans:gauche* $\approx$ 3:1). Radial distribution curves can also afford information concerning potential barriers hindering internal rotation.

The molecular intensity function to be inserted into Eq. (5) in order to obtain the form of $D(r)$ defined by Eq. (13) may be written to a good approximation

$$I_m(s) = \sum_{i=1}^{n} \sum_{j=1}^{n} c_{ij} \exp(-\langle l_{ij}^2\rangle s^2/2) \sin sr_{ij}/sr_{ij} \quad (i \neq j) \quad (15)$$

a function which is also defined in terms of the structural parameters, r_{ij} and $\langle l_{ij}^2\rangle^{1/2}$. Sometimes s is replaced in computations by the quantity q where $q = (10/\pi)s$. The experimental molecular intensity function is shown for 1,1-difluoroethylene in Fig. 5 and for 1,2-dichloroethane in Fig. 6. The radial distribution functions in Figs. 3 and 4 were computed from these functions using Eq. (12). The theoretical intensity curves in Figs. 5 and 6 were computed from Eq. (15) employing the structural parameters determined for the molecules except for the *trans* Cl—Cl distance for which Eq. (45) of Karle and Hauptman[18] was employed. Anharmonicity effects are taken into account by a theoretical function which alters the argument of the sine function in Eq. (15) to give $\sin(sr_{ij} - \kappa s^3)$. It is not often possible to evaluate κ with any degree of precision from electron diffraction data, even for simple molecules. When it is employed as part of the theoretical function, a value is estimated from other considerations such as spectroscopic analysis. The value of κ is generally quite small and is of the order of 10^{-5}–10^{-6} Å^3.

The key to accurate structure determination resides in having an accurate experimental background function to carry out the operation defined in

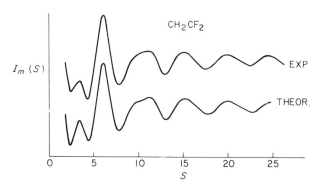

FIG. 5. The experimental and the theoretical molecular intensity curve computed from Eq. (15) for 1,1-difluoroethylene. (Karle and Karle.[9])

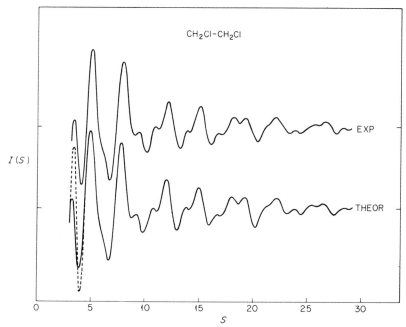

FIG. 6. The experimental and the theoretical molecular intensity curve computed from Eq. (15) for 1,2-dichloroethane. (Ainsworth and Karle.[42])

Eq. (9). This is facilitated by applying the positivity principle. In the application, the smoothness of the background line is maintained while its values are altered to optimize the positivity of the radial distribution function computed from Eq. (12). The data reduction leading to the radial distribution curves shown in Figs. 3 and 4 included such an optimization. It is seen that the distance distributions are positive and the regions outside these distributions are close to zero. Kimura and Iijima[38] have shown that the principle can be appropriately applied even when considerable differences occur in the atomic numbers of the ith and jth atoms resulting in a data reduction treatment which incorporates a significant $\cos(\eta_i(s) - \eta_j(s))$ term into the Q_{ij} function of Eq. (1).

The implementation of the theory of Debye as outlined in Eqs. (1)–(6) employing experimental data may now be summarized. Item (i) above may be handled by employing Eq. (9) to obtain a molecular scattering intensity with essentially constant coefficients. This is supplemented by the attachment of a theoretical function in the unmeasured region of very small s values and additional minor adjustments when required. Item (ii) concerning the limited range of data is handled by the introduction of a suitable damping function exp $(-as^2)$ into the integrand of the expression for the Fourier transform, Eq. (12), followed by a theoretical interpretation, e.g., Eq. (14), which ultimately eliminates the effect of the damping function on the final results. Item (iii) concerning the optimization of the background function $I_b(s)$ is taken care of by adjusting the theoretical values by involving the criteria of smoothness and the positivity of the resulting radial distribution function. Finally, for item (iv), concerned with the measurement of the relatively weak signal from the molecular scattering in the diffraction pattern, the sector device removes the steeply falling background, and the spinning of the photographic plates while they are being scanned removes the effect of graininess and averages out other photographic defects. Since the sector affects the total intensity $I_t(s)$ and the background intensity $I_b(s)$ similarly, their ratio in Eq. (9) is unaffected, and the data reduction proceeds as described.

IV. READING AN ELECTRON DIFFRACTION PAPER

The molecular scattering information available from an electron diffraction experiment is illustrated in Figs. 5 and 6. Often the data are usable out to 40 or more s units, although the vibrational motion tends to weaken the signal to background ratio considerably as s increases. In many cases the data deteriorate significantly in the region $s > 30$. As the number of parameters required to define a structure increases, the total amount of diffraction information remains essentially unaffected, putting limitations on the number of parameters which can be determined and their accuracy. Structural parameters associated with the atoms in a molecule which scatter more strongly, i.e., those with the larger atomic numbers, will be generally more readily and more accurately determined. Closely spaced distances are often difficult to resolve, and the structure determination may afford only an average value for the distances. By way of reducing the number of parameters, it is customary to assume that certain distances in a molecule are equal. In addition, symmetry characteristics are often assumed to prevail. For example, CCl_4 has been assumed to form a perfect tetrahedron. In this case there are two different distances $r(C-Cl)$ and $r(Cl-Cl)$ to determine among a total of 10 interatomic distances. If the structure of benzene is assumed to be based on two concentric regular hexagons, two different distances $r(C-C)$ and $r(C-H)$ need to be determined among a total of 66 interatomic distances. If some distances are slightly disparate, a recognizable broadening may occur in the corresponding radial distribution peak, which implies too large a vibrational amplitude to be attributable to distances of equal value. The assumption of a reasonable value for the amplitudes permits an estimation of the magnitudes of the differences in the distances comprising an unresolved peak. An additional opportunity for resolving closely spaced bonded and second nearest neighbor distances arises from known values of the longer distances in a molecule. It is sometimes necessary to permit distinct values with small differences to occur in bonded distances, for example, in order to account for the values of certain of the longer ones which are found.

A valuable source of additional structural information is the field of spectroscopy. The combination of spectroscopy and electron diffraction to enhance the results of each technique has been under continuing development for many years. With increasing experimental precision, a detailed comparison of the nature of the structural information derivable from spectroscopy and electron diffraction reveals subtle differences in the types of parameters which are obtainable. The differences arise from the presence of a variety of molecular motions, their effects on the structural parameters, and the nature of the quantities which are measured by the different techniques. In short, the types of parameters, as they are immediately revealed by electron diffraction and spectroscopy, need to be precisely defined in terms of the effects of the rotational and vibrational motions. Suitable corrections are required if detailed comparisons are to be carried out.

A. Distance Parameters

The main types of distances which occur in current discussions are as follows:

(i) r_e is the equilibrium internuclear distance defined by the minimum of the potential energy function.

(ii) r_m is the position of the maximum value of the probability function for an interatomic distance $P_{ij}(r)$ occurring, for example, in Eq. (3). Such values could be found from a calculation of $rf(r)$ employing Eq. (12).

(iii) r_a is the distance parameter r_{ij} obtained from Eq. (15) by fitting this theoretical function by least squares, for example, to experimental molecular intensity data. It has also been denoted by the symbol $r_g(1)$ [43] which represents the position of the center of gravity of $P_{ij}(r)/r$. The Fourier transform of Eq. (15) can be expressed in terms of the functions $P_{ij}(r)/r$.

(iv) r_g is the center of gravity position of the probability distribution function. It is the average value of the interatomic distance which may be directly calculated from $P_{ij}(r)$ and is therefore obtainable from an electron diffraction experiment. The theoretical expression for r_g in terms of r_a is

$$r_g = r_a + \langle l^2 \rangle / r_e \tag{16}$$

where $\langle l^2 \rangle$ is the mean squared amplitude of vibration, and in terms of r_e it is [43,44]

$$r_g = r_e + \delta r + \langle \Delta z \rangle + (\langle \Delta x^2 \rangle + \langle \Delta y^2 \rangle)/2r_e + \cdots \qquad (17)$$

where the z axis of the Cartesian coordinates is taken along the equilibrium positions of the nuclei, δr represents a correction for centrifugal distortion, and Δz, Δx, and Δy are the differences of the displacements of the pair of atoms in the directions of the three axes: $\Delta z = \Delta z_i - \Delta z_j$, etc. Averages are taken over the probability distribution function. The quantity $\langle \Delta z \rangle$ is called the mean parallel amplitude and depends on the cubic potential constants. It should vanish if the molecular vibrations are harmonic. The quantities $\langle \Delta x^2 \rangle$ and $\langle \Delta y^2 \rangle$ are called the mean square perpendicular amplitudes and are primarily dependent upon the quadratic potential constants. They remain finite when the vibrations are harmonic.

(v) r_α is an interatomic distance parameter and is defined in terms of the equilibrium internuclear distance and a term $\langle \Delta z \rangle$ which represents the variation in the mean positions of the nuclei due to anharmonicity in the vibrational motion,

$$r_\alpha = r_g - \delta r - (\langle \Delta x^2 \rangle + \langle \Delta y^2 \rangle)/2r_e + \cdots \qquad (18)$$

or

$$r_\alpha = r_e + \langle \Delta z \rangle + \cdots \qquad (19)$$

It is possible to calculate the harmonic contributions fairly readily from the theory of molecular vibrations given by Morino and Hirota.[45] A method for evaluating the centrifugal distortion has been described by Iwasaki and Hedberg.[46] These two types of calculation afford an evaluation of r_α from the experimentally obtained r_g by use of Eq. (17).

(vi) r_α^0 is the limit approached by r_α as the temperature approaches absolute zero,

$$r_\alpha^0 = \lim (T \to 0^\circ K) r_\alpha \qquad (20)$$

A method for computing r_α^0 from r_α has been described by Kuchitsu and Konaka.[47] If all the normal frequencies of a molecule are much higher than kT/hc at $T \approx 300^\circ K$, the contribution to the parameter r_α is mainly from the zero point vibrations, and therefore r_α and r_α^0 are essentially indistinguishable.

It is apparent from the above definitions that r_g is always larger than r_α. It is also expected that r_α and r_α^0 will be larger than r_e for bonded distances since the differences among them depend principally on the anharmonicity of the bond stretching vibrations. If the anharmonic term $\langle \Delta z \rangle$ is known, it is possible to derive r_e from r_α. Usually $\langle \Delta z \rangle$ is not known, but a reasonable estimate of r_e can be obtained nevertheless from a knowledge of r_α.

(vii) r_0 is an "effective" distance parameter obtained from measurement of the "effective" rotational constants (A_0, B_0, C_0) in the ground vibrational state from microwave or infrared spectra.[48]

(viii) r_s is a distance parameter obtained from rotational constants in the ground vibrational state, making use of the additional information available from isotopic substitution as described by Kraitchman[49] and Costain.[50]

(ix) r_z is a distance parameter derived from rotational constants (A_z, B_z, C_z) obtained from the "effective" rotational constants corrected for the contribution from harmonic vibrations.[51–54] Since no correction is made for the anharmonic contribution, it is expected that r_z will be comparable, to a good approximation, to r_α^0 derived from electron diffraction analysis.

There have been several investigations within the past 10 years devoted to the computation and comparison of the variety of distances defined in (i)–(ix) above. Some examples are investigations of CH_4 and CD_4 by Bartell et al.,[55] of CS_2 by Morino and Iijima,[56] of ethylene,[57] and of ethane and diborane[58] by Kuchitsu, of butadiene, acrolein, and glyoxal by Kuchitsu et al.,[59] and of formaldehyde, acetaldehyde, and acetone by Kato et al.,[60] and by Iijima and Kimura.[61] Kato et al.[60] list many estimates of the anharmonicity factor κ. It has been found that for bonded distances involving atoms such as carbon and oxygen, the various distance parameters defined above vary by a few thousandths of an Ångstrom unit. If one of the bonded atoms is a hydrogen atom, the various types of distances may vary by 0.01 Å or more. Such differences play an important role in optimizing the comparison between structural results from electron diffraction and spectroscopy.

Another subject in which small differences are important is the study of primary and secondary isotope effects. Secondary isotope effects concern changes in parts of a structure other than the bond including the isotopic substituent. The secondary isotope effect has been studied in ethane and deuteroethane by Bartell and Higginbotham,[62] employing the $r_g(1)$ structure. They reported that the C—C bond length in ethane exceeded that in deuteroethane by $\Delta r_g(1) = 0.0016$ Å with a standard deviation of ± 0.0007 Å. The authors pointed out that the significance of the numbers involved is quite borderline.

For questions of molecular conformation, aromaticity, and broader trends in bonding, for example, the small differences that distinguish the various distance parameters do not play a significant role, and for more complex structures the accuracy attainable minimizes the significance of such differences.

B. Shrinkage

An effect of the vibrational motion on the distance parameters r_a or r_g for the longer distances in a molecule often gives rise to a phenomenon called shrinkage. This occurs when the value of r_g, for example, for a longer distance, calculated from a knowledge of the bond distances and angles, is longer than the observed value. An example of shrinkage is given by a linear molecule such as CO_2 in which the value of the O—O distance is observed to be smaller than twice the value of the C—O distance. An indication to this effect was observed for CO_2 in 1949.[4] This arises owing to the fact that for the bending mode, the atoms deviate from the straight line connecting them. The detailed nature of the motion leading to the physical interpretation depends upon the anharmonicity. This matter has been discussed by Morino et al.[63] In a theoretical treatment of internal rotation about a single bond,[18] it was pointed out that motions of this type having large amplitudes could lead to significant shrinkage for *trans* distances. The theory developed in this latter study was employed in the analysis of the structure of 1,2-dichloroethane[42] to account for the effect of the internal rotation in shifting the position of the observed *trans* Cl—Cl distance to shorter values and distorting the shape of the peak in the radial distribution curve, Fig. 4. It is apparent that the effect of internal rotation on a *cis* distance would be to cause "elongation" rather than shrinkage. After some hiatus, renewed interest in the shrinkage phenomenon arose as a consequence of further observations of the effect by Bastiansen and collaborators[64-66] and the availability of the theory of Morino and Hiroto[45,67] for interpreting the observations. Examples of additional studies of the shrinkage effect have been given by Morino et al.,[68] Freyland et al.,[69] and a detailed discussion has been presented by Cyvin.[70] The magnitude of the shrinkage can be quite significant, involving changes as large as 0.03–0.04 Å in some molecules, and it is important to take this effect into consideration. It may lead to ambiguous alternatives concerning whether a molecule is bent at equilibrium or whether it merely appears to be because the observed longer distance parameters are shortened by the molecular motion.

C. Errors

The estimation and reporting of errors in a structure determination has not been standardized in the literature. Before the advent of generally available high speed computing facilities, uncertainties were estimated by a variation of parameters technique in which structural parameters were altered and comparisons were made between molecular intensity curves computed from these parameters employing Eq. (15) and the experimental molecular intensity curve. When the deviation between the computed and experimental curve was excessive, a limit of uncertainty of the set of parameters was implied. Such calculations were quite time consuming, and in complex structures the number of calculations required to probe thoroughly the multidimensional parameter space was quite prohibitive. Nevertheless the variation of parameters procedure had the virtue of keeping in the forefront the interdependence of the values of the structural parameters. This means, for example, that a change in one distance might be compensated for by a change in another in such a way that the effect on the resulting molecular intensity curves would not be noticeable. The values for the distance parameters may be said to be correlated. When there are several unresolved peaks in a radial distribution curve, as always occurs for complex structures, a proper analysis generally reveals that the distances and their associated vibrational amplitudes, corresponding to an unresolved peak, are highly correlated. An illustration of the treatment of the uncertainties and the dependent errors by the method of variation of parameters can be found in the investigation of benzene and cyclooctatetraene by I. L. Karle.[71] The correlation of overlapped peaks is well illustrated by these examples. In such investigations, the attempt was made to report the uncertainties as limits of error, implying that variations beyond the given bounds would lead to unacceptable comparisons with the molecular intensity curves.

Least-squares methods can be applied to intensity curves or radial distribution curves to determine structural parameters. As these methods have developed, the technique of variation of parameters has been employed quite infrequently, although, in some instances, variation of parameters has been combined with the least-squares technique. This is accomplished by holding some parameters at a succession of fixed values, each time permitting the remaining variable parameters to converge to their best values in the least-squares sense. The standard deviation obtained from a least-squares analysis measures only the internal consistency of the result. It is necessary to introduce an additional uncertainty which is a measure of the reproducibility of the curve being fit. Systematic errors which arise, for example, from the measurement of the accelerating voltage and the specimen to plate distance, and the behavior of the microdensitometer must be included. The extent to which a featureless background line and a nonnegative radial distribution curve with accurate areas have been obtained must also be taken into account. There are also minor limitations in the theoretical functions employed. In addition to these sources of uncertainty there is the important question of the dependency of the errors which is often concealed by the manner in which it is necessary to carry out a calculation in order to reach convergence. The general formulation of the least-squares problem often does not converge since the data are insufficient to resolve some of the structural parameters. Convergence is attained then by constraining the values of certain of the parameters. However the consequence of the latter can very well be to yield a least-squares standard deviation which is small only because the constraints have been imposed, bypassing the high correlation among the parameters. A suitable standard deviation should include all sources of error, both random and systematic and an estimate of the effect of correlation among the structural parameters. It is then possible to use the standard deviation σ to make an approximate estimate of a limit of error. For example $\pm 2\sigma$ gives a confidence level of about 95%, i.e., the probability of finding a result outside $\pm 2\sigma$ is about one chance in 20. It can be quite a long step from a least-squares standard deviation to a reasonable value for a limit of error in even moderately complex structures, and it is therefore worthwhile for the reader to determine, if possible, the extent to which the various sources of error have been taken into account in the reporting of final results. Discussions of the application of least-squares to electron diffraction analysis have been given by Hamilton,[72] Bonham and Bartell,[73] Hedberg and Iwasaki,[74] Morino et al.,[75] Murato and Morino,[76] Hilderbrandt and Bauer,[77] and Seip et al.[78]

The following is a list of questions which can facilitate the assimilation of the structural information in a manuscript.

1. What structural assumptions have been made?
2. What auxiliary structural information has been introduced?
3. What types of interatomic distance parameters are being quoted? When unstated, the parameter is usually r_a.
4. Have shrinkage effects been properly accounted for in molecular configurations that appear to be bent?
5. Have all sources of error been considered and limits of error reported? Has proper account been taken of correlation among the structural parameters?
6. Are the differences among the results of several structure determinations on the same substance contained within the limits of error, are they a matter of scale, or are there significant differences in the configuration?

V. REGULARITY OF BOND LENGTHS

As structural information has accumulated, a considerable amount of regularity concerning configuration and interatomic distance has been observed. It is therefore possible to anticipate the one or more possible configurations for a large variety of molecular structures, including approximate estimates of interatomic distances. Empirical rules help to organize the structural information and are also useful in suggesting additional research, often leading to a deeper understanding of bonding and configuration. Quite accurate empirical rules have been found to apply to carbon–carbon bonds in aliphatic hydrocarbons. They arose from data on such bonds which were mainly obtained from accurate spectroscopic investigations. The rules are based on the observation, as pointed out by Herzberg and Stoicheff,[79] that the bond lengths change systematically with changes in their environment. Further discussion of this subject, as more information became available, was presented by Costain and Stoicheff[80] and by Stoicheff.[81] It was found that the length of the bonds obeys an accurate linear relation[80,81] as a function of the number of adjacent bonds when C, H, N, or O are the attached atoms. Adjacent single, double, and triple bonds are each counted as one bond. A plot of the relation as published by Stoicheff[81] in 1962 is reproduced in Fig. 7. The equations for the lines are

$$r(\text{C—C}) = 1.299 + 0.040n, \qquad n = 2, 3, \cdots 6 \qquad (21)$$

and

$$r(\text{C}=\text{C}) = 1.226 + 0.028n, \qquad n = 2, 3, 4 \qquad (22)$$

where n is the number of adjacent bonds. The attachment of atoms such as halogens can cause wider deviations from the straight line than indicated in Fig. 7. Recent investigations have placed $r(\text{C—C})$ for 1,3-butadiene closer to

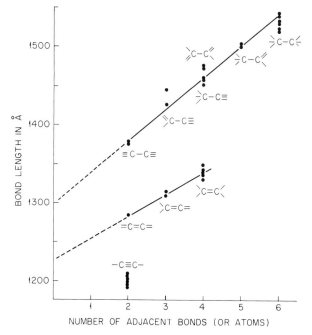

FIG. 7. The relationship between single and double CC bond lengths and the number of attached bonds. (Stoicheff.[81])

the line. However some exceptional values have also appeared. For example, Kuchitsu et al.[59] have reported $r(C—C)$ in glyoxal to be 1.525 Å, significantly different from the value of 1.46 Å predicted by the empirical rule.

A suggestive correlation for cyclic hydrocarbons has been found by Dallinga and Ros[82] between bond distance and overlap population derived from quantum mechanical calculations. A plot of the results from computations on several molecules is shown in Fig. 8. One-electron molecular orbitals were constructed by means of the iterative extended Hückel method as, for example, described by Basch et al.[83] and Schachtschneider et al.[84] A Mulliken population analysis was then applied to the results in order to obtain the overlap population. It was found worthwhile to include nonbonded interactions. In the structural models used for carrying out the calculations, the carbon–carbon single bond distances were all assumed to have the same value as were the carbon–hydrogen distances. The bond angles were chosen to agree with the available experimental values. Figure 8 shows a distinct trend correlating a decrease in interatomic distance with increasing population overlap.

The remainder of this article will be devoted to a review of the results of a number of recent structure investigations, often among related structures. It should be informative to bear in mind the implications of Figs. 7 and 8 in connection with the discussions of many of the structures.

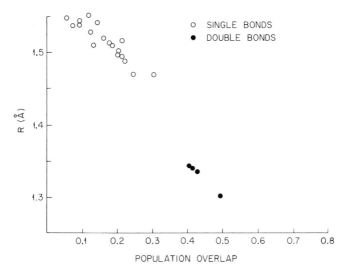

FIG. 8. The relationship between bond length (R) and calculated population overlap (O.P.) for cyclic hydrocarbons. (Dallinga and Ros.[82])

VI. SINGLE BONDS

It had long been considered that the single C—C bond length was properly represented by the value 1.5445 Å found in diamond by Lonsdale.[85] The *n* alkanes represent a system in which the C—C bonds might be expected to have the value found in diamond. The value for ethane determined by Bartell and Higginbotham,[62] however, was 1.534 Å, which is the same as the average C—C bond length in a series of *n*-hydrocarbons, butane through heptane, investigated by Bonham and collaborators.[86,87] The difference between the value for the C—C bond length in diamond and in the alkanes is significant and is an example of how the chemical environment of a bond affects its length.

It was suggested by Zeil *et al.*[88] that the C—C bond in 2,2-dimethylpropane (neopentane) might have a value intermediate between that for ethane, 1.534 Å, and that for diamond, 1.5445 Å, owing to the tetrahedral configuration of carbon atoms about the central carbon atom in neopentane. The result that they obtained had such a value, namely, 1.540 Å, agreeing with the value of 1.54 Å reported in an early study by Pauling and Brockway[89] using the visual method.

The structure of cyclopropane was investigated by Bastiansen *et al.*[90] employing data from a preliminary study in a thesis by P. N. Skancke. The structural parameters are shown in Fig. 9.

A generalization of the cyclopropane molecule is the molecule of spiropentane investigated by Dallinga *et al.*[91] It is shown in Fig. 10. In this figure, the plane through H-8—C-2—H-9 does not coincide with the bisectrix plane of angle C-1—C-2—C-3, but is rotated through an angle of 2.3° in the direction of the neighboring CH$_2$ group. The difference of 0.05 Å between C-1—C-2 and C-1—C-3 in spiropentane has been subjected to a theoretical analysis by P. Ros employing a Mulliken population analysis based on the results of an it-

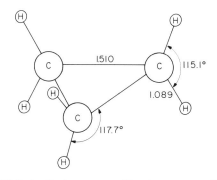

FIG. 9. Cyclopropane. (Bastiansen *et al.*[90])

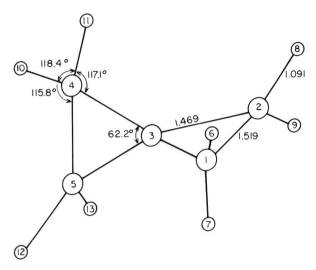

FIG. 10. Spiropentane. (Dallinga et al.[91])

erative extended Hückel calculation.[82,91] In the model used for the calculation all bonds were assumed to be equal. Dallinga et al.[91] state that from their experience with this type of calculation for cyclic hydrocarbons, they expect a difference of 0.01 in calculated C—C overlap population to correspond to a difference in bond length of about −0.01 Å. In the calculation for spiropentane, the difference of 0.050 Å between the bond lengths corresponds to a difference of −0.045 in overlap population. Corroborative results[91] are also obtained for cyclopropane and cyclopropene.

The structure of cyclobutane has been investigated by Almenningen and Bastiansen.[92] The C—C bond length at 1.548 Å is close to the value in diamond (1.5445 Å), differing somewhat from the average value of 1.533 Å found in the n-alkanes. Bicyclo[1,1,1]pentane shown in Fig. 11 is seen to be composed of cyclobutane rings. Its structure was investigated by Chiang and Bauer[93] who found the C—C bond length to be essentially the same as that in cyclobutane at 1.545 Å. The molecule has D_{3h} symmetry, however, which requires that the dihedral angle for any four-membered ring be 120° rather than the value of about 145° found for cyclobutane by Meiboom and Snyder[94] and by Dows and Rich.[95] A slight increase in the C—C single bond length over that in cyclobutane has been reported for perfluorocyclobutane by Chang et al.[96] The value found was 1.566 Å. The C—F distance was reported to be 1.333 Å and the dihedral angle increased to 162.6°.

The structures of silacyclobutane, 1,1-dichlorosilacyclobutane, and 1,1,3,3-tetrachlorosilacyclobutane have been investigated by Vilkov and

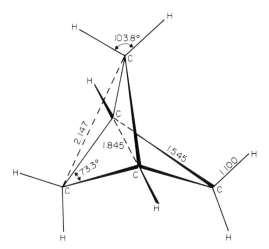

FIG. 11. Bicyclo[1,1,1]pentane. (Chiang and Bauer.[93])

collaborators.[97,98] In silacyclobutane, a carbon atom of cyclobutane is replaced by a silicon atom. The silicon atom is labeled as the 1 position. In none of these compounds was the ring found to be planar. The dihedral angle for the ring structures in the first two silacyclobutanes listed above was reported to be 150° and for the tetrachlorodisilacyclobutane, it was reported to be 166°. The C—C bond distance in silacyclobutane and in 1,1-dichlorosilacyclobutane was found to be fairly large with a value of 1.59 Å.

There have been some recent studies of the polycyclic hydrocarbons, bicyclo[2,1,1]hexane[99] (Fig. 12) bicyclo[2,2,1]heptane[100-102] (Fig. 13) and bicyclo[3,1,1]heptane (Fig. 14).[103] These materials show strain which may be correlated with unexpected reaction mechanisms and often very high reaction rates. Of interest also are the differences in the lengths of the carbon–carbon bonds. Detecting differences of a few hundredths of an angstrom in polycyclic hydrocarbons presents a difficult challenge for the electron diffraction method, not only because there are a large number of structural parameters, but also because the interpretation of the data is ambiguous. Changes in one part of a structure may be compensated for by changes in another part in fitting the diffraction data with a theoretical model. In such circumstances, it is often not possible to make a decision concerning the relative lengths of the bonds to within a few hundredths of an angstrom from the diffraction data alone. As pointed out previously, in order to proceed, auxiliary structural information is often introduced. This may take the form of parameter information obtained from comparable investigations by other means, e.g., X-ray analysis, limitations on the range of variation of structural parameters based on experience, and also theoretical calculations which may indicate roughly the

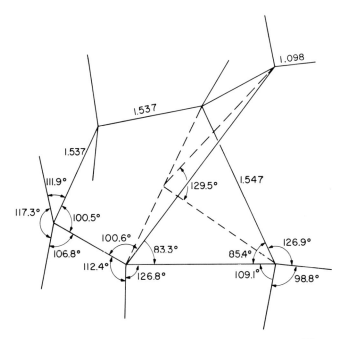

FIG. 12. Bicyclo[2,1,1]hexane. (Dallinga and Toneman.[99])

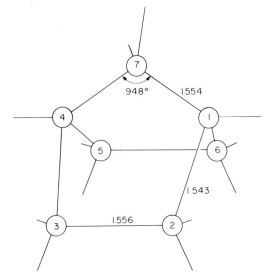

FIG. 13. Bicyclo[2,2,1]heptane. The values are averages from three structure determinations. (Yokozeki and Kuchitsu[100]; Dallinga and Toneman[101]; Chiang et al.[102])

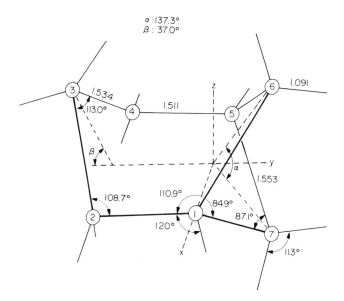

FIG. 14. Bicyclo[3,1,1]heptane. (Dallinga and Toneman.[103])

quality or relative ordering of a pair of distances. Additional assumptions concerning molecular symmetry or the equality of certain distances may also be made. The auxiliary structural information, although helpful, is not entirely satisfactory because its applicability to a particular investigation may be subject to some uncertainty. It is readily possible, however, to obtain average distances and the overall molecular configuration directly from the electron diffraction data. The values for the parameters in Figs. 12–14 are influenced by the many assumptions introduced into the analyses of the diffraction data. Three different sets of values[100-102] have been averaged for Fig. 13. Dallinga and Toneman[101] report that their calculations of Mulliken population overlap are in qualitative agreement with their results for the relative ordering of the C—C distances in bicyclo[2,2,1]heptane, r(C-1—C-2) = 1.534 Å, r(C-2—C-3) = 1.578 Å, and r(C-1—C-7) = 1.535 Å, and in bicyclo[3,1,1]heptane as shown in Fig. 14.

If a bond is formed with the loss of two hydrogen atoms at the 2,6 positions of bicyclo[2,2,1]heptane (norbornane), a substance called nortricyclene is formed. The structure of 4-chloronortricyclene has been investigated by Chiang et al.[104] They found that the three-membered ring formed by the C-1—C-2, C-2—C-6, and C-1—C-6 bonds has interatomic distances equal to 1.510 Å, the same as those for cyclopropane, a reduction from the value of about 1.543 Å for C-1—C-2 and C-1—C-6 in norbornane. The other C—C bond distances were reported to have a value of about 1.536 Å.

VII. SINGLE AND DOUBLE BONDS

Single carbon–carbon bonds are affected by the presence of multiple bonds. They will now be discussed with respect to double bonds. Alternate single and double bonds forming conjugated systems will be discussed separately. The simplest single and double bond combination is represented by propene which has been investigated by the visual method for electron diffraction and by microwave spectroscopy. McHugh and Schomaker employing electron diffraction[105] reported values of 1.33 Å and 1.49 Å for the double and single bond, respectively. Lide and Mann, employing microwave spectroscopy,[106] reported 1.353 and 1.488 Å for the corresponding distances. The double bond value is not greatly different than that for ethylene,[107] 1.337 Å, but the single bond value is about 0.045 Å shorter than the average value for n-alkanes.[86,87]

The structure of cyclopropene has been investigated by Kasai et al.[108] by means of microwave spectroscopy. The results are shown in Fig. 15. The C—C single bond distance and the structure of the CH_2 group are essentially the same as that found for cyclopropane (Fig. 9). A significant change occurs in the value of the C=C double bond decreasing almost 0.04 Å from the value in ethylene, 1.337 Å. An early electron diffraction investigation of cyclopropene by the visual method had been carried out by Dunitz et al.[108] and is in agreement with the above result within the quoted limits of error. A more recent electron diffraction investigation of cyclopropene has been

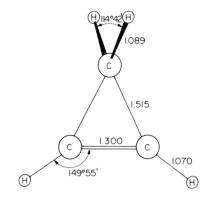

FIG. 15. Cyclopropene. (Kasai et al.[108])

carried out by Chiang,[110] obtaining comparable results for the carbon–carbon bonds. The reported values were 1.519 and 1.304 Å for the single and double bonds, respectively. The corresponding angles were reported at 132.9° for the C=CH angle and 117.6° for the HCH angle.

The replacement of the four hydrogen atoms in cyclopropene by chlorine atoms affects the carbon–carbon bonds and also shows differences among the carbon–chlorine bonds. The results of an electron diffraction investigation of tetrachlorocyclopropene by Mair and Bauer[111] is shown in Fig. 16.

Cyclobutene, the next molecule in the homologous series initiated by cyclopropene, has been investigated by microwave spectroscopy. The results found by Bak et al.[112] are illustrated in Fig. 17. In this molecule the C=C double bond distance is normal, and the C—C single bond distances increase in value as their separation from the double bond increases. The long C—C bond distance is comparable to that found in cyclobutane.[92]

The structure of perfluorocyclobutene has been investigated by Chang et al.[96] Its structure is shown in Fig. 18. Comparison with Fig. 17 indicates that the C=C bond length is not changed by fluorine substitution, the =C—C single bond is shortened by 0.01 Å, and the other C—C single bond is lengthened by 0.03 Å.

The structure of cis-3,4-dichlorocyclobutene-1 has been investigated by Bastiansen and Derissen.[113] They reported a structure with carbon–carbon

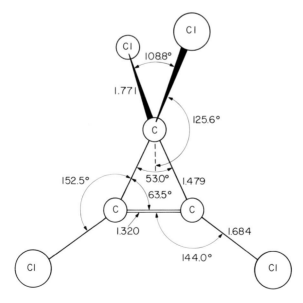

FIG. 16. Tetrachlorocyclopropene. (Mair and Bauer.[111])

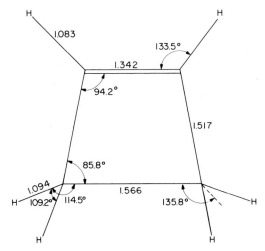

FIG. 17. Cyclobutene. (Bak et al.[112])

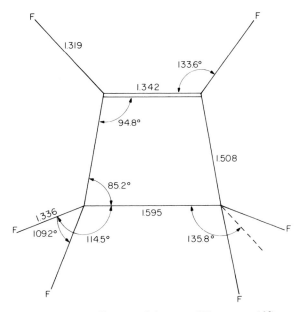

FIG. 18. Perfluorocyclobutene. (Chang et al.[96])

distances quite comparable to those found for perfluorocyclobutene (Fig. 18). The corresponding carbon–carbon distances were reported to have values at 1.349, 1.505, and 1.583 Å. The value for the C—Cl bonded distance was found to be normal at 1.771 Å. The angle C—C—Cl which compares to angle C—C—F at 114.5° was found to be 117.2°.

The configuration of cyclohexene was determined by Chiang and Bauer[114] to be in the half-chair conformation with C_2 symmetry. Its structure is illustrated in Fig. 19. The double bond distance is about the same as that in ethylene, and the single bond distances increase in value with increasing spacing from the double bond. The twofold axis intercepts the C-1=C-2 and C-4—C-5 bonds.

The structure of 1,4-cyclohexadiene has been investigated by Dallinga and Toneman[115] and by Oberhammer and Bauer.[116] Although the possibility that the molecule might be in the chair, boat, or skew form was considered in the analysis, the structure reported by Dallinga and Toneman was determined to be essentially planar with small uncertainty. The C=C double bond length was found to be 1.334 Å, and the C—C single bond length was found to be 1.496 Å. Both the C=C—C and C=C—H angles were reported to be 123.4°. The C—C—C and C—C—H angles, where the H atom is attached to a C atom with a double bond, were reported to be 113.3°. The values for these angles are in agreement with a simple empirical theory discussed by Wilson.[117] The H—C—H angle was found to be 103°. The result for 1,4-cyclohexadiene found by Oberhammer and Bauer[116] was different. Instead of finding a planar molecule, they found a boat shape with a dihedral angle of 159°

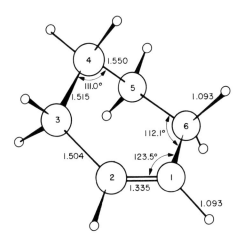

FIG. 19. Cyclohexene. (Chiang and Bauer.[114])

instead of 180°. Several of the values for the angles differ, but the values for the distances C=C = 1.347 Å and C—C = 1.511 Å are in fair agreement.

The molecular structure of bicyclo[2,2,1]hepta-2,5-diene shown in Fig. 20 has been the subject of four independent electron diffraction investigations.[100,118-120] The results are listed in Table I. As pointed out by Dallinga

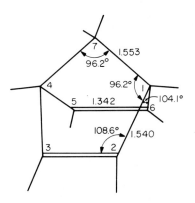

FIG. 20. Bicyclo[2,2,1]hepta-2,5-diene. The values are averages from four structure determinations. (Yokozeki and Kuchitsu[100]; Schomaker and Hamilton[118]; Muecke and Davis[119]; Dallinga and Toneman.[120])

and Toneman,[120] the results differ significantly at the 95% confidence level but the differences between Schomaker and Hamilton[118] and Muecke and Davis[119] are largely a matter of scale. The parameters given in Fig. 20 are an average of the four results in Table I. There have been two X-ray crystallographic studies of derivatives of bicyclo[2,2,1]hepta-2,5-diene. One is a $PdCl_2$ adduct investigated by Baenziger et al.[121] and the other is anti-7-norbornenyl-p-bromobenzoate investigated by MacDonald and Trotter.[122] The values found for the parameters were similar to those found by electron diffraction, but the X-ray investigations do not clarify which of the electron diffraction studies is the more correct. This is expected since the results of the

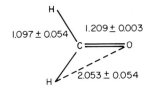

FIG. 21. Formaldehyde. Error limits at 99% confidence level. (Kato et al.[60])

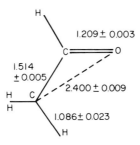

FIG. 22. Acetaldehyde. Error limits at 99% confidence level. (Kato et al.[60])

X-ray investigations of crystals vary somewhat from those of the vapor state even when identical materials are being investigated.

An electron diffraction investigation of formaldehyde, acetaldehyde, and acetone has been carried out by Kato et al.[60] These compounds are the simplest examples of organic molecules containing carbonyl groups. The results are shown in Figs. 21–23, where the errors are estimated by the authors as appropriate to a 99% confidence level. The evaluation of errors in this investigation by Kato et al.[60] is a good example of the proper treatment of the subject of experimental error in electron diffraction. Despite some differences in their environment, the C=O distances have essentially the same value in all three compounds. The C—C single bond lengths in acetaldehyde and acetone are also about the same, the values being intermediate between about 1.54 Å for normal C—C single bonds and about 1.47 Å for central C—C single bonds in conjugated systems.

It is of interest to compare the structure of acetone with that of isobutylene shown in Fig. 24, the latter being related to acetone by the replacement of the

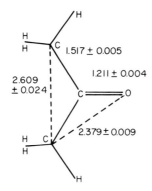

FIG. 23. Acetone. Error limits at 99% confidence level. (Kato et al.[60])

TABLE I

Results of Four Independent Electron Diffraction Investigations of Bicyclo[2,2,1]hepta-2,5-diene. The Error Estimates are Reported at about the 95% Confidence Level

Reference	Bond Lengths (Å)			Bond Angles (degrees)			
	C-2=C-3	C-1—C-2	C-1—C-7	1—7—4	2—1—7	2—1—6	1—2—3
Schomaker and Hamilton[118]	1.333 ± 0.008	1.522 ± 0.011	1.558 ± 0.016	96.7 ± 2.3	96.4 ± 0.7	102.2 ± 5.5	109.1 ± 0.4
Muecke and Davis[119]	1.357 ± 0.005	1.549 ± 0.008	1.568 ± 0.015	96.5	95.9	106.6 ± 0.8	108.5 ± 0.5
Yokozeki and Kuchitsu[100]	1.339 ± 0.005	1.533 ± 0.005	1.573 ± 0.010	92.0 ± 0.8			
Dallinga and Toneman[120]	1.341 ± 0.002	1.554 ± 0.002	1.514 ± 0.004	99.5 ± 0.6	96.2 ± 0.3	103.4 ± 0.4	108.2 ± 0.2

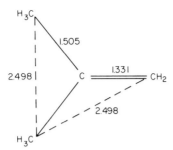

FIG. 24. Isobutylene. (Bartell and Bonham.[123])

oxygen atom by a CH_2 group. The molecular structure of isobutylene has been investigated by Bartell and Bonham.[123] It was found that the C—C single bonds were reduced to 1.505 Å from the normal value of about 1.54 Å. This compares to a value of 1.517 Å reported for acetone (Fig. 23). It is seen from Fig. 24 that the C=C double bond distance in isobutylene is close to that for ethylene.

The structures of acetone and isobutylene may be compared to those of the fluorinated derivatives $(CF_3)_2C$=O and $(CF_3)_2C$=CH_2 which were investigated by Hilderbrandt et al.[124] The authors also investigated the structure of $(CF_3)_2CNH$. They found that the replacement of CH_3 by CF_3 gave rise to a longer C—C bond accompanied by a lengthening of the C=X bond as indicated in Table II. The C—F bond distances are similar to those found in other studies, e.g., for hexafluoroethane the value is 1.32 ± 0.01 Å.[125]

It was pointed out by Hilderbrandt et al.[124] that inspection of fluorine substituted compounds for which structures have been determined show two

TABLE II

Comparison of Bond Distances in Acetone, Related Molecules, and Fluorinated Derivatives

Substance	Reference	r(C—C)	r(C=X)	r(C—F)
$(CH_3)_2C$=O	Kato et al.[60]	1.517	1.211	
	Hildebrandt et al.[124]	1.507	1.210	
$(CH_3)_2C$=CH_2	Bartell and Bonham[123]	1.505	1.331	
$(CF_3)_2C$=O	Hildebrandt et al.[124]	1.549	1.246	1.335
$(CF_3)_2C$=CH_2	Hildebrandt et al.[124]	1.533	1.373	1.327
$(CF_3)_2C$=NH	Hildebrandt et al.[124]	1.549	1.294	1.324

TABLE III

Structural Parameters for the Monomers and Dimers of Formic, Acetic, and Propionic Acids. Different Types of Errors are Listed (See Text).

	Formic Acid Monomer			Dimer	
	a	b	c	b	c
C=O	1.23 ± 0.01	1.216 ± 0.001	1.217 ± 0.003	1.216 ± 0.002	1.220 ± 0.003
C—O	1.36 ± 0.01	1.343 ± 0.001	1.361 ± 0.003	1.338 ± 0.001	1.323 ± 0.003
O—H	0.97 ± 0.05	0.973 ± 0.001	0.984 ± 0.024	0.978 ± 0.001	1.036 ± 0.017
C—H	1.09 (assumed)	1.103 ± 0.001	1.106 ± 0.024		1.082 ± 0.021
OH—O				2.764 ± 0.001	2.703 ± 0.007
O—C=O	122.4° ± 1°	125.5° ± 0.7°	123.4° ± 0.5°	121.2° ± 1°	126.2° ± 0.5°

	Acetic acid[d]		Propionic acid[e]	
	Monomer	Dimer	Monomer	Dimer
C=O	1.214 ± 0.003	1.231 ± 0.003	1.211 ± 0.003	1.232 ± 0.006
C—O	1.364 ± 0.003	1.334 ± 0.004	1.367 ± 0.004	1.329 ± 0.008
C—C=	1.520 ± 0.005	1.506 ± 0.005	1.518 ± 0.010	1.518 ± 0.015
C—C			1.543 ± 0.010	1.547 ± 0.015
O—H	0.97 (assumed)	1.03 (Assumed)	0.97 (Assumed)	1.03 (Assumed)
C—H	1.102 ± 0.012	1.102 ± 0.015	1.125 ± 0.012	1.127 ± 0.018
OH—O		2.684 ± 0.010		2.711 ± 0.014
C—C=O	126.6° ± 0.6°	123.6° ± 0.8°	126.7° ± 0.8°	122.9° ± 1.8°
C—C—O	110.6° ± 0.6°	113.0° ± 0.8°	111.2° ± 0.8°	113.4° ± 1.6°
O—C=O	122.8°	123.4°	122.1°	123.7°
C—C—C			112.8° ± 1.0°	112.0° ± 1.5°

[a] Karle and Karle.[126]
[b] Bonham and Su.[127]
[c] Almenningen et al.[128]
[d] Derissen.[129]
[e] Derissen.[130]

types of trends, in general corresponding to two types of bonding. The authors describe the two cases:

Case 1 is represented by: F_3C—X, F_2C=X, or FC=X, where X is N, C, O, or a halogen, but not double bonded to a third atom. Here the substitution of fluorine for hydrogen decreases the CX bond length. An exception appears to be hexafluoroethane[125] where the C—C bond length increases to 1.56 Å.

Case 2 is represented by: F_3C—X=Y where X is C or N and Y is C, N, or O. In this case the examples investigated indicate a lengthening of the CX bond.

There have been some recent investigations of the vapor phase structures of formic, acetic, and propionic acids. The molecules occur as monomers and dimers in equilibrium. The latter are formed through cyclic hydrogen bond bridging. As the temperature increases, the concentration of monomer increases. Some of the structural parameters for the acids are listed in Table III. In Table III, the errors are estimated limits from variation of parameters, Karle and Karle,[126] least-squares standard deviations, Bonham and Su,[127] and standard deviations with scale and correlation errors included, Almenningen et al.,[128] and Derissen.[129,130] If the standard deviations for the monomer of formic acid given by Almenningen et al.[128] are multiplied by a factor to give limits of error, e.g., a factor of 2 or 2.5 for confidence levels of 95 or 99%, respectively, it is seen that the limits of error so obtained are in very good agreement with those obtained by Karle and Karle[126] from variation of parameters. The two determinations are in agreement well within the error limits. The results in Table III indicate that there is no great change in forming the dimer from the monomer. With regard to the dimer of formic acid, it is apparent that there are some discrepancies which are significantly beyond the implications of the estimated standard deviations.

The structure of hexafluoropropene is under investigation by Bauer and Chang[131] and Karle et al.[131a] The electron diffraction data imply that the ethylene carbon atoms and their bonded attachments do not lie in a plane, in agreement with an earlier investigation by Buck and Livingston.[132] Two structural models which are consistent with the experimental data are shown in Figs. 25(a) and (b). In Fig. 25(a), the deviation from planarity of the bonded attachments to the ethylene carbons arises from a rotation of the CF_2 group about the C=C axis by about 38°. In Fig. 25(b), the deviation from planarity arises from a rotation of the CF_3 group out of the plane formed by the remaining five atoms. In this model, the C—C single bond makes an angle of about 25° with respect to the plane. These nonplanar structures are at variance with the microwave results of Jacob and Lide[131b] and the Raman measurements of Harvey and Nelson,[131c] which are compelling and indicate that the ethylene carbons have planar attachments and that the molecule possesses C_s symmetry. In this instance, there may be a fundamental process that is being overlooked in the analysis of the electron diffraction patterns.

Short C=C bond lengths have been found in several molecules, e.g., cyclopropene,[108–110] described above; 1,1-difluoroethylene[9,133]; tetrafluoroethylene,[9] and butatriene.[134]. The r_a distance value for the C=C bond length was found to be 1.311 Å for CH_2CF_2[9] and 1.313 Å for CF_2CF_2.[9] Transformation to r_g distances would increase these values by about 0.001 Å. In a microwave investigation of CH_2CF_2, Laurie and Pence[133] report an r_s value of 1.315 Å. Almenningen et al.[134] found a value of 1.318 Å for the end

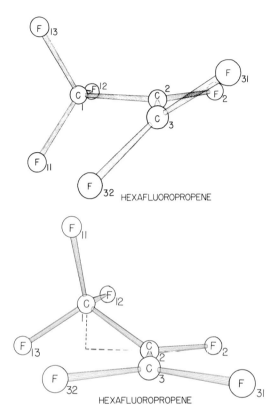

FIG. 25. Hexafluoropropene. In model (a) the dihedral angle between the CF_2 and CCF planes at both ends of the C=C double bond is about 38°. In model (b) the C—C single bond is raised about 25° out of the plane formed by two carbon and three fluorine atoms. (Chiang and Bauer[131]; Karle et al.[131a]) Both models are incompatible with spectroscopic measurements (see text).

C=C bond in butatriene and a value of 1.283 Å for the central C=C bond. The four carbon atoms in butatriene have a linear configuration.

A structure analysis of acetic anhydride has been carried out by Romers and collaborators.[135] The structure of the molecule which has C_2 symmetry is shown in Fig. 26. The two acetyl moieties are planar, except for hydrogen atoms, and are relatively rotated with a dihedral angle of 78.8°.

A molecule which is similar to acetic anhydride is the keto form of acetylacetone. The structures of the molecules in the keto–enol equilibrium of acetylacetone have been investigated by Lowrey et al.[136] The results are shown in Fig. 27. In the keto form of acetylacetone a CH_2 group replaces the

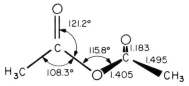

FIG. 26. Acetic Anhydride. The planar portions of the two acetyl moieties are relatively rotated with a dihedral angle of 78.8°. (Romers et al.[135])

central oxygen atom in acetic anhydride. The structures of the two molecules are quite similar. In the keto form of acetylacetone the dihedral angle between the two planar acetonyl moieties is approximately 46°. A significant structural change takes place as the enol form of acetylacetone is formed from the keto form by the migration of a hydrogen atom. Except for methyl hydrogens, the molecule is planar and the internal hydrogen bridge has the small O—O separation of 2.38 Å. The C—C bond distances also change, being smaller in the enol form. The enol form was found to occur to about 66% at $T = 378°K$, in agreement with the value found by Reeves[137] by spectroscopic analysis of the liquid. The structure of the enol form agrees well with that found for the corresponding portion of tetraacetylethane investigated by means of X-ray analysis by Schaefer and Wheatley.[138] They found an internal hydrogen bridge, for example, of 2.424 Å, a ring C—C distance of 1.403 Å, and an end C—C distance of 1.489 Å. It is of interest that the ring C—C distance is close to that for an aromatic system.

VIII. CONJUGATED SYSTEMS

A considerable number of structure investigations have been carried out by electron diffraction on conjugated systems, affording information concerning

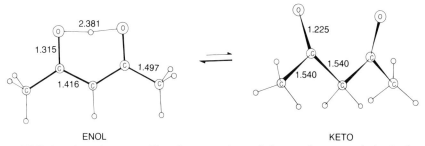

FIG. 27. Acetylacetone. The planar portions of the two ketone moieties in the keto form are relatively rotated with a dihedral angle of $\approx 46°$. The enol form is planar except for the methyl hydrogen atoms. (Lowrey et al.[136])

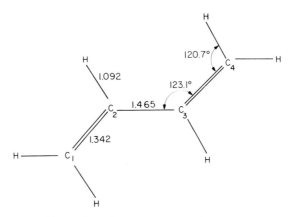

FIG. 28. 1,3-Butadiene. The values are averages from two structure determinations. (Haugen and Traetteberg[139]; Kuchitsu et al.[59])

the molecular configuration and the detailed structural parameters. A simple conjugated system is represented by the molecule 1,3-butadiene shown in Fig. 28. Its structure has been investigated by Haugen and Traetteberg[139] and by Kuchitsu et al.[59] The values in Fig. 28 are an average of the results of the two investigations which agree quite closely. As can be seen from the figure, the single and double bonded carbon–carbon distances were found to be 1.465 and 1.342 Å, respectively. The value for the double bond is close to that of 1.337 Å for ethylene.[107] The value for the single bond is considerably shorter than that reported for ethane, 1.534 Å.[62]

The structure of isoprene, the 2-methyl derivative of 1,3-butadiene, has been investigated by Vilkov and Sadova.[140] They reported comparable values for the single and double bonded carbon–carbon distances in the butadiene portion, 1.470 and 1.338 Å, respectively. The C—C single bond involving the methyl carbon was found to be 1.510 Å, somewhat reduced from that in ethane. In their analysis the authors found agreement with a planar model in the *trans* configuration and also with a model twisted through 40° from the *trans* position, if some small changes were made in the bond angles.

The structure of the 2,3-dimethyl derivative of butadiene shown in Fig. 29 has been investigated by Aten et al.[141] There appears to be a significant increase in the C—C single bond value in the conjugated system compared to that for butadiene. The C—C single bonds involving the methyl carbons have the same value as that found for isoprene.[140] The 2,3-dimethyl derivative of butadiene is in the *trans* configuration, and it was reported that there is no evidence for significant deviation from heavy atom coplanarity.

A more complex example of a conjugated system is the molecule of 1,3,5-*trans*-hexatriene illustrated in Fig. 30. Its structure has been investigated by

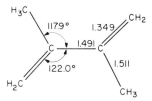

FIG. 29. 2,3-Dimethylbutadiene. (Aten et al.[141])

Traetteberg.[142] It is seen from Fig. 30 that the values of the terminal and central C=C double bonds in 1,3,5-*trans*-hexatriene are 1.337 and 1.368 Å, respectively. This difference is consistent with the indications from a theoretical calculation employing the Pariser–Parr–Pople approximation carried out by Roos and Skancke.[143] The calculated values for the terminal and central C=C double bonds were found to differ, having the values 1.347 and 1.354 Å, respectively. The C—C single bond length was calculated to be 1.461 Å. The experimental value for the terminal C=C bond is the same as that in ethylene.[107] Conjugation has reduced considerably the value of the C—C single bond which generally averages about 1.54 Å. In benzene where all the bonds are equivalent, the value for the carbon–carbon bond is 1.396 Å.[71,144,145]

An additional indication that the terminal C=C double bonds are shorter than the central one was reported by Traetteberg[146] for 1,3,5-*cis*-hexatriene shown in Fig. 31. The values determined were 1.336 and 1.362 Å, respectively. A theoretical calculation[143] gave the same values as stated above for the *trans*-hexatriene. The calculated difference between the terminal and central double bonds amounts to 0.007 Å, whereas the differences reported for the *trans* and *cis* isomers are 0.031 and 0.026 Å, respectively. All the differences are in the same direction, however.

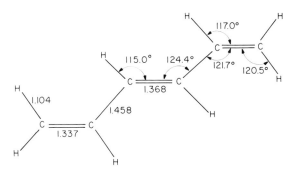

FIG. 30. 1,3,5-*trans*-Hexatriene. (Traetteberg.[142])

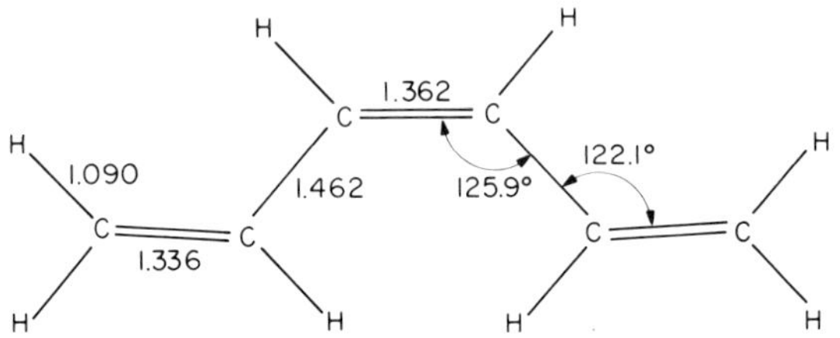

FIG. 31. 1,3,5-*cis*-Hexatriene. (Traetteberg.[146])

The values for the longer interatomic distances in *trans*- and *cis*-hexatriene, as obtained from the electron diffraction investigation, are affected by the molecular motion. The general motion, including torsional motion in the *trans* molecule, would cause shrinkage in the longer nonbonded distances which could not be readily distinguished from a twist about the central double bond. It was concluded[142] that the shortening of the longer internuclear distances could be accounted for by shrinkage effects and that the *trans* isomer was probably planar. The situation is somewhat different for the *cis* isomer owing to the fact that a twist about the central double bond increases the values of the longer nonbonded distances. Torsional oscillation would also increase the distances, whereas other types of motion would cause shrinkage. For the *cis*-hexatriene[146] it was concluded that a torsional twist of 10° about the central double bond was probably indicated causing the molecule to be nonplanar. It is not apparent that torsional oscillations about a planar molecule could be ruled out, however, and the author does not imply that the conclusion is definitive. The bond lengths in the two isomers of hexatriene are about the same. The *cis* isomer, however, is subject to a certain amount of crowding by the proximity of the two ends of the molecule. Traetteberg[146] suggests that this may account for the small increase observed for the value of the angle C=C—C from 124.4° to 125.9° in going from the *trans* to the *cis* isomer and may also influence a real twisting about the central double bond.

The molecule of 1,3-cyclohexadiene shown in Fig. 32 has three different kinds of single bonds. Its structure has been investigated in two microwave[147,148] and three electron diffraction investigations.[116,149,150] The microwave investigations established a nonplanar conformation for the molecule with a 17.5° torsional angle between two planar ethylene groups. Since the microwave investigation afforded information concerning only the moments of inertia, it was necessary to assume values for bond lengths and valence angles in order to establish the torsional angle. The detailed parameters were determined by electron diffraction. In the investigation by Dallinga and Toneman,[150] the moments of inertia from the microwave investigations were incorporated into the electron diffraction analysis. A similar type of calculation has been described by Kuchitsu *et al.*[59] The results of Dallinga

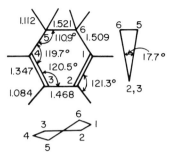

FIG. 32. 1,3-Cyclohexadiene. The values are averages from four structure determinations. (Oberhammer and Bauer[116]; Butcher[147]; Traetteberg[149]; Dallinga and Toneman.[150])

nd Toneman[150] and the microwave results of Butcher[147] are quite close, whereas the results of Oberhammer and Bauer[116] and those of Traetteberg[149] re quite close with small differences among the two sets. The average values re shown in Fig. 32. The estimated error limits[116] for C-4—C-5 and C-5—C-6 re fairly large, ≈0.02 Å, owing to a high correlation with the values for the corresponding vibrational amplitudes. It is interesting to note that the lengths f the single bonds vary from a value characteristic of a conjugated system to value close to that for ethane, 1.534 Å, the value increasing the further the ond is displaced from the influence of the double bonds.

Investigations have been made of several conjugated 1,3-cycloalkadienes.[116,147,149-152] Table IV, taken partly from Traetteberg,[152] shows some comparative results. The lengths of the single bonds listed in Table IV are seen to

TABLE IV

Parameters for Conjugated 1,3-Cycloalkadienes

Parameter	C_6H_8[a]	C_7H_{10}[b]	C_8H_{12}[c]
C-1=C-2	1.347 Å	1.35 Å	1.347 Å
C-2—C-3	1.468 Å	1.48 Å	1.475 Å
C-4—C-5	1.509 Å	1.54 Å	1.509 Å
C-5—C-6	1.521 Å	1.55 Å	1.542 Å
∠C=C—C	120.5°	129°	129.0°
β[d]	17.7°	0°	37.8°

[a] Oberhammer and Bauer,[116] Butcher,[147] Traetteberg,[149] and Dalinga and Toneman.[150]
[b] Chiang and Bauer.[151]
[c] Traetteberg.[152]
[d] Dihedral angle between the two planar ethylene groups.

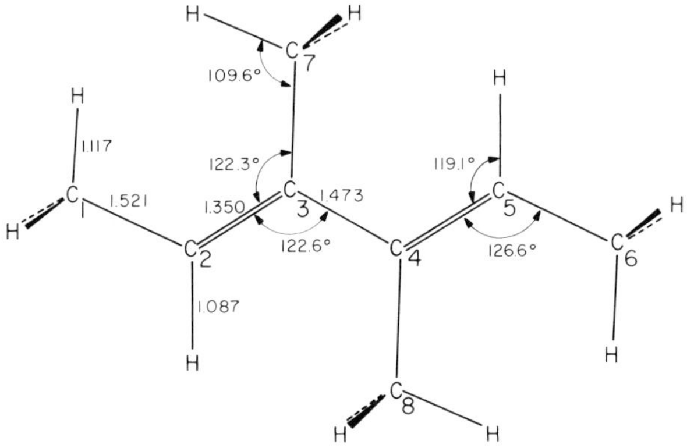

FIG. 33. *cis,cis*-3,4-Dimethyl-2,4-hexadiene. The molecule deviates from planarity (see text). (Traetteberg.[153])

increase as the bonds occur farther from the influence of the double bond. The results for the bond lengths in the various materials are fairly comparable.

The structures of three isomers of 3,4-dimethyl-2,4-hexadienes (*cis,cis-*, *cis,trans-* and *trans,trans-*) have been investigated by Traetteberg.[153] The structures of these isomers involve deviations from planarity. The *cis,cis* isomer shown in Fig. 33 was found to be roughly planar with a 27° angle of torsion around the C-3—C-4 bond, a 16° deviation from planarity at the C=C double bonds and about a 33° angle of torsion at the methyl groups. The other two isomers were found to have approximately *gauche* conformations at the C-3—C-4 bond. Table V shows that the values for the bonded distances in the three isomers have the usual values and do not differ significantly.

The structure of the molecule of acrolein has recently been determined by microwave investigation [154] and two electron diffraction investigations,[155,1] The relatively flat molecule with a possible minor torsion about the C-1—C-

TABLE V

Some Bonded Distances for Three Isomers of 3,4-Dimethyl-2,4-hexadiene

Distance	*cis,cis*	*trans,trans*	*cis,trans*
C-1—C-2	1.521	1.521	1.528
C-2=C-3	1.350	1.349	1.359
C-3—C-4	1.473	1.479	1.460

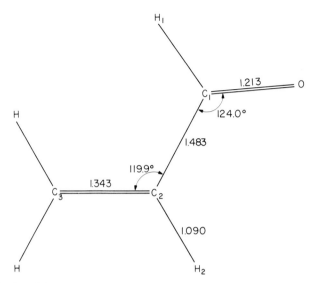

FIG. 34. Acrolein. The values are averages from two structure determinations. Traetteberg[155]; Kuchitsu et al.[156])

ond of the order of 10° is shown in Fig. 34. The values given are averages of the results from the two electron diffraction investigations. Table VI shows a comparison of the structural results. The values for the distances from the electron diffraction analyses are r_g values and those from the microwave analysis are r_s values which are best compared with r_e values. However, it is seen that the results are nevertheless quite comparable, and the distances assume normal values for a conjugated system. The uncertainties in the second column are standard deviations. The uncertainties in the third and fourth columns represent estimated limits of error.

TABLE VI

Comparison of Results for Acrolein

Parameter	E.D.[155]	E.D.[156]	M.W.[154]
C-1—H-1	1.099 Å ass.	1.128 ± 0.03 Å	1.108 ± 0.003 Å
C-2—H-2	1.079 ± 0.002	1.100 ± 0.010	1.086 ± 0.005
C=C	1.340 ± 0.001	1.345 ± 0.003	1.345 ± 0.003
C—C	1.481 ± 0.001	1.484 ± 0.004	1.470 ± 0.003
C=O	1.209 ± 0.001	1.217 ± 0.003	1.219 ± 0.005
∠CCC	119.9° ± 0.4°	119.8° ± 0.2°	119.8° ± 0.2°
∠CCO	124.7° ± 0.2°	123.3° ± 0.3°	123.3° ± 0.3°

An example of a conjugated hydrocarbon possessing a five-membered ring is the molecule dimethylfulvene illustrated in Fig. 35. Its structure was investigated by Chiang and Bauer[114] who found that the values for the single and double bond lengths were quite similar to those in 1,3-butadiene[59,139] except for the single bonds to the methyl carbons. As can be seen from Fig. 35, these lengths are 1.510 Å, a value intermediate between that for butadiene and the average value of 1.533 Å,[86,87] for n-hydrocarbons.

The molecule of trimethylenecyclopropane shows the significant effect of double bonds attached to each of the carbon atoms on the value of the single bonds in the cyclopropane ring. In an investigation of trimethylenecyclopropane by Dorko et al.,[157] it was found that the C=C double bond length was 1.343 Å, close to the value in ethylene, whereas the C—C single bond length was found to be 1.453 Å, a considerable decrease from the value of 1.510 Å for cyclopropane. The distances are, in fact, quite similar to those in butadiene. Both trimethylenecyclopropane and dimethylfulvene show little aromatic character in terms of the observed bond lengths.

The molecule 1,6-methano-1,3,5,7,9-cyclodecapentaene shown in Fig. 36, possesses ten π electrons associated with a cyclic array. It was of interest to determine whether the 10 carbon atoms of the ring are planar and the extent of the aromaticity. Accordingly, an electron diffraction investigation was carried out by Montgomery and Coetzer.[158] As seen in Fig. 36 the distances in the symmetric 10-membered ring show small deviations from the value in benzene,[71,144,145] 1.396 Å, and the bonded distance to the methylene carbon is comparable to that found for a single bond in a conjugated system. The 10 atoms in the ring were found to be roughly planar with carbon atoms 3, 4, 8, 9, and the bridgehead carbons 1 and 6 slightly above the plane (by approximately 0.25 Å) described by atoms 2, 5, 7, and 10.

The structure of thiophene illustrates the effect of a sulfur atom on a conjugated system. An electron diffraction investigation of thiophene has

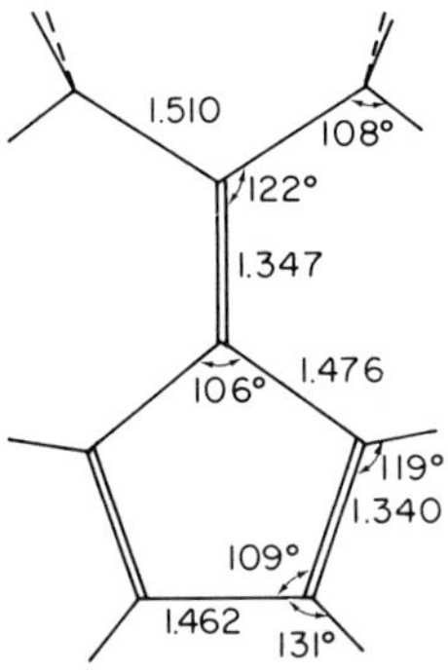

FIG. 35. Dimethylfulvene. (Chiang and Bauer.[114])

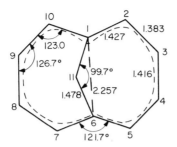

FIG. 36. 1,6-Methano-1,3,5,7,9-cyclodecapentaene. (Montgomery and Coetzer.[158])

been carried out by Bonham and Momany,[159] giving results which are in very good agreement with a microwave study by Bak et al.[160] The electron diffraction results are illustrated in Fig. 37. The S atom may be regarded as having replaced an ethylene group in benzene. In that sense it is seen to have decreased the aromaticity since the carbon–carbon bonds are not equal but lie on either side of the value in benzene.

In the sense that an NCH_3 group in N-methylpyrrole may be regarded as replacing an ethylene group in benzene, a structure investigation by Vilkov et al.[161] has shown that the NCH_3 group decreases the aromaticity even more than a sulfur atom. The larger difference between the values of the carbon–carbon bonds is apparent in comparing Figs. 37 and 38.

A molecule comparable to thiophene is 1,2,5-thiadiazole shown in Fig. 39. It was investigated by electron diffraction by Momany and Bonham[162] and by microwave spectroscopy by Dobyns and Pierce,[163] obtaining very good agreement. The electron diffraction results are given in Fig. 39. From computations of double bond character, Momany and Bonham[162] infer that the introduction of two nitrogen atoms into the thiophene ring at the 2,5 positions appears to have little effect on the overall aromaticity. They point out that

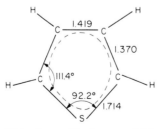

FIG. 37. Thiophene. (Bonham and Momany.[159])

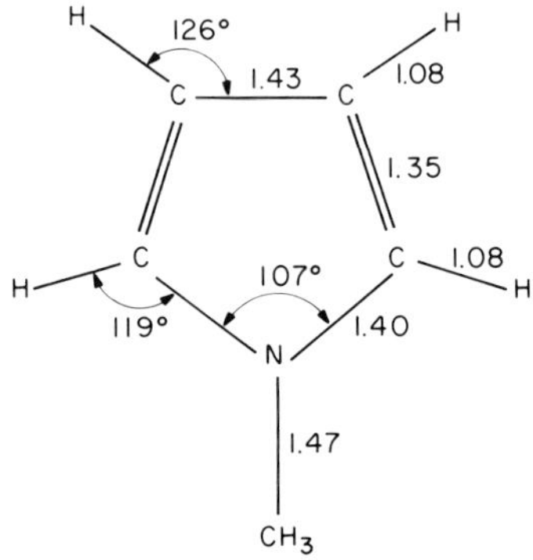

48   J. Karle

FIG. 38. N-methylpyrrole. (Vilkov et al.[161])

this conclusion is compatible with the comparable values for the length of the C—C bond opposite the S atom for thiophene and 1,2,5-thiadiazole.

A more recent investigation of thiophene, 2-chlorothiophene, and 2-bromothiophene has been carried out by Harshbarger and Bauer.[164] Their results for thiophene are essentially the same as those shown in Fig. 37 with the exception that the C—C distance opposite the S atom was found to be 1.442 instead of 1.419 Å. The halothiophenes were investigated[164] in order to determine whether the thiophene ring is as resistant to distortion on substitution as benzene. The indication from many structure investigations of substituted benzene rings is that the carbon skeleton does not deviate greatly from D_{6h} symmetry or the C—C bond distance of 1.396 Å.[165] The electron diffraction investigation did not lead to a unique model for either 2-chlorothiophene or 2-bromothiophene. Two indistinguishable possibilities are shown for each compound in Figs. 40 and 41. Harshbarger and Bauer[164] indicate that chemical and physical evidence favors model A somewhat for each compound. In both cases the thiophene rings are significantly distorted by the halogen substitution in the 2 position, suggesting that the thiophene ring is less resistant to distortion from this type of substitution than is the

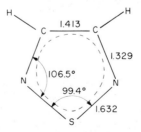

FIG. 39. 1,2,5-Thiadiazole. (Momany and Bonham.[162])

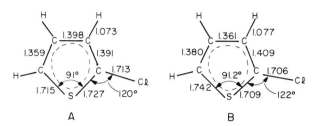

FIG. 40. 2-Chlorothiophene. Model A is favored (see text). (Harshbarger and Bauer.[164])

benzene ring. This observation would be consistent with the conclusion of Bonham and Momany,[159] based on the structure of thiophene, that the thiophene ring is less aromatic than the benzene ring. However, in a reinvestigation of these halothiophenes by Derissen et al.,[164a] it is reported that because of the very high correlation among the geometric and vibrational parameters, it is not possible to determine from the electron diffraction data whether the thiophene ring deviates from C_{2v} symmetry or whether it is distorted by the halogen substitution.

IX. TRIPLE BONDS

The lengths of ordinary C≡C triple bonds are relatively insensitive to their environment. Very small differences appear to have been found to occur as shown in Table VII.[166-168] The C—C single bond length in dimethylacetylene is the same as that in 1,3-butadiene.[59,139] The single bond length in vinylacetylene is significantly smaller. The C=C double bond length in vinylacetylene is the same as that in 1,3-butadiene. The C≡C triple bond distance in perfluorodimethylacetylene has been reported by Chang et al.[169] to be 1.201 Å. The C—C single bond distance in this molecule was found to be

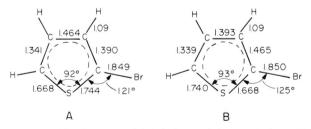

FIG. 41. 2-Bromothiophene. Model A is favored (see text). (Harshbarger and Bauer.[164])

TABLE VII

Values for Carbon–Carbon Bonds (r_g) (Measured in Angstroms) for Several Aliphatic Hydrocarbons Containing a Triple Bond

Compound	Reference	$r_g(C{\equiv}C)$	$r_g(C{=}C)$	$r_g(C{-}C)$
HC≡CH	Tanimoto et al.[166]	1.212		
H₃C—C≡C—CH₃	Tanimoto et al.[167]	1.214		1.468
H₂C=C(H)—C≡CH	Fukuyama et al.[168]	1.215	1.344	1.434
HC≡C—C≡CH	Tanimoto et al.[166]	1.218		1.384

1.475 Å and the C—F distance was found to be 1.333 Å. It is of interest that the substitution of fluorine atoms for hydrogen atoms in dimethylacetylene has so little effect on the carbon–carbon distances.

The structure of acrylonitrile investigated by Fukuyama and Kuchitsu[170] may be compared with that of vinylacetylene.[168] The lengths of the C=C double bonds in both compounds were found to be the same. The C—C single bond length in acrylonitrile was found to be slightly larger than that in vinylacetylene, 1.438 Å instead of 1.434 Å, a difference of low statistical significance. Fukuyama and Kuchitsu[170] point out that several recent investigations have indicated that a C—C single bond in a conjugated system increases its length as one of the atoms adjacent to the C—C bond is substituted by a more electronegative atom or group. Some of these effects are illustrated in Table VIII.[59,139,155,156,166,171–173]

TABLE VIII

Some Examples Showing the Increase in C—C Single Bond Length as an Atom Adjacent to the C—C Bond is Substituted by a More Electronegative Atom or Group

Compound	Reference	$r(C{-}C)$
O=CH—C≡CH	Costain and Morton[171]	1.445 (r_s)
H₂C=CH—C≡N	Costain and Stoicheff[172]	1.426 (r_s)
O=CH—CH=O	Kuchitsu et al.[59]	1.526 (r_g)
O=CH—CH=CH₂	Traetteberg,[155] Kuchitsu et al.[156]	1.483 (r_g)
H₂C=CH—CH=CH₂	Kuchitsu et al.,[59] Haugen and Traetteberg[139]	1.467 (r_g)
N≡C—C≡N	Morino et al.[173]	1.391 (r_g)
HC≡C—C≡CH	Tanimoto et al.[166]	1.384 (r_g)

The relative insensitivity of the C≡C bond length to its environment has been studied further by Zeil and his collaborators in a series of acetylene derivatives. The materials studied were $(CH_3)_3$—C—C≡C—Cl,[174,175] $(CH_3)_3$Si—C≡C—H, $(CH_3)_3$Si—C≡C—Cl, and $(CH_3)_3$Ge—C≡C—Cl.[176] In the latter work, Zeil et al. have collected together additional structural work on related compounds by microwave spectroscopy. They point out that from values of nuclear quadrupole coupling constants measured by nuclear quadrupole resonance spectroscopy, there is strong evidence of a bond type in the halogenated acetylenes which can be described by

$(CH_3)_3C$—C≡C—Cl $(CH_3)_3C$—$\overset{-}{C}$=C=$\overset{+}{Cl}$
$(CH_3)_3X$—C≡C—Cl $(CH_3)_3X$—$\overset{-}{C}$=C=$\overset{+}{Cl}$ (X = Si, Ge, Sn)
                          $(CH_3)_3\overset{-}{X}$=C=C=$\overset{+}{Cl}$
                          $(CH_3)_3\overset{-}{X}$=C=$\overset{+}{C}$—Cl

Zeil et al.[176] state that they would expect an increase in C≡C bond length and a decrease in force constant of the C≡C bond with an increase in the number of mesomeric structures. In an as yet unpublished investigation by Hüttner and Zeil the dependence of the force constant has been verified. The question arises concerning whether there is a correspondence between the C≡C bond distances and the physical behavior leading to the mesomeric structures proposed by the chemists. Zeil et al.[176] find that there are no significant differences within the limits of error of the C≡C bond length between molecules with two and four mesomeric structures. They conclude that if the force constants and nuclear quadrupole coupling constants indicate the existence of the pd-π-bond, then the geometry of molecules of the type $(CH_3)_3X$—C≡C—Y will not be affected within the error limits of electron diffraction experiments (± 0.005–0.008 Å) by this bond type.

X. CONFORMERS

Molecular conformation is an important field of investigation for which the electron diffraction technique is well suited. This is already apparent from the discussions in the previous sections. Many problems arise concerning, for example, *cis* or *trans* stabilization or the relative incidence of either form in a sample of gas. Population estimates can lead to estimates of the energy difference between the *trans* and *gauche* forms, and the detailed study of the vibrational motion can lead to estimates of the potential barriers hindering internal rotation.

An early investigation of the rotational isomers of 1,2-dichloroethane was carried out by Ainsworth and Karle.[42] They found that the predominant isomer was in the *trans* form with the remainder in the *gauche* positions at

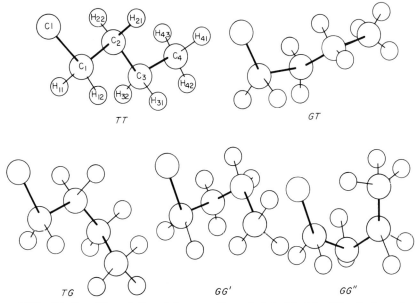

FIG. 42. Possible rotational isomers for *n*-butyl chloride. The first symbol refers to chlorine. (Ukaji and Bonham.[180])

109 ± 5° from the *trans* equilibrium position. The amount of *gauche* isomer was found to be 27 ± 5% at $T \approx 295°K$. From the change of mean moment with temperature, Mizushima *et al.*[177] obtained 25% *gauche* form at $T \approx 356°K$. Bernstein[178] reported 22% *gauche* at $T \approx 298°$ and 27% at $T \approx 356°K$.

In an investigation of symmetric tetrabromoethane by George *et al.*,[179] it has been found that the molecule is predominantly, if not totally, in the *gauche* form at $T \approx 378°K$.

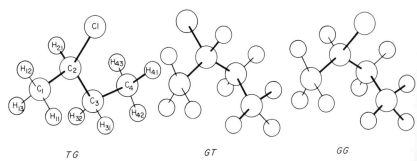

FIG. 43. Possible rotational isomers for *sec*-butyl chloride. The first symbol refers to chlorine. (Ukaji and Bonham.[180])

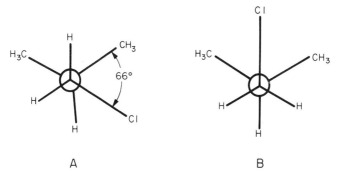

FIG. 44. Possible rotational isomers for isobutyl chloride. (Pauli et al.[181]). In A, one methyl group is *trans* and the other is *gauche* to the Cl atom. In B, both methyl groups are *gauche* to the Cl atom.

Investigations have been carried out on the conformations of *n*-butyl chloride,[180] *sec*-butyl chloride,[180] and isobutyl chloride[181] by Bonham and his collaborators. Figures 42–44 show structural models for rotational isomers for *n*-butyl chloride, *sec*-butyl chloride, and isobutyl chloride, respectively. In the labeling for the first two figures, the first symbol refers to chlorine. It was found that the *trans* form at the methyl end of *n*-butyl chloride was more stable than the *gauche* form by about 415 cal/mole, and the *gauche* form at the chlorine end was more stable than the *trans* form at the same end of the molecule by about 318 cal/mole ($T \approx 300°K$). Ukaji and Bonham[180] assign an uncertainty of about 150 cal/mole. For *sec*-butyl chloride, the energy difference between the more stable isomer with the Cl atom *gauche* and the methyl group *trans* and the other two isomers Cl-*trans*, methyl-*gauche*, and Cl-*gauche*, methyl-*gauche* was estimated to be 415 cal/mol ($T \approx 300°K$), with an uncertainty of about -200 and $+300$ cal/mole. The results for *sec*-butyl chloride imply that a methyl group is attracted more to a Cl atom than another methyl group. This is consistent with the findings of Morino and Kuchitsu[182] for *n*-propyl chloride in which the ratio of *gauche* form to *trans* form is 4:1. (Note that the *gauche* form has a twofold multiplicity as compared to a single *trans* form.) In Table IX which is taken from Ukaji and Bonham,[180] the occurrence of the various rotational isomers is given for several compounds.

Since a methyl atom is attracted more to a Cl atom than another methyl group, there exists the possibility that a strong enough attraction between a Cl atom and a methyl group exists to make conformer B of isobutyl chloride in Fig. 44 the more stable of the two. This turned out not to be the case, and it was found[181] that the ratio of conformer A to conformer B was 4:1 ($T \approx 300°K$).

54   J. Karle

TABLE IX

The Occurrence of Rotational Isomers in Several Related Substances.

% of isomers[a]	n-Butyl chloride	sec-Butyl chloride	n-Butane	n-Propyl chloride
TT	11	—	60	—
TG	11	25	40	20
GT	37	48	—	80
GG'	17	27	—	—
GG"	24	—	—	—

[a] In the labeling of the isomers, the first symbol refers to the chlorine atom and the second to the methyl group. Only the second symbol in the pair is to be used for n-butane and only the first for n-propyl chloride.

In the investigation of the conformation of n-butyl bromide by Momany et al.,[183] it was found that the distribution of conformers was 60% *trans* and 40% *gauche* at each end of the molecule. The possible conformers are the same as those for n-butyl chloride illustrated in Fig. 42. The free energy change in going from the *trans* to the *gauche* conformer was found to be 650 ± 150 cal/mole ($T \approx 300°K$). The conformer percentages were quite similar to those found for n-pentane.[184]

The investigations thus far indicate that a Cl atom stabilizes *gauche* conformations if the opposite substituent is a methyl or methylene group, and a Br atom or a methyl group stabilizes *trans* conformations when the opposite group is a methyl or methylene. The Cl atom in isobutyl chloride, however, stabilizes a *trans,gauche* form rather than a *gauche,gauche* form (Fig. 44). A microwave study of n-propyl fluoride by Hirota[185] indicates that the F atom stabilizes the *gauche* conformation.

A Cl atom stabilizes an approximately *gauche* conformation in monochlorodimethyl ether, investigated by Planje et al.[186] The dihedral angle which the Cl atom makes with respect to the O—C bond was found to be 74°, as compared to an ideal *gauche* value of 60°. It is also of interest that the bond between the chloromethyl carbon atom and the oxygen atom was found to be 1.368 Å, whereas the other C—O bond was found to be 1.414 Å. The C—O bonds in dimethyl ether were reported by Kimura and Kubo[187] to be 1.416 Å. Some possible rotational isomers for this molecule are shown in Fig. 45. Isomer I having C_{2v} symmetry is the observed structure. The hydrogen atoms undergo restricted rotation about the axes formed by the C—O bonds.

The structure of 2-methyl-1-butene was investigated by Shimanouchi et al.[188] as an example of rotational isomerism about a C—C axis adjacent to a

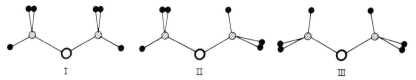

FIG. 45. Possible rotational isomers for dimethyl ether. (Kirmura and Kubo.[187])

C=C bond. An infrared investigation[189] indicated that there were two rotational isomers in the vapor state, and the electron diffraction investigation[188] was undertaken to determine their structures and relative composition. The possible rotational isomers can be seen in Fig. 46 where X represents a CH_2 group. It was found that the two rotational isomers were the *trans* and *gauche* forms with the *trans* form present to 46 ± 5%. The *trans* form is more stable than the *gauche* by 300 ± 120 cal/mole ($T \approx 300°K$). The dihedral angle for the *gauche* form was found to be 72.7 ± 5.0°, similar to the value found for monochlorodimethyl ether.[186]

An investigation comparable to that for 2-methyl-1-butene was carried out by Abe et al.[190] on methylethyl ketone in which rational isomerism may be studied about a C—C axis adjacent to a C=O bond. It had been found from an infrared investigation[191] that the molecule has two isomers in the vapor phase. Figure 46 shows possible rotational isomers, where X represents an oxygen atom. It was found that the *trans* form occurred to 95 ± 3% and was more stable than the presumed *gauche* form for the minor constituent by 2.1 ± 0.4 kcal/mole.

The conformational isomers of cyclopropyl carboxaldehyde are different than those for simple alkyl aldehydes as shown in an investigation by Bartell and Guillory.[192] Possible isomers are shown in Fig. 47. It was found that *trans* and *cis* isomers occurred, implying a twofold barrier exceeding 2.5 kcal/mole, rather than the usual threefold barrier. The isomers occurred in almost equal concentration.

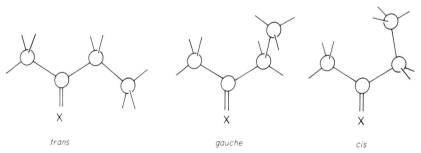

FIG. 46. Possible rotational isomers for 2-methyl-1-butene (X = CH_2) and methylethyl ketone (X = O). (Shimanouchi et al.[188]; Abe et al.[190])

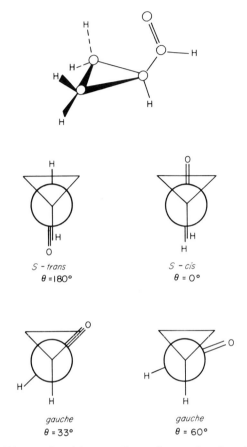

FIG. 47. Possible rotational isomers for cyclopropyl carboxaldehyde. (Bartell and Guillory.[192])

Isopropyl carboxaldehyde is an alkyl aldehyde which closely resembles cyclopropyl carboxaldehyde. It was of interest to compare the corresponding rotational barriers, and such an investigation was carried out by Guillory and Bartell.[193] Possible rotational isomers for isopropyl carboxaldehyde are shown in Fig. 48. It was found that the barrier had a threefold component exceeding 2 kcal/mole and the predominant isomer was the *gauche* form occurring to about 90%. The other form was the *trans*. Figure 48 shows that the *gauche* form involves an eclipsing of C=O and CH_3 groups.

A further study of the conformational behavior of cyclopropane derivatives was carried out by Bartell *et al.*[194] They investigated cyclopropylmethyl ketone and cyclopropanecarboxylic acid chloride. Possible rotational

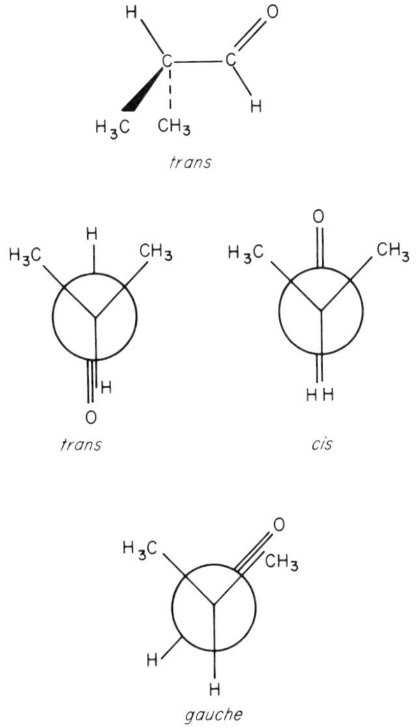

FIG. 48. Possible rotational isomers for isopropyl carboxaldehyde. (Guillory and Bartell.[193])

isomers are shown in Fig. 49. The isomers were found to exist in conformations similar to those for cyclopropyl carboxaldehyde. Again *cis* and *trans* rather than *trans* and *gauche* rotational isomers were observed. In cyclopropylmethyl ketone, the *cis* form was reported to occur to 80 ± 15%, and the remainder was in the *trans* form. Likewise in cyclopropanecarboxylic acid chloride, the *cis* form was reported to occur to 85 ± 15% and the remainder in the *trans* form.

The molecule of dimethyl diselenide contains a bonded pair of Group VI atoms which characteristically form compounds in which the attachments occur with a dihedral angle close to 90°. In a study of this molecule by D'Antonio et al.,[195] it was found that it possessed C_2 symmetry with a dihedral angle of 87.5°. This value agrees well with that of 82° estimated from infrared and Raman spectroscopy by Green and Harvey.[196] The value of the dihedral angle for dimethyldisulfide was determined by Sutler et al.[197] from microwave spectroscopy to be 85°.

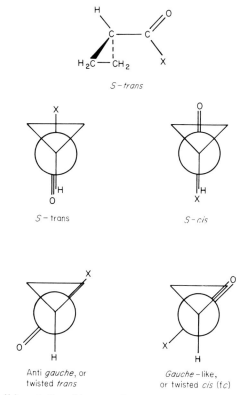

FIG. 49. Possible rotational isomers for cyclopropylmethyl ketone (X = CH_3) and cyclopropane carboxylic acid chloride (X = Cl). (Bartell et al.[194])

In its attachment to other atoms, the relative orientation of the plane of a phenyl group often comes into question. Several such molecules have been investigated by Vilkov and his collaborators.[198–201] For C_6H_5—$CH(CH_3)_2$, cumene,[198] and C_6H_5—$CH(CH_2)_2$, phenylcyclopropane,[199] it was reported that the predominant isomer is the one in which the plane of the benzene ring is perpendicular to the plane of the three attached carbon atoms. On the contrary, for C_6H_5—$N(CH_3)_2$, N-dimethylaniline,[200] the predominant isomer is the one in which the plane perpendicular to the plane of the benzene ring is also perpendicular to the plane formed by the attached nitrogen and two carbon atoms. In phenyl-$POCl_2$,[201] the P atom has an approximately tetrahedral configuration and the benzene ring is oriented so that the P=O bond is in the plane of the ring.

An interesting conformational problem arises in the case of substituted cyclobutanes resulting from the fact that the cyclobutane ring is usually non-planar. Structure investigations, have shown that the dihedral angle is often found to be between 145°–155°,[93,202] except in some derivatives for which the molecule has a center of symmetry, thereby requiring a planar ring. When the cyclobutane ring is bent, there are conformational problems quite similar to those for cyclohexane[203–205] in which atoms attached to the ring can adopt two types of positions termed "equatorial" (e) and "axial" (a). This terminology is illustrated in Fig. 50 for 1,3-dihalocyclobutane derivatives whose conformations have been investigated by Almenningen et al.[206] In Fig. 50 the e and a attachments are labeled for the 1,3 positions. The equatorial attachments are directed closer to the average plane of the cyclobutane ring than the axial ones are. There is only one possible *trans* molecule if the two substituents are identical. This was found for *trans*-1,3-dibromocyclobutane,[206] which had to be in the (ae) conformation. For *trans*-1,3-chlorobromocyclobutane a mixture was found,[206] Cl(e)Br(a) and Cl(a)Br(e). For *cis*-1,3-dibromocyclobutane and *cis*-1,3-chlorobromocyclobutane both the (aa) and (ee) conformations are possible, but only the (ee) form has been observed.[206] In all four of these compounds the dihedral angle in the cyclobutane ring was found to be close to 147°.

The potential barriers hindering rotation about single bonds have been studied with continuing interest over the years since an understanding of the characteristics of such barriers can afford a deep insight into intramolecular forces. A review article describing many ways of determining barriers hindering internal rotation has been written by Wilson.[207] An approximate method employing electron diffraction has been developed for evaluating these barriers.[208] The procedure is based on separating the effect of the torsional motion on the total amplitude of vibration for an interatomic distance from that of the frame vibrations. It is the longer distances between off-axis atoms, for example, the distances between chlorine atoms on different carbon atoms in hexachloroethane, whose vibrational amplitudes receive a contribution from the torsional motion. The remaining contribution derives from the vibration of the molecule frame. If the contribution from the torsional oscillation alone were known, information concerning the rotational barrier could be derived. The separation of the contribution of the torsional oscillation from that of the other vibrations may be accomplished from a knowledge

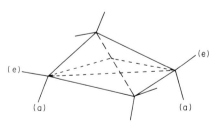

FIG. 50. 1,3-Dihalocyclobutane showing equatorial (e) and axial (a) orientations at the 1 and 3 positions. (Almenningen et al.[206])

of the functional form of the contribution of the frame vibrations as a function of the relative position of a pair of atoms on the circle of rotation described by the torsional motion. In several instances it was found that the determination of the height of the potential barrier was not sensitively dependent upon the nature of the functional dependence of the contribution from the frame vibrations on the angle of rotation.[208] In special instances it is possible to determine the functional dependence of the frame vibrations.[209] This requires at least three different distances affected by the torsional motion to occur in a molecule. Ideally, it would be desirable to compute the contributions of the frame vibrations to the total amplitude from spectroscopic data and molecular force models. A considerable amount of early work in this regard has been carried out by Morino and his collaborators[20,21,45,210] and continues to the present in connection with the vibrational aspects of electron diffraction research. A review[211] and a book[70] concerning the field of vibrational analysis in electron diffraction and spectroscopy have been written by Cyvin.

The molecule of chloropicrin, Cl_3CNO_2,[209] serves as an example of the type of information available from the study of hindered internal rotation in a molecule having three different distances affected by the torsional motion. For a threefold barrier, a barrier height of 3.4 kcal/mole ($T \approx 318°K$) was found. The agreement between the calculated and experimentally observed amplitudes of vibration for the barrier height of 3.4 kcal/mole was found to be very good. In addition it was possible to determine the different contributions (0.05–0.01 Å) from the frame vibrations to the root mean square amplitudes of each of the three different O—Cl distances. Another molecule whose potential barrier has been investigated is hexachloroethane. Using the data from three different structure investigations,[210,212,213] it was found[208] that the barrier height was 11.0, 14.7, and 12.8 kcal/mole, respectively. The temperature for the first experiment is not certain, but the latter two were carried out at 347° and 480°K, respectively. For hexafluoroethane,[214] a barrier of 4.3 kcal/mole ($T \approx 295°K$)[208] was calculated. This may be compared to two thermodynamic calculations, one by Pace and Aston[215] and the second by Mann and Plyler[216] in which the values of 4.35 and 3.92 kcal/mole, respectively, were obtained.

XI. A VARIETY OF TOPICS

A. Steric Effects

The crowding of large atoms or groups in molecules can lead to distortions in the structure from commonly found configurations. In the case of hexachlorobenzene, the molecule was found by Strand and Cox[217] to be planar with D_{6h} symmetry. In contrast, there is the indication, though not completely definitive, that the bromine atoms in hexabromobenzene investigated by Strand[218] occur alternately above and below the plane of the benzene ring giving a distorted structure with S_6 symmetry. The smallest Cl—Cl distance in hexachlorobenzene was reported to be 3.11 Å,[217] while twice the van der Waals radius for Cl is about 3.6 Å. For hexabromobenzene, the smallest Br—Br distance was reported to be 3.31 Å,[218] whereas twice the van der Waals radius is about 3.9 Å. The authors also investigated the structures of 1,2,4,5-tetrachlorobenzene[217] and *ortho*-dibromobenzene.[218] It was found that the shortest Cl—Cl and Br—Br distances increased by 0.08 and 0.07 Å, respectively, distorting the angles between the ortho carbon–halogen bonds from the value of 60°. The results were 62.8° between the *ortho* C—Cl bonds and 63.6° between the ortho C—Br bonds.

Another example of the effect of crowding occurs in the molecule $(CF_3)_2NN(CF_3)_2$, tetrakis (trifluoromethyl) hydrazine, investigated by Bartell and Higginbotham.[219] Owing to the crowding of the trifluoromethyl groups, the usual pyramidal configuration around the nitrogen atoms was found to be flattened, i.e., angle NNC was 119 ± 1.5° and angle CNC was 121 ± 1.5° with the three angles around the N atom adding up to 359° instead of 360° for a perfectly flat model. In contrast, electron diffraction investigations of $N(CH_3)_3$,[220] $N(CF_3)_3$,[221] and $N_2H_2(CH_3)_2$,[222] employing visually estimated intensity data, reported a CNC angle of 108 ± 4°, 114 ± 3°, and 110 ± 4°, respectively. An additional unusual structural feature in $(CF_3)_2NN(CF_3)_2$ is the relative shortness of the N—N bond length. It was found to be 1.40 Å, about 0.05 Å less than the N—N bond length in hydrazine.[223] The shortening of a bond length is a structural feature which is well correlated with an increase of stability. It was pointed out by Bartell and Higginbotham[219] that $(CF_3)_2NN(CF_3)_2$ has been reported by Young and Dresdner[224] to be considerably more stable to thermal rupture of the N—N bond than a number of other hydrazine derivatives.

B. Free Radicals

Electron diffraction offers the opportunity to study the structure of free radicals in the gaseous state. Some are stable, but even if they are not, a half-life of a second or less should suffice if the radicals are prepared, for example by heating, close to the jet opening where diffraction of the electron beam takes place.

The structure of a stable free radical di-*t*-butylnitroxide has been investigated by Andersen and Andersen.[225] It was proposed[226] that the stability of this molecule was due to a three electron nitrogen–oxygen bond and that additional stability might arise from steric interference with the formation of a dimer. The results of the structure investigation are shown in Fig. 51. The C—C bond length of 1.536 Å and that of 1.512 Å for the C—N bond are normal. The value of 1.280 Å for the N—O bond is consistent with a three electron bond, as pointed out by Andersen and Andersen.[225] For example, N—O bonds in nitryls and nitric acid have values ranging 1.19–1.25 Å, whereas N—OH bonds in a variety of compounds have values ranging from 1.41–1.51 Å.[227] The N—OH bond in solid HNO_3 appears to be an exception at 1.29 Å. However in the gas phase it has a value of 1.41 Å. The molecule of di-*t*-butylnitroxide was found to have a twofold axis of symmetry. The two butyl groups are relatively staggered with respect to each other, each group having a carbon atom rotated 42° about a C—N axis. A twist of 0° is defined as the position in which a butyl group has a methyl carbon *trans* to the oxygen atom.

The structure of the free radical triphenylmethyl has been investigated by Andersen.[228] Electron diffraction photographs were obtained from the vapor as hexaphenylethane was melted. As melting proceeded, the vapor pressure remained constant and was suitable for obtaining photographs. When melting was completed, the vapor pressure dropped to a negligible value. The structure of the molecule may be seen in Fig. 52. It has a propellor shape with a threefold axis and is only slightly pyramidal at the central carbon

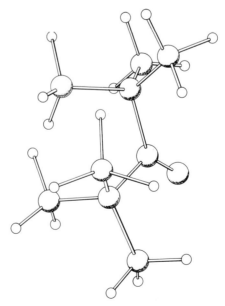

FIG. 51. Di-*t*-butylnitroxide free radical. (Andersen and Andersen.[225])

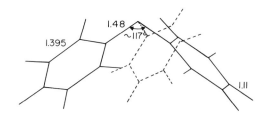

FIG. 52. Triphenylmethyl free radical. (Andersen.[228])

atom. Each phenyl group is rotated by 40°–45°. The angle of rotation is defined as zero when the ring plane normal lies in the plane defined by the threefold axis and the central bond to the ring. A comparison of this structure can be made with that of triphenylmethane, also investigated by Andersen.[229] It was observed that the molecule has C_3 symmetry; the C—C distance in the ring is 1.40 Å, whereas that from the central carbon is 1.53 Å; the C—H distance is 1.08 Å; the CCC angle from the central carbon is 112°; and the angle of rotation of the phenyl groups is 45°. The changes taking place on free radical formation are apparent from reference to Fig. 52.

Electron diffraction photographs of NF_2 were obtained by Bohn and Bauer[230] by heating N_2F_4. It was found that the N—F bond distance was 1.363 Å which may be compared to 1.371 Å in NF_3 and 1.393 Å in N_2F_4. The FNF angles in both compounds are comparable, 102.5° and 103.7° for NF_2 and N_2F_4, respectively. The N—N distance in N_2F_4 was found to be 1.53 Å, considerably longer than the value 1.40 Å reported by Bartell and Higgenbotham[219] for $N_2(CF_3)_4$.

The structure of indenyl free radical, a material which is quite unstable, has been investigated by Schäfer.[231] The free radical was obtained by thermal decomposition of diindenylcobalt in a purity of at least 80%. Average parameters were obtained for the radical which was assumed to have C_{2v} symmetry (Fig. 53). The average C—C bond length was found to be 1.415 Å with the indication that the actual distances may range 1.37–1.46 Å. An SCF—MO calculation has been carried out by Schäfer and Jensen[232] on the indenyl radical, and the theory was found to be in reasonable agreement

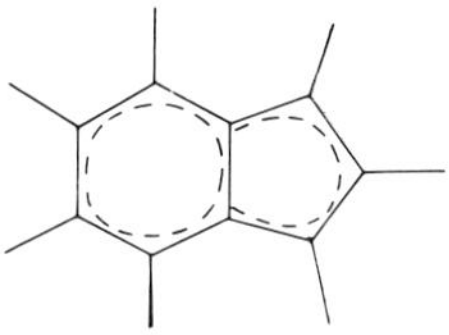

FIG. 53. Indenyl free radical. (Schäfer.[231])

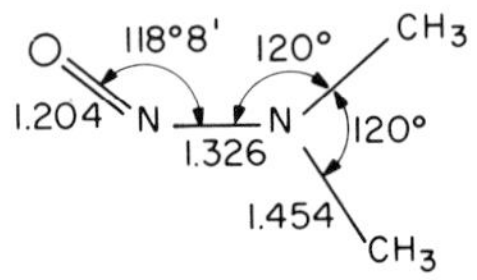

64   J. Karle

FIG. 54. Dimethylnitrosoamine. (Vilkov and Nazarenko.[234])

with the range of C—C bond distances suggested by the electron diffraction experiment.

C. Other Structural Types

There are many structural types which do not fit readily into the categories of the previous sections. A few examples will be described.

Trivalent nitrogen may vary in configuration from the pyramidal to the essentially planar form. A listing of both types of configuration for molecules in the vapor phase has been given by Vilkov and Hargittai.[233] The differences in structure are induced by the substituents attached to the nitrogen atom. A planar configuration about a nitrogen atom has been reported in dimethylnitrosoamine by Vilkov and Nazarenko.[234] In this molecule, the entire heavy atom structure was found to be planar and is shown in Fig. 54.

In the molecule of $(CH_3)_2NSO_2Cl$, both the geometrical orientation of the nitrogen bonds and the nature of the hexavalent sulfur bonds are of interest. Its structure has been investigated by Vilkov and Hargittai.[233] The configuration found for $(CH_3)_2NSO_2Cl$ is illustrated in Fig. 55. The bonding configuration about the nitrogen atom was found to be pyramidal with all angles $112 \pm 2°$. The value of 1.47 Å for the C—N bond distance was assumed. Angle NSCl was reported to be $112 \pm 3°$, and angle OSO was reported to be $120 \pm 5°$.

The structure of dimethylsulfodiimine has been reported by Oberhammer and Zeil.[235] It was found to have the configuration illustrated in Fig. 56. It

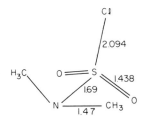

FIG. 55. The molecule of $(CH_3)_2NSO_2Cl$. (Vilkov and Hargittai.[233])

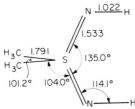

FIG. 56. Dimethylsulfodiimine. (Oberhammer and Zeil.[235])

possesses C_{2v} symmetry. The configuration is distorted tetrahedral, and the sulfur–nitrogen bonds were found to have values corresponding to a double bond distance.

Organo-substituted phosphorus fluorides are represented by the molecules of CH_3PF_4 and $(CH_3)_2PF_3$. Their structures have been investigated by Bartell and Hansen[236] and are shown in Figs. 57 and 58. The molecules are distorted

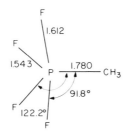

FIG. 57. The molecule of CH_3PF_4. (Bartell and Hansen.[236])

trigonal bipyramids in which the methyl groups occupy equatorial positions. The axial P—F bonds are longer than the equatorial ones. Apparently the substitution of methyl groups for fluorine atoms distorts the bipyramid so that the P—F bonds are bent away from the CH_3 groups.

Trivalent phosphorus forms pyramidal bonds, as does nitrogen. Two trivalent phosphorus compounds, *tris*-(dimethylamino)phosphene,

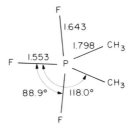

FIG. 58. The molecule of $(CH_3)_2PF_3$. (Bartell and Hansen.[236])

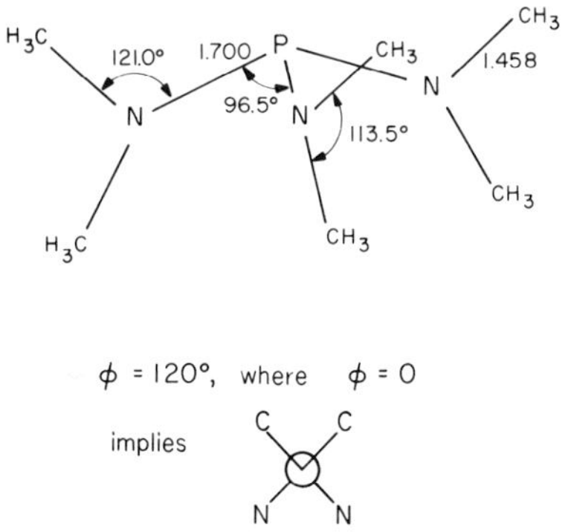

FIG. 59. Tris(dimethylamino)phosphene. (Vilkov et al.[237])

[(CH$_3$)$_2$N]$_3$P (Fig. 59) and *tris*-(ethyleneimino)phosphine, [(CH$_2$)$_2$N]$_3$P, (Fig. 60) have been investigated by Vilkov et al.[237] The molecules have C_3 symmetry. The N—P—N angle is seen to be about 97°. There are significant differences between corresponding bonded distances in the two molecules.

The molecule of tetramethyldiborane has an interesting structural feature in that, similarly to diborane, the boron atoms are joined through two hydrogen bridges. A determination of the structure of tetramethyldiborane has been carried out by Carroll and Bartell.[238] The molecule is illustrated in Fig. 61. Although the molecule of diborane has a similar configuration, the B—B distance is shorter, having a value of 1.775 Å.[239] The methyl groups in tetramethyldiborane are staggered with respect to the bonds radiating from the boron atoms, and the barrier to rotation was estimated at 1 kcal/mole.

Cyclobutadiene iron tricarbonyl has a sufficient vapor pressure to make it suitable for an electron diffraction investigation. Its structure has been investigated by Oberhammer and Brune[240] from a sample warmed to 65°C.

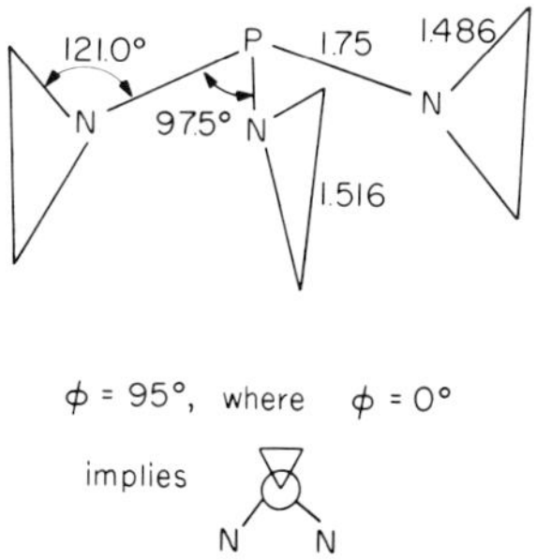

FIG. 60. Tris(ethyleneimino)phosphene. (Vilkov et al.[237])

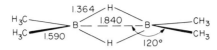

FIG. 61. Tetramethyldiborane. (Carroll and Bartell.[238])

The structure is shown in Fig. 62. The cyclobutadiene ring was found to be square, in agreement with NMR measurements. It was not possible from the electron diffraction data to determine whether free rotation or restricted rotation about an equilibrium position takes place. The NMR measurements in solution indicated free rotation of the carbonyl groups.[241] The hydrogen atoms of the cyclobutadiene rings are bent toward the iron atom by 8.6 ± 6.3°.

The structures of vinyltrichloro-, phenyltrichloro-, and phenylmonochlorosilane have been investigated by Vilkov and his collaborators.[242-244] The arrangement of the bonds around the silicon atom were found to be approximately tetrahedral with angle Cl—Si—Cl 107° in the trichlorosilanes and angles Cl—Si—C and H—Si—H 111° and 110°, respectively, in the monochlorosilane. In vinyltrichlorosilane the double bond is *cis* to a chlorine atom. The other two compounds appear to have a low barrier hindering rotation. The C—C bond distances in the benzene rings show no effect from attachment to the silicon atom.

The attachment of a Hg atom to a benzene ring appears to increase somewhat the carbon–carbon bond distance. For phenyl mercuric bromide, C_6H_5HgBr, Vilkov and Akishin[245] report a C—C distance of 1.42 ± 0.01 Å, in contrast to 1.396 Å for benzene. For diphenylmercury, Vilkov *et al.*[246] report a value of 1.415 ± 0.006 Å for the C—C distance. In both molecules the two bonds to the mercury atom form a linear configuration. The C—Hg bonded distance was found to be 2.068 ± 0.02 Å in C_6H_5HgBr, whereas it was found to increase to 2.092 ± 0.005 Å in $(C_6H_5)_2Hg$.

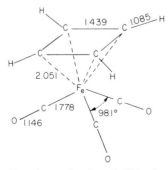

FIG. 62. Cyclobutadiene iron tricarbonyl. (Oberhammer and Brune.[240])

XII. CONCLUSIONS

It is quite apparent that much useful information concerning configuration and conformation has been made available to the organic chemist by the electron diffraction technique. The facility with which analyses can be carried out has been benefited by improvements in instrumentation and densitometry and the advances in computer technology. To a certain extent, trends in bonding as a function of chemical environment have been determined sufficiently reliably to encourage quantum chemical studies and afford a deeper understanding of molecular structure. This is particularly true when results from electron diffraction are combined with those from spectroscopy, an activity which has already had a good start and, in time, should increase in importance, to the benefit of both fields and to the users of structural information.

References

[1] J. Karle and I. L. Karle, *in* "Determination of Organic Structures by Physical Methods" (E. A. Braude and F. C. Nachod, eds.), Vol. 1, p. 427. Academic Press, New York, 1955.
[2] P. Debye, *J. Chem. Phys.* **9**, 55 (1941).
[3] J. Karle and I. L. Karle, *J. Chem. Phys.* **15**, 764 (1947).
[4] I. L. Karle and J. Karle, *J. Chem. Phys.* **17**, 1052 (1949).
[5] J. Karle and I. L. Karle, *J. Chem. Phys.* **18**, 957 (1950).
[6] C. Finbak and O. Hassel, *Arch. Math. Naturvidensk.* **45**, No. 3 (1941).
[7] H. Viervoll, *Acta Chem. Scand.* **1**, 120 (1947).
[8] O. Hassel and H. Viervoll, *Acta Chem. Scand.* **1**, 149 (1947).
[9] I. L. Karle and J. Karle, *J. Chem. Phys.* **18**, 963 (1950).
[10] C. Finbak, *Avh. Nor. Vidensk.-Akad. Oslo, Mat.-Naturvidensk. Kl.* No. 13 (1937).
[11] P. P. Debye, *Phys. Z.* **40**, 66 and 404 (1939).
[12] I. L. Karle, D. Hoober, and J. Karle, *J. Chem. Phys.* **15**, 756 (1947).
[13] S. H. Bauer, F. A. Keidel, and R. B. Harvey, "An Evaluation of Quantitative Procedures for the Estimation of Intensities of Diffracted Electrons," Off. Nav. Res. Rep. NRO52-040 (1949).
[14] K. P. Coffin, Doctoral Thesis, Cornell University (1951).
[15] J. Karle and I. L. Karle, *Amer. Mineral.* **33**, 767 (1948).
[16] J. Karle, *J. Chem. Phys.* **13**, 155 (1945).
[17] J. Karle, *J. Chem. Phys.* **15**, 202 (1947).
[18] J. Karle and H. Hauptman, *J. Chem. Phys.* **18**, 875 (1950).
[19] J. Karle, *J. Chem. Phys.* **22**, 1246 (1954).
[20] Y. Morino, K. Kuchitsu, and T. Shimanouchi, *J. Chem. Phys.* **20**, 726 (1952).
[21] Y. Morino, K. Kuchitsu, A. Takahashi, and K. Maeda, *J. Chem. Phys.* **21**, 1927 (1953).
[22] L. O. Brockway, *Rev. Mod. Phys.* **8**, 231 (1936).
[23] V. Schomaker and R. Glauber, *Nature (London)* **170**, 290 (1952).

[24] R. Glauber and V. Schomaker, *Phys. Rev.* **89**, 667 (1953).
[25] H. L. Cox, Jr. and R. A. Bonham, *J. Chem. Phys.* **47**, 2599 (1967).
[26] C. Tavard, D. Nicolas, and M. Rouault, *J. Chim. Phys.* **64**, 540 (1967).
[27] O. Bastiansen and P. N. Skancke, *Advan. Chem. Phys.* **3**, 323 (1961).
[28] L. S. Bartell, *in* "Physical Methods of Chemistry" (A. Weissberger and B. W. Rossiter, eds.), 4th ed. Wiley (Interscience), New York, 1972, p. 125.
[29] S. H. Bauer, *in* "Physical Chemistry" (D. Henderson, ed.), Vol. 4, p. 741. Academic Press, New York, 1970.
[30] O. Bastiansen, O. Hassel, and E. Risberg, *Acta Chem. Scand.* **9**, 232 (1955).
[31] L. O. Brockway and L. S. Bartell, *Rev. Sci. Instrum.* **25**, 569 (1954).
[32] H. C. Corbet, G. Dallinga, F. Oltmans, and L. H. Toneman, *Rec. Trav. Chim. Pays-Bas* **83**, 789 (1964).
[33] L. V. Vilkov, N. G. Rambidi, and V. P. Spiridonov, *J. Struct. Chem. (USSR)* **8**, 786 (1967).
[34] W. Zeil, J. Haase, and L. Wegmann, *Z. Instrumenten K.* **74**, 84 (1966).
[35] Y. Murata, K. Kuchitsu, and M. Kimura, *J. Appl. Phys. Jap.* **9**, 591 (1970).
[36] S. H. Bauer and K. Kimura, *J. Phys. Soc. Jap.* **17**, Part BII, 300 (1961).
[37] R. A. Bonham and T. Ukaji, *J. Chem. Phys.* **36**, 72 (1962).
[38] M. Kimura and T. Iijima, *J. Chem. Phys.* **43**, 2157 (1965).
[39] L. S. Bartell, L. O. Brockway, and R. H. Schwendeman, *J. Chem. Phys.* **23**, 1854 (1955).
[40] C. Degard, *Bull. Soc. Roy. Sci. Liege* **12**, 383 (1937).
[41] V. Schomaker, *Meet. Amer. Chem. Soc., Baltimore* (1939).
[42] J. Ainsworth and J. Karle, *J. Chem. Phys.* **20**, 425 (1952).
[43] L. S. Bartell, *J. Chem. Phys.* **23**, 1219 (1955).
[44] Y. Morino, K. Kuchitsu, and T. Oka, *J. Chem. Phys.* **36**, 1108 (1962).
[45] Y. Morino and E. Hirota, *J. Chem. Phys.* **23**, 737 (1955).
[46] M. Iwasaki and K. Hedberg, *J. Chem. Phys.* **36**, 2961 (1962).
[47] K. Kuchitsu and S. Konaka, *J. Chem. Phys.* **45**, 4342 (1966).
[48] C. H. Townes and A. L. Schawlow, "Microwave Spectroscopy". McGraw-Hill, New York, 1955.
[49] J. Kraitchman, *Amer. J. Phys.* **21**, 17 (1953).
[50] C. C. Costain, *J. Chem. Phys.* **29**, 864 (1958).
[51] T. Oka, *J. Phys. Soc. Jap.* **15**, 2274 (1960).
[52] T. Oka and Y. Morino, *J. Mol. Spectrosc.* **6**, 472 (1961).
[53] V. W. Laurie and D. R. Herschbach, *Bull. Amer. Phys. Soc.* [2] **5**, 500 (1960).
[54] D. R. Herschbach and V. W. Laurie, *J. Chem. Phys.* **37**, 1668 (1962).
[55] L. S. Bartell, K. Kuchitsu, and R. J. deNeui, *J. Chem. Phys.* **35**, 1211 (1961).
[56] Y. Morino and T. Iijima, *Bull. Chem. Soc. Jap.* **35**, 1661 (1962).
[57] K. Kuchitsu, *J. Chem. Phys.* **44**, 906 (1966).
[58] K. Kuchitsu, *J. Chem. Phys.* **49**, 4456 (1968).
[59] K. Kuchitsu, T. Fukuyama, and Y. Morino, *J. Mol. Struct.* **1**, 463 (1967–1968).
[60] C. Kato, S. Konaka, T. Iijima, and M. Kimura, *Bull. Chem. Soc. Jap.* **42**, 2148 (1969).
[61] T. Iijima and M. Kimura, *Bull. Chem. Soc. Jap.* **42**, 2159 (1969).
[62] L. S. Bartell and H. K. Higginbotham, *J. Chem. Phys.* **42**, 851 (1965).
[63] Y. Morino, J. Nakamura, and P. W. Moore, *J. Chem. Phys.* **36**, 1050 (1962).
[64] A. Almenningen, O. Bastiansen, and T. Munthe-Kaas, *Acta Chem. Scand.* **10**, 261 (1956).
[65] A. Almenningen, O. Bastiansen, and M. Traetteberg, *Acta Chem. Scand.* **13**, 1699 (1959).

[66] O. Bastiansen and M. Traetteberg, *Acta Crystallogr.* **13**, 1108 (1960).
[67] Y. Morino, *Acta Crystallogr.* **13**, 1107 (1960).
[68] Y. Morino, S. J. Cyvin, K. Kuchitsu, and T. Iijima, *J. Chem. Phys.* **36**, 1109 (1962).
[69] W. Freyland, J. Haase, and W. Zeil, *Z. Naturforsch. A* **21**, 1945 (1966).
[70] S. J. Cyvin, "Molecular Vibrations and Mean Square Amplitudes." Elsevier, Amsterdam, 1968.
[71] I. L. Karle, *J. Chem. Phys.* **20**, 65 (1952).
[72] W. C. Hamilton, Ph.D. Thesis, California Institute of Technology (1954).
[73] R. A. Bonham and L. S. Bartell, *J. Chem. Phys.* **31**, 702 (1959).
[74] K. Hedberg and M. Iwasaki, *Acta Crystallogr.* **17**, 529 (1964).
[75] Y. Morino, K. Kuchitsu, and Y. Murata, *Acta Crystallogr.* **18**, 549 (1965).
[76] Y. Murata and Y. Morino, *Acta Crystallogr.* **20**, 605 (1966).
[77] R. Hilderbrandt and S. H. Bauer, *J. Mol. Struct.* **3**, 825 (1969).
[78] H. M. Seip, T. G. Strand, and R. Stolevik, *Chem. Phys. Lett.* **3**, 617 (1969).
[79] G. Herzberg and B. P. Stoicheff, *Nature (London)* **175**, 79 (1955).
[80] C. C. Costain and B. P. Stoicheff, *J. Chem. Phys.* **30**, 777 (1959).
[81] B. P. Stoicheff, *Tetrahedron* **17**, 135 (1962).
[82] G. Dallinga and P. Ros, *Rec. Trav. Chim. Pays-Bas* **87**, 906 (1968).
[83] H. Basch, A. Viste, and H. B. Gray, *J. Chem. Phys.* **44**, 10 (1966).
[84] J. H. Schachtschneider, R. Prins, and P. Ros, *Inorg. Chim. Acta* **1**, 462 (1967).
[85] K. Lonsdale, *Phil. Trans. Roy. Soc. London, Ser. A* **240**, 219 (1946).
[86] R. A. Bonham and L. S. Bartell, *J. Amer. Chem. Soc.* **81**, 3491 (1959).
[87] R. A. Bonham, L. S. Bartell, and D. A. Kohl, *J. Amer. Chem. Soc.* **81**, 4765 (1959).
[88] W. Zeil, J. Haase, and M. Dakkouri, *Z. Naturforsch. A* **22**, 1644 (1967).
[89] L. Pauling and L. O. Brockway, *J. Amer. Chem. Soc.* **59**, 1223 (1937).
[90] O. Bastiansen, F. N. Fritsch, and K. Hedberg, *Acta Crystallogr.* **17**, 538 (1964).
[91] G. Dallinga, R. K. van der Draai, and L. H. Toneman, *Rec. Trav. Chim. Pays-Bas* **87**, 897 (1968).
[92] A. Almenningen and O. Bastiansen, *Acta Chem. Scand.* **15**, 711 (1961).
[93] J. F. Chiang and S. H. Bauer, "The Molecular Structure of Bicyclo [1,1,1] Pentane," *J. Amer. Chem. Soc.* **92**, 1614 (1970).
[94] S. Meiboom and L. C. Snyder, *J. Amer. Chem. Soc.* **89**, 1038 (1967).
[95] D. A. Dows and N. Rich, *J. Chem. Phys.* **47**, 333 (1967).
[96] C. H. Chang, R. F. Porter, and S. H. Bauer, *J. Mol. Struct.* **7**, 89 (1971).
[97] L. V. Vilkov, V. S. Mastryukov, Yu. V. Baurova, V. M. Vdovin, and P. L. Grinberg, *Dokl. Akad. Nauk SSSR* **117**, 1084 (1967).
[98] L. V. Vilkov, M. M. Kysakov, N. S. Nametkin, and V. D. Oppenheim, *Dokl. Akad. Nauk SSSR* **183**, 94 (1968).
[99] G. Dallinga and L. H. Toneman, *Rec. Trav. Chim. Pays-Bas* **86**, 171 (1967).
[100] A. Yokozeki and K. Kuchitsu, *Bull. Chem. Soc. Jap.* **44**, 2356 (1971).
[101] G. Dallinga and L. H. Toneman, *Rec. Trav. Chim. Pays-Bas* **87**, 795 (1968).
[102] J. F. Chiang, C. F. Wilcox, Jr., and S. H. Bauer, *J. Amer. Chem. Soc.* **90**, 3149 (1968).
[103] G. Dallinga and L. H. Toneman, *Rec. Trav. Chim. Pays-Bas* **88**, 185 (1969).
[104] J. F. Chiang, C. F. Wilcox, Jr., and S. H. Bauer, *Tetrahedron* **25**, 369 (1969).
[105] J. P. McHugh and V. Schomaker, "A Report for the Year 1954–55 for the Division of Chemistry and Chemical Engineering," California Institute of Technology (1955).

[106] D. R. Lide, Jr. and D. E. Mann, *J. Chem. Phys.* **27**, 868 (1957).
[107] L. S. Bartell, E. A. Roth, C. D. Hollowell, K. Kuchitsu, and J. E. Young, *J. Chem. Phys.* **42**, 2683 (1965).
[108] P. H. Kasai, R. J. Myers, D. F. Eggers, Jr., and K. B. Wiberg, *J. Chem. Phys.* **30**, 512 (1959).
[109] J. D. Dunitz, H. G. Feldman, and V. Schomaker, *J. Chem. Phys.* **20**, 1708 (1952).
[110] J. F. Chiang, *J. Chin. Chem. Soc.* **17**, 65 (1970).
[111] H. J. Mair and S. H. Bauer, *J. Phys. Chem.* **75**, 1681 (1971).
[112] B. Bak, J. J. Led, L. Nygaard, J. Rastrup-Andersen, and G. O. Sorensen, *J. Mol. Struct.* **3**, 369 (1969).
[113] O. Bastiansen and J. L. Derissen, *Acta Chem. Scand.* **20**, 1089 (1966).
[114] J. F. Chiang and S. H. Bauer, *J. Amer. Chem. Soc.* **91**, 1898 (1969).
[115] G. Dallinga and L. H. Toneman, *J. Mol. Struct.* **1**, 117 (1967–1968).
[116] H. Oberhammer and S. H. Bauer, *J. Amer. Chem. Soc.* **91**, 10 (1969).
[117] E. B. Wilson, Jr., *Tetrahedron* **17**, 191 (1962).
[118] V. Schomaker and W. Hamilton, cited by W. G. Woods, R. A. Carboni, and J. D. Roberts, *J. Amer. Chem. Soc.* **78**, 5653 (1956).
[119] T. W. Muecke and M. I. Davis, *Trans. Amer. Crystallogr. Ass.* **2**, 173 (1966).
[120] G. Dallinga and M. H. Toneman, *Rec. Trav. Chim. Pays-Bas* **87**, 805 (1968).
[121] M. C. Baenziger, G. F. Richards, and J. R. Doyle, *Acta Crystallogr.* **18**, 924 (1965).
[122] A. C. MacDonald and J. Trotter, *Acta Crystallogr.* **19**, 456 (1965).
[123] L. S. Bartell and R. A. Bonham, *J. Chem. Phys.* **32**, 824 (1960).
[124] R. L. Hilderbrandt, A. L. Andreassen, and S. H. Bauer, *J. Phys. Chem.* **74**, 1586 (1970).
[125] D. A. Swick and I. L. Karle, *J. Chem. Phys.* **23**, 1499 (1955).
[126] I. L. Karle and J. Karle, *J. Chem. Phys.* **22**, 43 (1954).
[127] R. A. Bonham and L. S. Su, *Symp. Gas Phase Mol. Struct., 2nd, 1968* Abstract M-1 (1968).
[128] A. Almenningen, O. Bastiansen, and T. Motzfeldt, *Acta Chem. Scand.* **23**, 2848 (1969).
[129] J. L. Derissen, *J. Mol. Struct.* **7**, 67 (1971).
[130] J. L. Derissen, *J. Mol. Struct.* **7**, 81 (1971).
[131] S. H. Bauer and C. H. Chang, *Abstracts 161st Meeting American Chemical Society*, Los Angeles, Calif., March 28, 1971, Paper Phys. 14., and private communication.
[131a] J. Karle, A. H. Lowrey, C. F. George, and P. D'Antonio (in progress).
[131b] E. J. Jacob and D. R. Lide, Jr., *Abstracts 26th Symp. Mol. Struct. Spectry.*, Columbus, Ohio, June 14, 1971, Paper B2, and private communication.
[131c] A. B. Harvey and L. Y. Nelson, *J. Chem. Phys.*, **55**, 4145 (1971).
[132] F. A. M. Buck and R. L. Livingston, *J. Amer. Chem. Soc.* **70**, 2817 (1948).
[133] V. W. Laurie and D. T. Pence, *J. Chem. Phys.* **38**, 2693 (1963).
[134] A. Almenningen, O. Bastiansen, and M. Traetteberg, *Acta Chem. Scand.* **15**, 1557 (1961).
[135] H. J. Vledder, F. C. Mijlhoff, J. C. Leyte, and C. Romers, *J. Mol. Struct.* **7**, 421, (1971).
[136] A.H. Lowrey, C. F. George, P. D'Antonio, and J. Karle, *J. Amer. Chem. Soc.* **93**, 6399 (1971).
[137] L. W. Reeves, *Can. J. Chem.* **35**, 1351 (1957).
[138] J. P. Schaefer and P. J. Wheatley, *J. Chem. Soc.*, A p. 528 (1966).

[139] W. Haugen and M. Traetteberg, *Selec. Top. Struct. Chem.*, p. 113 (1967).
[140] L. V. Vilkov and N. I. Sadova, *J. Struct. Chem.* (*USSR*) **8**, 398 (1967).
[141] C. F. Aten, L. Hedberg, and K. Hedberg, *J. Amer. Chem. Soc.* **90**, 2463 (1968).
[142] M. Traetteberg, *Acta Chem. Scand.* **22**, 628 (1968).
[143] B. Roos and P. N. Skancke, *Acta Chem. Scand.* **21**, 233 (1967).
[144] A. Almenningen, O. Bastiansen, and L. Fernholt, *Kgl. Nor. Vidensk. Selsk. Skr.* [N.S.] No. 3 (1958).
[145] K. Kimura and M. Kubo, *J. Chem. Phys.* **32**, 1776 (1960).
[146] M. Traetteberg, *Acta Chem. Scand.* **22**, 2294 (1968).
[147] S. S. Butcher, *J. Chem. Phys.* **42**, 1830 (1965).
[148] G. Luss and M. D. Harmony, *J. Chem. Phys.* **43**, 3768 (1965).
[149] M. Traetteberg, *Acta Chem. Scand.* **22**, 2305 (1968).
[150] G. Dallinga and L. H. Toneman, *J. Mol. Struct.* **1**, 11 (1967–1968).
[151] J. F. Chiang and S. H. Bauer, *J. Amer. Chem. Soc.* **88**, 420 (1966).
[152] M. Traetteberg, *Acta Chem. Scand.* **24**, 2285 (1970).
[153] M. Traetteberg, *Acta Chem. Scand.* **24**, 2295 (1970).
[154] E. A. Cherniak and C. C. Costain, *J. Chem. Phys.* **45**, 104 (1966).
[155] M. Traetteberg, *Acta Chem. Scand.* **24**, 373 (1970).
[156] K. Kuchitsu, T. Fukuyama, and Y. Morino, *J. Mol. Struct.* **4**, 41 (1949).
[157] E. A. Dorko, J. L. Hencher, and S. H. Bauer, *Tetrahedron* **24**, 2425 (1968).
[158] L. K. Montgomery and J. Coetzer, *Symp. Gas Phase Mol. Struct., 2nd, 1968* Abstract M–13 (1968); private communication (1970).
[159] R. A. Bonham and F. A. Momany, *J. Phys. Chem.* **67**, 2474 (1963).
[160] B. Bak, D. Christensen, L. Hansen-Nygaard, and J. Rastrup-Anderson, *J. Mol. Spectrosc.* **7**, 58 (1961).
[161] L. V. Vilkov, P. A. Akishin, and B. M. Presnjakova, *J. Struct. Chem.* (*USSR*) **3**, 6 (1962).
[162] F. A. Momany and R. A. Bonham, *J. Amer. Chem. Soc.* **86**, 162 (1964).
[163] Sr. V. Dobyns and L. Pierce, *J. Amer. Chem. Soc.* **85**, 3553 (1963).
[164] W. R. Harshbarger and S. H. Bauer, *Acta Crystallogr., Sect. B* **26**, 1010 (1970).
[164a] J. L. Derissen, J. W. M. Kocken, and R. H. van Weelden, *Acta Crystallogr., Sect. B* **27**, 1692 (1971).
[165] S. H. Bauer, K. Katada, and K. Kimura, in "Structural Chemistry and Molecular Biology" (A. Rich and N. Davidson, eds.), p. 653, Freeman, San Francisco, California, 1968.
[166] M. Tanimoto, K. Kuchitsu, and Y. Morino, *Bull. Chem. Soc. Jap.* **44**, 386 (1971); M. Tanimoto, University of Tokyo (1969).
[167] M. Tanimoto, K. Kuchitsu, and Y. Morino, *Bull. Chem. Soc. Jap.* **42**, 2519 (1969).
[168] T. Fukuyama, K. Kuchitsu, and Y. Morino, *Bull. Chem. Soc. Jap.* **42**, 379 (1969).
[169] C. H. Chang, A. L. Andreassen, and S. H. Bauer, *J. Org. Chem.* **36**, 920 (1971).
[170] T. Fukuyama and K. Kuchitsu, *J. Mol. Struct.* **5**, 131 (1970).
[171] C. C. Costain and J. R. Morton, *J. Chem. Phys.* **31**, 389 (1959).
[172] C. C. Costain and B. P. Stoicheff, *J. Chem. Phys.* **30**, 777 (1959).
[173] Y. Morino, K. Kuchitsu, Y. Hori, and M. Tanimoto, *Bull. Chem. Soc. Jap.* **41**, 2349 (1968).
[174] J. Haase, W. Steingross, and W. Zeil, *Z. Naturforsch. A* **22**, 195 (1967).

1. Electron Diffraction   73

[175] H.-K. Bodenseh, R. Gegenheimer, J. Mennicke, and W. Zeil, *Z. Naturforsch. A* **22**, 523 (1967).
[176] W. Zeil, J. Haase, and M. Dakkouri, *Discuss. Faraday Soc.*, 149 (1969).
[177] S. Mizushima, Y. Morino, I. Watanabe, T. Simanouti, and S. Yamaguchi, *J. Chem. Phys.* **17**, 591 (1949).
[178] H. J. Bernstein, *J. Chem. Phys.* **17**, 258 (1949). (Structural model corrected for the result reported by Ainsworth and Karle.[42])
[179] C. F. George, A. H. Lowrey, P. D'Antonio, and J. Karle, in progress.
[180] T. Ukaji and R. A. Bonham, *J. Amer. Chem. Soc.* **84**, 363 (1962).
[181] G. H. Pauli, F. A. Momany, and R. A. Bonham, *J. Amer. Chem. Soc.* **86**, 1286 (1964).
[182] Y. Morino and K. Kuchitsu, *J. Chem. Phys.* **28**, 175 (1958).
[183] F. A. Momany, R. A. Bonham, and W. H. McCoy, *J. Amer. Chem. Soc.* **85**, 3077 (1963).
[184] R. A. Bonham, L. S. Bartell, and D. A. Kohl, *J. Amer. Chem. Soc.* **81**, 4765 (1959).
[185] E. Hirota, *J. Chem. Phys.* **37**, 283 (1962).
[186] M. C. Planje, L. H. Toneman, and G. Dallinga, *Rec. Trav. Chim. Pays-Bas* **84**, 232 (1965).
[187] K. Kimura and M. Kubo, *J. Chem. Phys.* **30**, 151 (1959).
[188] T. Shimanouchi, Y. Abe, and K. Kuchitsu, *J. Mol. Struct.* **2**, 82 (1968).
[189] T. Shimanouchi, Y. Abe, and M. Mikami, *Spectrochim. Acta, Part A* **24**, 1037 (1968).
[190] M. M. Abe, K. Kuchitsu, and T. Shimanouchi, *J. Mol. Struct.* **4**, 245 (1969).
[191] T. Shimanouchi, Y. Abe, and M. Mikami, *Spectrochim. Acta, Part A* **24**, 1037 (1968).
[192] L. S. Bartell and J. P. Guillory, *J. Chem. Phys.* **43**, 647 (1965).
[193] J. P. Guillory and L. S. Bartell, *J. Chem. Phys.* **43**, 654 (1965).
[194] L. S. Bartell, J. P. Guillory, and A. T. Parks, *J. Phys. Chem.* **69**, 3043 (1965).
[195] P. D'Antonio, C. F. George, A. H. Lowrey, and J. Karle, *J. Chem. Phys.* **55**, 1071 (1971).
[196] W. H. Green and A. B. Harvey, *J. Chem. Phys.* **49**, 3586 (1968).
[197] D. Sutler, H. Dreizler, and H. D. Rudolph, *Z. Naturforsch. A* **20**, 1676 (1965).
[198] L. V. Vilkov, N. I. Sadova, and S. S. Mochalov, *Dokl. Akad. Nauk SSSR* **179**, 896 (1968).
[199] L. V. Vilkov and N. I. Sadova, *Dokl. Akad. Nauk SSSR* **162**, 565 (1965).
[200] L. V. Vilkov and T. P. Timasheva, *Dokl. Akad. Nauk SSSR* **161**, 351 (1965).
[201] L. V. Vilkov, N. I. Sadova, and I. Yu. Zilberg, *J. Struct. Chem. (USSR)* **8**, 528 (1967).
[202] E. Adman and T. N. Margulis, *J. Amer. Chem. Soc.* **90**, 4517 (1968); B. Greenberg and B. Post, *Acta Crystallogr., Sect. B* **24**, 918 (1968).
[203] O. Hassel, *Tidsskr. Kjemi, Bergv. Met.* **3**, 32 (1943).
[204] O. Hassel, *Research (London)* **3**, 504 (1950).
[205] O. Hassel, *Quart. Rev., Chem. Soc.* **7**, 221 (1953).
[206] A. Almenningen, O. Bastiansen, and L. Walloe, *Selec. Top. Struct. Chem.*, p. 91 (1967).
[207] E. B. Wilson, Jr., *Advan. Chem. Phys.* **2**, 367 (1959).
[208] J. Karle, *J. Chem. Phys.* **45**, 4149 (1966).
[209] R. E. Knudsen, C. F. George, and J. Karle, *J. Chem. Phys.* **44**, 2334 (1966).
[210] Y. Morino and E. Hirota, *J. Chem. Phys.* **28**, 185 (1958).

[211] S. J. Cyvin, *Acta Polytech. Scand. Phys. Nucl. Ser.* No. 8 (1960).
[212] A. Almenningen, B. Andersen, and M. Traetteberg, *Acta Chem. Scand.* **18**, 603 (1964).
[213] D. A. Swick, I. L. Karle, and J. Karle, *J. Chem. Phys.* **22**, 1242 (1954).
[214] D. A. Swick and I. L. Karle, *J. Chem. Phys.* **23**, 1499 (1955).
[215] E. L. Pace and J. G. Aston, *J. Amer. Chem. Soc.* **70**, 566 (1948).
[216] D. E. Mann and E. K. Plyler, *J. Chem. Phys.* **21**, 1116 (1953).
[217] T. G. Strand and H. L. Cox, Jr., *J. Chem. Phys.* **44**, 2426 (1966).
[218] T. G. Strand, *J. Chem. Phys.* **44**, 1611 (1966).
[219] L. S. Bartell and H. K. Higginbotham, *Inorg. Chem.* **4**, 1346 (1965).
[220] L. O. Brockway and H. O. Jenkins, *J. Amer. Chem. Soc.* **58**, 2036 (1936).
[221] R. L. Livingston and G. Vaughan, *J. Amer. Chem. Soc.* **78**, 4866 (1956).
[222] W. Beamer, *J. Amer. Chem. Soc.* **70**, 2979 (1948).
[223] Y. Morino, T. Iijima, and Y. Murata, *Bull. Chem. Soc. Jap.* **33**, 46 (1959).
[224] J. A. Young and R. D. Dresdner, *J. Org. Chem.* **28**, 833 (1963).
[225] B. Andersen and P. Andersen, *Trans. Amer. Crystallogr. Ass.* **2**, 193 (1966).
[226] A. K. Hoffmann and A. T. Henderson, *J. Amer. Chem. Soc.* **83**, 4671 (1961).
[227] "International Tables for X-Ray Crystallography" (C. H. MacGillavry and G. D. Rieck, eds.), Vol. III, p. 270. Kynoch Press, Birmingham, England, 1968.
[228] P. Andersen, *Acta Chem. Scand.* **19**, 629 (1965).
[229] P. Andersen, *Acta Chem. Scand.* **19**, 622 (1965).
[230] R. K. Bohn and S. H. Bauer, *Inorg. Chem.* **6**, 304 (1967).
[231] L. Schäfer, *J. Amer. Chem. Soc.* **90**, 3919 (1968).
[232] L. Schäfer and H. H. Jensen, *Acta Chem. Scand.* **24**, 1125 (1970).
[233] L. V. Vilkov and I. Hargittai, *Acta Chim. Acad. Sci. Hung.* **52**, 423 (1967).
[234] L. V. Vilkov and I. I. Nazarenko, *J. Struct. Chem. (USSR)* **9**, 887 (1968).
[235] H. Oberhammer and W. Zeil, *Z. Naturforsch. A* **24**, 1612 (1969).
[236] L. S. Bartell and K. W. Hansen, *Inorg. Chem.* **4**, 1777 (1965).
[237] L. V. Vilkov, L. C. Haikin, and V. V. Evdokimov, *J. Struct. Chem. (USSR)* **10**, 1101 (1969).
[238] B. L. Carroll and L. S. Bartell, *Inorg. Chem.* **7**, 219 (1968).
[239] L. S. Bartell and B. L. Carroll, *J. Chem. Phys.* **42**, 1135 (1965).
[240] H. Oberhammer and H. A. Brune, *Z. Naturforsch. A* **24**, 607 (1969).
[241] H. A. Brune, H. P. Wolff, and H. Hüther, *Chem. Ber.* **101**, 1485 (1968); *Z. Naturforsch. B* **23**, 1184 (1968).
[242] L. V. Vilkov, V. C. Mastryukov, and P. A. Akishin, *J. Struct. Chem. (USSR)* **5**, 183 (1964).
[243] L. V. Vilkov, V. C. Mastryukov, and P. A. Akishin, *J. Struct. Chem. (USSR)* **5**, 906 (1964).
[244] L. V. Vilkov and V. C. Mastryukov, *Dokl. Akad. Nauk SSSR* **162**, 1306 (1965).
[245] L. V. Vilkov and M. G. Akishin, *J. Struct. Chem. (USSR)* **9**, 690 (1968).
[246] L. V. Vilkov, M. G. Akishin, and G. I. Mamaeva, *J. Struct. Chem. (USSR)* **9**, 372 (1968).

Spin Saturation Labeling 2

J. W. FALLER

I. Introduction	75
II. Theory	76
III. Applications and Examples	81
A. Two Sites	81
B. Three Sites	85
C. Four Sites	89
IV. Measurement Techniques	90
A. Degree of Saturation	90
B. Relaxation Times	91
V. Sources of Error	93
VI. Practical Aspects	95
References	96

I. INTRODUCTION

An extremely useful technique for the study of reaction mechanisms has developed over the past few years; it is the purpose of this chapter to point out its utility, to provide a source of references, and to present an adequate background to understand its practical applications. The relevant theory was initially developed by McConnell and Thompson[1] and subsequently extended by Hoffman and Forsen.[2] A sufficient number of studies on simple systems have been completed to verify that the theoretical models are adequate. Proton exchange in ammonia[3] and carbon-13 exchange[4] in the carbonic acid system ($HCO_3^- \rightleftharpoons H_2CO_3 \rightleftharpoons CO_2$) provided the original tests of the method. Initially, chemical systems were chosen specifically to illustrate the potential of the approach and its application to the study of chemical exchange rates; however, the great power of this technique, which will be denoted here as *spin saturation labeling* or *spin saturation transfer*,* in the elucidation of reaction mechanisms has not been exploited until recently.

---

* The following terms are synonymous and have been used previously: molecular transfer of nonequilibrium nuclear spin magnetization; indirect spin saturation; and spin depolarization transfer.

The spin saturation method represents a unique way of obtaining kinetic data on systems that are initially at equilibrium. In many ways it is similar to the chemical relaxation techniques developed by Eigen and others,[5-9] in which the equilibrium is suddenly disturbed and the concentrations of the components are followed as a function of time as the system adjusts to the new set of equilibrium conditions. At chemical equilibrium the relative population of spin states by magnetic nuclei may be described by a Boltzmann factor. In this NMR experiment, the equilibrium conditions are altered and the time dependences of the populations of the various spin states are followed by intensity changes in the NMR spectrum. Interpretation of these intensity changes often allows one to elucidate some of the details of rearrangement mechanisms by effectively labeling migrating nuclei or groups.

When a reaction occurs in the appropriate rate range (first-order rate constants between 0.005 and 5 second^{-1}), it is feasible to measure rates with the method. Although it tends to be useful over a narrower range of rates than the analysis of NMR line shapes, it may be used for slower reactions, and thus complements the line shape technique as a kinetic tool. Furthermore, it is more sensitive to small differences in rates, which makes it preferable in the determination of mechanisms. In many cases where one might use deuterium labeling as a method of following a rearrangement or an exchange process, the spin saturation labeling technique is an adequate, if not superior, substitute. For instance, with the aid of an NMR instrument with a spin decoupler, any proton having a chemical shift different from the others in a compound can be specifically "labeled" with very little effort. Furthermore, because of the time intervals involved, one might not be able to detect an intermediate stage in a deuterium scrambling experiment, whereas it could be identified easily in the spin saturation experiment.

II. THEORY

For the sake of clarity, a simple system in which protons might occupy only two different chemical environments will be considered initially. These sites, A and B, would presumably have different electronic surroundings; hence different magnetic fields should result at the nuclei and give rise to a characteristic frequency for each site. Ordinarily, a thermal equilibrium obtains, and a Boltzmann factor determines the populations of the spin states. The difference in population of spin levels may be considered as a difference in the number of nuclei aligned with the external magnetic field and those aligned against it. In terms of macroscopic properties, this gives rise to a net magnetic moment in the solution. The z magnetization, i.e., the component in the external field direction (z), is proportional to the difference in spin populations. The contribution to the total magnetization in the z direction from nuclei in site A at time t is given by $M_z^A(t)$. Ordinarily, a weak radiofrequency magnetic field at the resonance frequency does not appreciably alter the population of the spin states at a particular site (the states are denoted as $|\alpha\rangle_A$ and $|\beta\rangle_A$). The intensity of an observed resonance signal depends upon the strength of the rf field and the z magnetization of the nuclei at that site. However, if a strong rf field is applied at the resonance frequency of nuclei in site A, both spin states will become essentially equally populated. Only a small fraction of the original intensity would be observed if the resonance position were swept through with a weak rf field, and the resonance would be said to be saturated. Nevertheless, over a short period of time thermal equilibrium would be restored, and the intensity would gradually recover. Hence saturation of the resonance for nuclei in site A corresponds to producing a population difference of zero and M_z^A of zero. In addition to the chemical shift, therefore, there is associated with each site a characteristic lifetime for each spin state and a time constant or half-life associated with the recovery of populations expected at thermal equilibrium. The time constant associated with the restoration of the resonance intensity is known as the longitudinal relaxation time or T_1. Values of T_1 between 1 and 30 seconds are fairly common for protons in organic and organometallic compounds. The relaxation rate is a characteristic of each site, and the appropriate time constants would be designated T_{1A} and T_{1B}.

The relaxation described above relates specifically to changes of nuclear spin states at a specific site; however, if chemical exchange occurs, then nuclei are also transferred from one site to another. The exchange is considered to take place without relaxation occurring in the transfer process, i.e., no spin flips occur during the exchange. Hence, if the exchange rate is comparable to the relaxation rate at both sites, complete saturation of one resonance will result in partial saturation of the other. The energy diagram for such a system with four states and four paths of interconversion is shown in Fig. 1. In some sense the system is like a laser in that "pumping" the levels in site A will cause a population change in the levels at site B. This technique relies on a form of Overhauser effect, and it is traditional to treat these exchange effects and relaxation in terms of average lifetimes rather than rate constants. Therefore, the lifetimes τ_A and τ_B can then be defined as the reciprocal of the first-order rate constant for leaving sites A and B, respectively. Consequently, the following lifetimes, which are the reciprocals of the first-order rate constants for leaving a particular spin state can then be defined*:

$$\tau_{1A} = (T_{1A}^{-1} + \tau_A^{-1})^{-1} \quad (1)$$

$$\tau_{1B} = (T_{1B}^{-1} + \tau_B^{-1})^{-1} \quad (2)$$

* τ_{1A} for the $|\alpha\rangle$ state is not exactly equal to τ_{1A} for the $|\beta\rangle$ state, due to the slight difference in upward and downward transition probabilities arising from the Boltzmann distribution between states from thermal population.

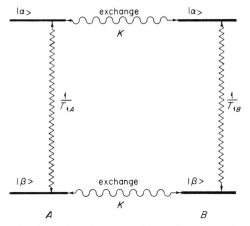

FIG. 1. Energy levels and exchange pathways for a two-site spin-½ case.

The McConnell modification of the Bloch equations are the starting point for all of the computations of the effect of saturating one resonance by saturating another.[10]

$$\frac{dM_z^A}{dt} = \frac{M_0^A - M_z^A}{T_{1A}} - \frac{M_z^A}{\tau_A} + \frac{M_z^B}{\tau_B} \qquad (3)$$

where M_0^A and M_0^B are the normal equilibrium magnetizations of the protons in sites A and B (or experimentally, the signal intensities for A and B at thermal equilibrium as determined by a weak nonsaturating magnetic field). If a saturating rf field at the resonance frequency of B is turned on at $t = 0$ and saturation takes place instantaneously,* one obtains the following expression for the magnetization of nuclei at A:

$$M_z^A(t) = M_0^A \left[\frac{\tau_{1A}}{\tau_A} \exp\left(-\frac{t}{\tau_{1A}}\right) + \frac{\tau_{1A}}{T_{1A}} \right] \qquad (4)$$

A new equilibrium value of the magnetization of nuclei at A is approached asymptotically such that at $t = \infty$.

$$\frac{M_z^A(\infty)}{M_0^A} = \frac{\tau_{1A}}{T_{1A}} \qquad (5)$$

Subtraction of Eq. (5) from (4) and rearrangement gives

$$\frac{M_z^A(t)}{M_0^A} - \frac{M_z^A(\infty)}{M_0^A} = \frac{\tau_{1A}}{\tau_A} \exp\left(-\frac{t}{\tau_{1A}}\right) \qquad (6)$$

* Due to the exponential nature of the result, the assumption of instantaneous saturation is not necessary. That is, the zero of time can be chosen arbitrarily.

Upon removal of the saturating rf field at the resonance corresponding to protons in site B, the initial thermal equilibrium magnetization at A recovers, but in a more complex fashion than in the saturation experiment. The complications arise from the spin states at A being pumped by chemical exchange at different rates because of the changing magnetization at B. In the saturation experiment, the magnetization at B remains constant, so that only the rate of exchange between $|\alpha\rangle$ and $|\beta\rangle$ changes with time, which leads to a simple exponential expression. Thus, one can obtain T_{1A} and τ_A from the ratio of the new equilibrium intensity to the original intensity of a resonance and the time constant (τ_{1A}) associated with the approach of the intensity to its new value. Several representative curves are shown in Fig. 2. Taking the case when the first-order rate constant for leaving site A is 0.1 second^{-1}, one can readily determine the ratio of τ_{1A}/T_{1A} to be 0.5 from the ratio of the intensities before and after saturation. The exponential approach to equilibrium gives τ_{1A} which can be obtained from the slope of $\ln [M_z^A(t) - M_z^A(\infty)]$ versus t, i.e., the natural logarithm of the difference between the intensity at any given time and that at the new equilibrium value is plotted versus time. In principle, it should be possible to restrict attention to the saturation curve only; in practice, however, it is often convenient to use the recovery curve, because the recovery time is determined by T_{1A} rather than τ_{1A} when the exchange rate is fast $(T_{1A} \gg \tau_A)$[11]; therefore, the time required for recovery is longer than that for saturation and is more easily measured.

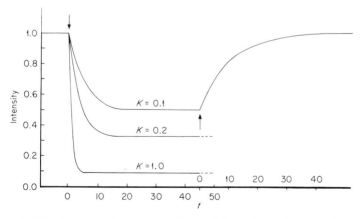

FIG. 2. The decay and recovery of signal intensity indicating changes in magnetization at site A upon application or removal of an irradiation field at site B. The arrow pointing downward indicates the time at which the field at B would be turned on; the time indicated by the arrow pointed upwards represents the point where the field would be turned off. k is given in second^{-1} and T_{1A} taken to be 10 seconds.

Furthermore, T_{1A} can be determined by a variety of methods, which will be discussed later, and one can thus readily obtain the ratio of τ_{1A}/T_{1A} from intensity ratios in the spectra such that τ_A can be readily determined.

It should be noted that no specific mention of relative populations of the nuclei in each site was necessary in the previous discussion. One is dealing with residence times of nuclei in particular sites, and the nature of the method is to produce first-order or pseudo-first-order rate constants for nuclei leaving a particular site; hence, relative populations are not needed. Nevertheless, there is frequently considerable variation in the conventions chosen by different investigators, and it is relatively easy to err in choosing the appropriate convention. Generally, the problems arise from neglecting to define carefully "exchange rate." Theoretical approaches employ a description of exchanging systems in terms of mean exchange lifetimes. It is essential to recognize that this term often refers to all sites considered as a group, and it represents the reciprocal of a sum of first-order rate constants, that is,

$$1/\tau = 1/\tau_A + 1/\tau_B \qquad (7)$$

This "mean lifetime" is operationally defined by the following relationship:

$$\tau = (1 - p_i)\tau_i \qquad (8)$$

where p_i is the fraction of nuclei at site i. Care must be taken to avoid confusion of the "mean lifetime" τ with the mean lifetime for nuclei in a particular site, τ_i, particularly when comparing data obtained from line shape analysis. Failure to take account of these definitions has resulted in an enormous number of rates being reported with large errors (often by a factor of 2).

Multiple-site problems can be treated readily by extension of Eq. (3) to include magnetization entering from other sites. The problem is particularly simple for saturation of a given site when one assumes that the other sites are completely saturated. In this case all of the $M_z^i = 0$, and Eqs. (4)–(6) still are correct. Thus, when all other exchanging sites can be saturated one can determine τ_A and T_{1A} from the amount of saturation and the time dependence of approach to equilibrium. For cases involving three sites these results can be achieved relatively easily in practice, and a specific example has been treated in detail by Forsen and Hoffman.[12] For larger numbers of sites, however, it is often more convenient to work only with equilibrium magnetizations, rather than to follow the time dependence of magnetizations.

It is often feasible to determine the value of T_1 for individual sites in separate experiments (see Section V), and in these cases it is only necessary to observe the degree of saturation for the resonances at equilibrium. The equilibrium measurements are quite straightforward, because an equation, such as (3), is available for each site. At equilibrium the derivative with respect to time is zero; the magnetization at any site being saturated is zero; and therefore, there are as many linear equations as there are unknowns for the number of sites involved. Examples are given in the following section which illustrate the applications of the various methods.

III. APPLICATIONS AND EXAMPLES

A. Two Sites

1. Rate Determinations

The activation energy for the ring inversion process in cyclohexane is a key problem in the field of conformational analysis. Lack of accurate kinetic data over a wide temperature range has led to substantial discrepancies between the activation parameters reported by various workers. The spin saturation transfer technique was applied to the cyclohexane system by Anet and Bourne[11] in order to extend the range of the kinetic measurements beyond those available from line shape and pulsed NMR methods. The

$$\text{(Ia)} \rightleftharpoons \text{(Ie)} \qquad (9)$$

inversion process [(Ia) $\rightleftharpoons$ (Ie)] becomes sufficiently slow at low temperatures such that a carbon disulfide solution of the compound at $-100°$ shows resonances for the equatorial and axial protons separated by about 28 Hz at 60 MHz. In order to avoid problems arising from coupling of the protons, cyclohexane-d_{11} was used and the deuterium resonances were decoupled. Hence, the lone proton in a given molecule is either axial or equatorial, and since coupling is eliminated, the PMR spectrum consists of two sharp resonances. The first resonance corresponds to the group of molecules (Ia) which have the proton in an axial position, whereas, the other resonance corresponds to the group of molecules (Ie) which have an equatorial proton.

If the resonance for the axial proton in (Ia) is saturated and the conversion to (Ie) is faster than the relaxation time of an equatorial proton, the resonance for the equatorial proton becomes partially saturated. At $-97°$ saturation of the low field resonance causes the intensity of the high field resonance to decrease to 12% of its original value. The slope of a plot of the natural logarithm of the approach to the new equilibrium value, $\ln [M_z(t) - M_z(\infty)]$, versus time gave $\tau_1 = 3.0$. Since $\tau_1/T_1 = 0.12$ from the degree of saturation, one obtains a value for T_1 of 25 seconds. From the definition of τ_1, one obtains a value for τ of 3.8 seconds, i.e., a value of $k = 0.26$ second^{-1}. Assuming the value of T_1 to be constant, Anet and Bourne were able to carry their rate studies to $-117°$ and measure rate constants as low as 0.004 second^{-1}.* The ability of this technique to measure the slower rates allowed these workers to obtain more accurate activation parameters than were previously available.

This capability of measuring rates occurring in the range between 10^{-3} and 1 second^{-1} is particularly useful when unstable species are studied, and the higher temperatures necessary for line shape analysis lead to rapid decomposition. The barrier to rotation in dimethylcyclopropylcarbinyl cation (II) was successfully measured[13] by saturating one methyl resonance and noting the saturation of the other between $-49°$ and $-21°$.

(II)

Conformational equilibria in [18]annulenes,[14,15] dehydro[16]annulenes,[16] and bisallylpalladium[17] have also been studied qualitatively by the saturation method. Furthermore, hindered rotation of the nitrosyl moiety in p-nitroso-N,N-dimethylaniline has been demonstrated by spin saturation.[14] Although the method is particularly suited to the detection of intramolecular interconversions, several studies of intermolecular proton exchange have been investigated. Qualitative studies of phenolic hydroxyl proton saturation upon irradiation of the resonance of trace quantities of water in chloroform have suggested a structural tool for the assignment of phenolic resonances.[18] A detailed study of exchange in alcohols and between salicylaldehyde and o-hydroxyacetophenone has been carried out by Fung[19] and by Forsen and Hoffman,[20] respectively. Since a pseudo-first-order rate constant is measured, it is possible to measure a fairly wide range of second-order rate constants by varying the concentrations of reactants. An interesting application was reported which involves a bimolecular kinetics study of proton exchange between fluorenyllithium and fluorene in dimethyl sulfoxide.[21]

This particular case is interesting because the protons which are transferred between molecules are not the ones which are observed in the saturation experiment. Saturation of the resonance assigned to the H_b proton causes saturation of the resonance due to H_a. The method, however, is only sensitive to changes in environment; hence, the proton which becomes an $H_{a'}$ proton after the transfer of an H_b proton has remained attached to the same fluorenyl

---

* In principle the relaxation times for the two sites need not be the same. In this case they are equal within experimental error.

moiety throughout the exchange process; whereas the proton which is transferred begins in an H_b environment and ends up in an identical environment.

2. Detection of Exchange with Weak Resonances

A particular advantage of the spin saturation method is that of specifically demonstrating which sites are involved in exchange. Furthermore, only the lifetime and relaxation time of one site determine the degree of saturation at that site; consequently, it is feasible to impart a high degree of saturation in a strong resonance at a site with a high population of nuclei by irradiating a weak resonance.

This technique has been applied in studies of the restricted rotation in sulfonium ylides which has been suggested in a fairly wide range of substituted derivatives.

Although this interpretation is correct in some instances, it has been shown that restricted rotation may not be the only source of the observed temperature dependence or of the interchange of isomers. Irradiation of the weak resonance associated with the trace of water remaining in the solvent showed that there was exchange between the water and the methine proton, thus suggesting that in some cases the mechanism responsible for the broadening of the resonances was an interconversion proceding via the following path[22]:

$$(CH_3)_2S^+CH=CO^-Ph + H_2O \rightleftarrows (CH_3)_2S^+-CH_2\overset{O}{\overset{\|}{C}}Ph + OH^- \quad (11)$$

Resonances which are very weak or are obscured can also be located, as has been shown in certain allyl complexes of molybdenum. Two isomeric configurations of π-cyclopentadienyldicarbonylmolybdenum-π-methallyl have been detected.[23] Each isomer should give rise to three resonances for the methallyl moiety.

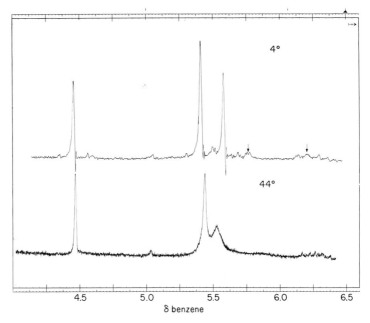

(VII)

The resonances for the major isomer are clearly visible in Fig. 3 at 4°. The broadening observed in the resonances at 44° suggested that a significant amount of the minor isomer was present, but the resonances of the isomer were obscured by impurities (the location of two of the resonances are indicated by arrows).

Confirmation of the location of the *anti*-proton resonance of the minor isomer was provided by a double irradiation experiment. At 8° the region

FIG. 3. The 100-MHz PMR spectrum of the methallyl moiety in π-cyclopentadienyldicarbonylmolybdenum-π-methallyl.

from 600 to 640 Hz upfield from benzene was irradiated in an experiment at 100 MHz while observing the major-component *anti*-proton resonance at 558 Hz. Assuming configurational interconversion occurs at a faster rate than spin–lattice relaxation of the *anti* protons, irradiation at the resonance position of the *anti* protons of the minor isomer should result in reduced intensity of the analogous resonance of the other isomer due to transfer of spin saturation. Irradiation between 620 and 622 Hz resulted in a minimum in the intensity of the resonance at 558 Hz, thus establishing the location of the *anti* resonance of the minor isomer at 6.21δ (see Fig. 4). A similar experiment confirmed the location of the methyl resonance of the isomer at 5.78δ.

B. Three Sites

Forsen and Hoffman[12] treated this case in detail in a study of the mechanism of keto–enol tautomerism in acetylacetone. Six rate constants are involved in the three site case, as indicated below:

(12)

Fortunately, the equilibrium constants determine the ratios of the rate constants between two sites so that the determination of six independent rate constants is not necessary. The T_1's for the sites can be determined by simultaneously saturating the other sites. Any individual rate constant can be determined by continuously saturating one resonance and then observing the decay of magnetization at the third site when the second site is saturated.

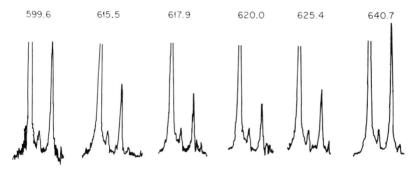

FIG. 4. The effect of irradiation in the region of the *anti* resonance of the minor isomer upon the *anti* resonance of the major isomer of

π-C$_5$H$_5$Mo(CO)$_2$(π-2-CH$_3$C$_3$H$_4$).

In this particular case, site C

$$CH_3\overset{O}{\underset{\|}{C}}-CH_2-\overset{O}{\underset{\|}{C}}-CH_3 \rightleftharpoons CH_3\overset{O\cdots H-O}{\underset{\|}{C}}-CH=\overset{}{\underset{|}{C}}-CH_3 \qquad (13)$$

is taken as the methylene protons in acetylacetone; site B, as the hydroxyl proton in the enol; and site A, as the methyne proton in the enol. When the methylene protons were saturated ($M_z^C = 0$), the intensity of the hydroxyl proton decreased to 62% of its original value ($M_z^B(\infty)/M_0^B = 0.62$). Simultaneous saturation of the methyne proton A ($M_z^A = 0$) had no significant effect on the magnetization at B (the saturation was observed to be 63%). This result requires that the direct transfer of magnetization from A to B be negligible, i.e., $k_{ab} \simeq 0$. Saturation of A alone produces a reduction to 57% in C and 78% in B, thus suggesting that saturation is effectively transferred to B via C. Crudely, it might be considered that nuclei in the irradiated site are completely saturated, and they gradually recover their normal magnetization as they migrate to other sites. As long as only one path is involved, each subsequent site along the journey will show a lower amount of saturation than the previous site. This evidence supports the mechanism which indicates that the primary process which is responsible for the rapid interchange of enolic hydroxyl and olefinic protons involves the intermediate conversion to the keto form.

A similar treatment has provided evidence for the principal pathway of *cis–trans* isomerization of π-crotylpalladium complexes.[24] The problem is to determine whether the mechanism involves the intramolecular interconversion of the isomers via a rotation of the crotyl moiety or the intermolecular exchange of ligands. Reaction of a chloro-bridged dimer with an amine produces the two isomers in almost equal proportions, and the reaction proceeds far to the right, leaving a negligible amount of free amine in solution. Reaction of the dimer with two-thirds of an equivalent of amine produces an equal number of *cis*, *trans*, and dimer sites.

$$\left\langle\left(Pd\overset{Cl}{\underset{Cl}{\diagup\diagdown}}Pd\right)\right\rangle + 2L \rightleftharpoons \left\langle\left(Pd\overset{Cl}{\underset{L}{}}\right)\right. \left\langle\left(Pd\overset{L}{\underset{Cl}{}}\right)\right. \qquad (14)$$

$$\qquad\qquad\text{(VIII)} \qquad\qquad\qquad \text{(IX)} \qquad \text{(X)}$$

$$\qquad\text{dimer} \qquad\qquad\qquad\qquad\qquad \textit{cis} \qquad \textit{trans}$$

The methyl resonances are considered for convenience. At low temperature, irradiation of the *cis* methyl resonance causes a high degree of saturation in the dimer resonance and smaller degree of saturation in the *trans* resonance. Following arguments similar to those in the acetylacetone case discussed previously, it is possible to conclude that the predominant pathway of isomerization involves the intermediacy of the dimer (or of a species which is rapidly converted to the dimer) rather than rotation. Thus under these conditions, a given *cis* complex is most likely to be converted first to the dimer, then subsequently to the *trans* complex rather than proceeding directly from *cis* to *trans*.

Effects of Spin Coupling

Figure 5 shows the results of the previous experiment but illustrates in certain respects a six-site case because of spin–spin coupling to the methyl group. A doublet with a separation of 7 Hz is observed for each methyl group due to coupling with the proton on the adjacent carbon atom. The upfield peak of each doublet corresponds to one spin state of the proton (β) on the adjacent carbon atom, while the lower peak would correspond to the other (α). If the isomerization occurs without spin flips of that proton occurring (which is the usual case), complete saturation of the upfield component will lead to partial saturation of only the upfield components of the other doublets; that is, there are two different sets of molecules, those with the adjacent

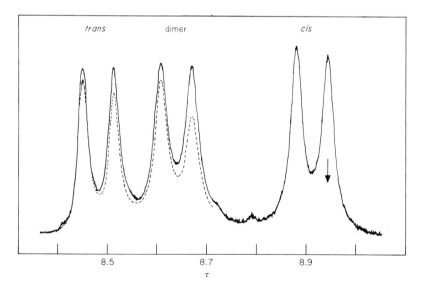

FIG. 5. The crotyl methyl region of the 100-MHz PMR spectrum of a mixture of *cis*- and *trans*-π-crotyl(2-picoline)palladium chloride and π-crotylpalladium chloride dimer at $-56°$. The broken line indicates the effect of saturation transfer upon the *trans* and *dimer* resonances upon saturation of the upfield component of the *cis* resonance.

proton α and those with the adjacent proton β, and they are not interconverting rapidly.*

Alternatively, in intramolecular exchanges it often is possible to determine relative signs of coupling constants. Aside from the *cis–trans* equilibria in the crotylpalladium complexes, there is pi–sigma rearrangement,[25] which interconverts *syn* and *anti* protons (see 15).

$$\begin{array}{c}\text{H}_1\\ \text{H}_2\\ \text{(PdXL)}\end{array} \rightleftarrows \begin{array}{c}\text{H}_1\ \text{H}_2\\ \diagup\diagdown\\ \text{PdXL}\end{array} \rightleftarrows \begin{array}{c}\text{H}_1\ \text{H}_2\\ \text{(XLPd)}\end{array} \quad (15)$$

In this case the central proton is coupled to the *syn* proton, giving a splitting of 7 Hz, and to the *anti* proton, giving a splitting of 12 Hz. Saturation of the upfield component of the *anti* doublet causes saturation in the upfield component of the *syn* doublet. Again it is convenient to consider the molecules in the solution to be divided into two groups based on the spin state of the central carbon atom. Irradiation of the upfield resonance of the *anti* doublet would correspond to "saturating the protons" in the *anti* position of the molecules which have the central proton in the β spin state. The rearrangement proceeds without relaxation of protons attached to the central carbon atom, and therefore, the saturation is transferred only to the *syn* sites in molecules with the central proton in the β spin state. Since the upfield component of the *syn* doublet is saturated, the sign of the coupling constant to the central proton must be the same as that for the *anti* proton–central proton coupling constants.

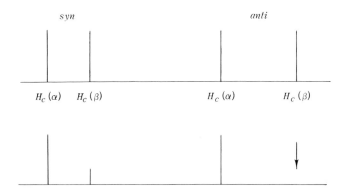

* The partial saturation of the low field components is due to inadvertent partial saturation of the low field component of the *cis* doublet.

C. Four Sites

Mechanisms

The heptamethylbenzenonium ion is a "fluxional" or stereochemically nonrigid structure, i.e., it undergoes a rapid intramolecular rearrangement to an ion of equivalent structure. It is of considerable interest to determine the nature of the methyl migration responsible for the reaction. In particular, it is of interest to determine whether the reaction proceeds via 1,2 shifts or if there is a random migration of the methyl group about the various carbon atoms in the ring.

At 28° the PMR spectrum appears as four fairly sharp resonances as indicated schematically in Fig. 6. As the temperature is raised the resonances begin to broaden, and appreciable coalescence of the peaks occurs about 60°. Calculation of the line shapes expected for the 1,2-shift model and the random rearrangement model reveal subtle differences in the expected line shapes,

$$
\begin{array}{c}
\text{(structure with labels 1,1 top; 2,2 sides; 3,3 sides; 4 bottom)} \\
\text{(XI)}
\end{array}
\rightleftarrows
\begin{array}{c}
\text{(structure with labels 2 top; 1,1 sides; 3,3 sides; 2,2; 4,3 bottom)}
\end{array}
\rightleftarrows \text{etc.} \quad (16)
$$

which suggest that the 1,2 shift is the most plausible mechanism.[26] Nevertheless, the differences in the spectra expected for the two models is such that the possibility of simultaneous operation of the two mechanisms was not

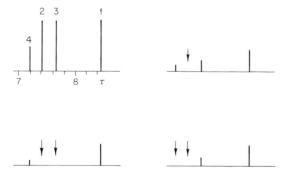

FIG. 6. The effects of saturating certain resonances in the PMR spectrum of heptamethylbenzenonium ion at 28°. The numbers refer to the assignment of the methyl groups.

ruled out, nor is the line shape method sensitive enough to eliminate other, more complicated reaction paths.

Spin saturation transfer, however, provides convincing evidence that the random migration of the methyl group does not play a significant role in the rearrangement of the ion.[27] With the 1,2-shift mechanism, methyl groups will be transferred into site 1 via site 2, whereas a random rearrangement would allow for the direct transfer of nuclei from sites 3 and 4, as well as from site 2. Saturation of the methyl protons in site 2 reduces the intensity of the resonance due to methyl groups at site 1; however, no additional decrease in intensity is noted upon saturation at 3 or 4. A further reduction of 22% would be expected upon irradiation at 3 or 4, if the random migration mechanism were in operation. Since one can observe a change greater than about 1%, it follows that the portion of the rearrangement represented by the random migration mechanism must be very minor, if it occurs at all.

IV. MEASUREMENT TECHNIQUES

Although many of the original experiments were performed using field-swept instruments, the recent availability of frequency-swept* instruments greatly facilitates spin saturation labeling experiments. Use of the frequency-swept instruments often allows effective use of the technique on resonances which are separated by as little as 15 Hz, whereas earlier experiments often required a minimum separation at 1 ppm. It will be assumed that a frequency-swept instrument is used in the following discussion. The review by Forsen and Hoffman[2] discusses many details of field sweep experiments.

A. Degree of Saturation

Measurements of the saturation alone often provide sufficient information to solve many mechanism problems. Accuracy is important in experiments of this nature, but in determinations of relaxation times, this ratio is used as an infinity value, and small errors in the ratio can produce catastrophic

---

* The principal advantage of frequency-swept instruments is the capability of the instrument to irradiate continuously a particular resonance, i.e., saturate a particular site. In practice "frequency-swept" spectra are generated in some instruments by variations in the field and simultaneous adjustment of locking frequencies. In any event the resonance is irradiated at all times, even though the field is varied. Although one speaks of irradiating with a second rf field, this field is often generated by modulation of the magnetic field rather than modulation of the radio frequency from the transmitter. Regardless of the technical aspects, the instruments can be considered as constant field instruments with the capability of irradiating at or sweeping several frequencies.

results in the τ_1 and T_1 evaluations. Furthermore, the ratio is very sensitive to rate constant, and small errors are magnified in the calculation of the rate constants.

In order to determine the ratio of τ_{1A} to T_{1A} it is necessary to measure an intensity ratio. Generally one can maintain acceptable accuracy by merely comparing the heights of a resonance before and after saturation at the other sites. It is often useful to measure the areas to check on the accuracy of using peak heights as measures of intensity. Although one can use the electronic integration provided by the instruments, it is frequently preferable to record the resonances on an expanded sweep scale and to determine the areas either using a planimeter or cutting and weighing the shapes. In either case the areas should be checked to be sure that an error is not introduced by the inadvertent removal of spin coupling by the fields used to saturate other sites. Assuming that the precautions have been taken to eliminate the errors to be discussed in Section V, ratios of peak heights or peak areas which represent accurate ratios of z magnetization can usually be obtained easily and rapidly.

B. Relaxation Times

Since the determination of the value of τ_{1A}/T_{1A} is obtained directly from equilibrium intensity measurements, a straightforward determination of T_1 is desirable for the measurement of rates. As indicated in Section II, following the decrease in intensity at a particular site after application of saturating fields at other sites allows the relaxation time to be obtained indirectly from τ_1. This is most easily accomplished by using several external oscillators to generate the fields (H_2, H_3, etc.) as in other double and triple resonance experiments. Since fairly low power levels are necessary, it is practical to perform multiple resonance experiments with very little difficulty. In this method, the magnetic field is held constant, and observing rf field is generated at the site for which the magnetization is to be determined. This latter step is often accomplished by merely moving the recorder pen of the instrument to the center of the resonance. The decay of intensity is most easily followed, using an external recorder with a reliable time base and a fast response time. When τ_{1A} is on the order of a few seconds, common laboratory recorders often have sufficiently fast response; however, one frequently encounters values between 0.2 and 2 seconds, and faster equipment is needed. The Sanborn 151–100 recorder which was supplied with field-swept Varian instruments is adequate; however, recorders, more often than not, are chosen because of their availability to the users rather than a need for highly sophisticated equipment. Consequently, high-speed oscillographic recorders or storage oscilloscopes are often used. Further sophistications are possible with small laboratory computers generating the frequencies within the computer and collecting the data via an analog–digital converter directly into the computer memory for analysis, which greatly facilitates the measurements.

In certain cases it may be inconvenient to determine the relaxation times by the method suggested above. Often it is possible to measure the relaxation time by other methods at lower temperatures (or slower exchange rates) and extrapolate the values to the region of interest. Usually this requires lowering the temperature 20°–40°; however, one often notes variations in T_1 of less than 15% over this range. Spin-echo or pulse methods[28] could be used but generally would prove inconvenient. Application of adiabatic-fast-passage techniques[29] allow for determinations of T_1, which are quite similar in principle to a 180°–90° pulse sequence and can be performed on more conventional spectrometers.

The field-swept adiabatic-fast-passage experiment consists of rapidly sweeping the field through a resonance with a high rf power and then subsequently sweeping the field with low rf power to observe the signal as it returns to the normal intensity. The first passage with high power is equivalent to a 180° pulse, i.e., it inverts the magnetization ($M_z \rightarrow -M_z$). Subsequent observation of the resonance at regular time intervals with low power, which does not significantly alter the value of M_z, will show it to be "upside down" as the negative magnetization decays to zero and then approaches its original value "rightside up." The time constant for the exponential approach to the original intensity is T_1. Modifications of this procedure, which allow the determination of relatively short relaxation times, have been reported.[30] In investigations of mechanisms, one often merely wishes to assure himself that the relaxation times for all of the sites are approximately the same. This can be readily determined by observations on an oscilloscope during a field sweep which passes through an intense center-band frequency and a weak side-band frequency. Each resonance in the spectrum passes sequentially through the center band, giving the magnetization at each site the equivalent of a 180° pulse. Each resonance is observed after the same time interval from the "pulse" by the side band. If the side band is moved or the sweep speed slowed until the resonances have nearly zero intensity in the side-band spectrum, the relative values of T_1 for each resonance can be ascertained.

Alternatively, one has the option of frequency-swept adiabatic-fast-passage methods. Generally, these require an additional piece of equipment, such as a function generator, to allow the sweeping of frequency. This method permits the use of field-locked spectrometers, and smaller differences in chemical shifts between sites are necessary. Finally, one can use the direct approach if T_1 is fairly long, i.e., with the instrument centered on a resonance, the resonance can be saturated with a high rf power, the power turned down quickly, and the recovery of intensity followed on an external recorder.[31]

V. SOURCES OF ERROR

The spin saturation labeling experiments, unlike spin decoupling experiments, require fairly low rf power levels; hence there is seldom a problem in achieving complete saturation of a given resonance. Nevertheless, the equations assume that the magnetization at an irradiated site is zero, and it is imperative that this condition be met within experimental error ($M(_2\infty)/M_0 < 0.005$). The degree of saturation can be checked by lowering the power level and quickly sweeping over the resonance as it recovers and extrapolating back to zero time.[11] In practice, however, it usually is more convenient to increase the power level of the saturating rf field (H_2) above that where no additional saturation is observed in the resonance being monitored (H_1).

On the other hand, care must be taken to avoid saturation of the resonance being observed by H_1. Lower levels of power are often necessary since the signal is usually monitored by maintaining H_1 at resonance during the entire experiment. The power level of H_2 must also be kept as low as possible if the separation of the resonance to be saturated and that to be observed is small, due to possibility of inadvertently partially saturating the observed resonance with the H_2 field.

The power levels are usually too low to cause spin decoupling; however, decoupling of splittings less than 1 Hz and the effects of spin tickling can alter the line shape of the observed resonance. Consequently, it is essential to determine the area of the resonance in order to ascertain if a measurement of the peak height is sufficiently accurate. In the case of intramolecular interchanges it is also necessary to lower the temperature or slow the exchange to the point where no spin saturation transfer should affect the spectrum and investigate possible increases in magnetization arising from nuclear Overhauser effects. The nuclear Overhauser effect (NOE) results in an increase in magnetization at a site via relaxation processes which are stimulated by magnetic transitions in other nuclei very near to that site. The NOE arises through dipole interactions and decreases as the cube of the distance between the irradiated and observed site. Generally one needs fairly high power levels to observe an NOE, so that it is unusual to run into this problem; nevertheless, the same conditions which give the largest NOE's are also those desired for spin saturation transfer experiments.

An unusual occurrence, but one which must be considered, is the propagation of demagnetization via spin–spin interactions.[32] When the nuclei in two sites are strongly coupled and the chemical shift difference between them is small, a given resonance cannot be exclusively assigned to a given site due to the mixing of spin functions. When saturation is transferred into one of these sites, partial collapse of both resonances is observed.

The most common sources of error arise from variations in rf power levels and the efficiency of the detection system when the additional fields are turned on or off. The most common technique for generating the H_2 and H_3 rf fields is to modulate a master frequency (e.g., 100 MHz) with audio oscillators operating in a range between 1 and 10 KHz. This reduces the power at the original radiofrequency (the center band) and produces sum and difference frequencies on either side of the center band. These sidebands are usually used for observing and irradiating rf fields. The distribution of power between the center band and the side bands is of particular importance because the alteration of the power level of one side band will influence the power level of the others. Consequently, when one turns on the H_2 field, the level of the observing field H_1 is reduced and the apparent value of the magnetization is decreased. Overloading of the phase sensitive detector or amplifiers can also cause similar problems. It is therefore imperative to determine the magnitude of the effects of raising and lowering the power levels of the various oscillators on the intensity of the observed resonance. This is most conveniently arranged by irradiating some region of the spectrum which is irrelevant to the saturation transfer experiment, i.e., a place where no resonances are observed. Obviously, one wishes to measure M_0 values with the saturating irradiation at an irrelevant position and at the same power level as when the H_2 field is at the resonance to be saturated and $M_z(\infty)$ is measured. (In magnetization recovery experiments M_0 should generally be determined with H_2 off.) One sometimes must return to the theory to determine the appropriate manner to measure M_0 or make appropriate corrections, but this problem can often be minimized or eliminated by using high center-band power levels and low audio oscillator power levels for the modulation producing the sidebands.

Not as many subtleties arise in the determination of relaxation times. Insufficient speed in recorder response is perhaps the most common difficulty. When using magnetization recovery curves, one must be careful that the limitations of the theory are not exceeded, i.e., the exchange rate must be sufficiently fast to satisfy the condition that $\tau_1 \ll T_1$. One kinetic problem, which is quite common in first-order approach to equilibrium, is the necessity of accuracy for the infinity value. A time of at least $5 \times T_1$ should be allowed before measuring limiting values of intensity. Sometimes it is convenient to take ratios of differences in intensity at certain time intervals (Guggenheim method [33]) to eliminate the undue importance attributed to the infinity value.

Detailed studies of the accuracy to be expected of this method have not been performed as in the case of the other NMR kinetic methods [31,34]; nevertheless, some comments on accuracy, although they are subjective estimates, are useful. It appears that one can expect the value of T_1 to be within 15% of the *true* value, and magnetization ratios are usually reproducible to within ± 0.01. The precision available in the determination of relative degrees of saturation is quite good. Within a given experiment the relative values of T_1 for the various sites appear to be good to several percent, and these ratios are the important factor in determining mechanisms. The accuracy of the rate constants determined by the method varies substantially over the range where the method is applicable (this is largely determined by the variation in rate constant attending the ± 0.01 uncertainty in magnetization ratio). In the regions where the method is most accurate, it appears unlikely that rates reliable to better than $\pm 20\%$ of those obtained by other methods are obtained.

VI. PRACTICAL ASPECTS

Spin saturation transfer experiments provide a more accurate way of determining probabilities of magnetization transfer between various sites than line shape analysis. On the other hand, line shape analysis produces more accurate rates (on the order of 10% error) and generally covers a wider range of rates. For investigations of many mechanisms and experiments similar in purpose to deuterium-labeling studies, measurements at a single temperature are sufficient. Spin-saturation transfer experiments are preferable in this case, but the two methods are complementary. Often it is useful to check the relative accuracy of the two methods in regions of overlap in rate. Initial line broadening can generally measure the rate for leaving a site by

$$k = \pi W \tag{9}$$

where W is the additional full-width at half-height observed in the line due to exchange. Hence, it is often more convenient to use line shape methods for rate studies* and saturation transfer experiments to determine site exchange probabilities. Nevertheless, the saturation transfer method is more suited to slow rates, and it is feasible to extend its range by increasing the value of T_1 relaxation times as much as possible. This can generally be achieved by following the same precautions as in the nuclear Overhauser effect experiments, i.e., using solvents which contain no hydrogen and degassing the solvent to remove dissolved oxygen.[2,39] In practice it is sometimes useful to modify T_1 at different temperatures to achieve the greatest precision in the measurements. When attempting to determine accurate activation parameters, it is desirable to cover as wide a range of rates and temperatures as possible, so the use of both methods is advantageous. In these experiments the temperature instability appears to be the major source of error; hence, wide

---

* Line shape analysis and the errors inherent in the method have been discussed elsewhere,[34-38] but familiarity with the technique and its problems aid in the choice of methods and in ascertaining sources of error. It is particularly important to consider carefully definitions of lifetimes (Section II), and it is occasionally useful in complex cases to consider examples which have been worked out for temperature-jump relaxation.[5-9]

temperature ranges tend to reduce systematic errors. In general, we have found that for activation energies of intramolecular reactions, in which it is possible to measure a wide range of temperatures, log A generally lies in the range 12.3–12.8.[40] Hence, intramolecular reactions which appear to have anomalously small values of log A, and by inference unusual values of ΔS^*,[34,41,42] are generally regarded with suspicion and have often resulted from systematic experimental errors.

Spin saturation transfer experiments complement NMR line shape analysis in kinetic and mechanistic studies. Further, the spin saturation labeling technique often provides a convenient alternative to deuterium labeling in the study of rearrangement and exchange mechanisms.

Note Added in Proof: Since the preparation of this manuscript an excellent review of nuclear Overhauser effects has appeared.[43]

References

[1] H. M. McConnell and D. D. Thompson, *J. Chem. Phys.* **26**, 958 (1957).
[2] R. A. Hoffman and S. Forsen, *Progr. Nucl. Magn. Resonance Spectrosc.* **1**, 15 (1966).
[3] H. M. McConnell and D. D. Thompson, *J. Chem. Phys.* **81**, 85 (1959).
[4] A. Patterson, Jr. and R. Ettinger, *Z. Elektrochem.* **64**, 98 (1960).
[5] M. Eigen and L. DeMaeyer, *Tech. Org. Chem.* **8b**, 895 (1963).
[6] M. Eigen, W. Kruse, G. Maass, and L. DeMaeyer, *Progr. React. Kinet.* **2**, 285 (1964).
[7] M. Eigen and R. G. Wilkins, *Advan. Chem. Ser.* **49**, 55 (1965).
[8] M. Eigen, *Angew. Chem., Int. Ed. Engl.* **3**, 1 (1964).
[9] G. H. Czerlinski, "Chemical Relaxation." Dekker, New York, 1966.
[10] H. M. McConnell, *Chem. Phys.* **28**, 430 (1958).
[11] F. A. L. Anet and A. J. R. Bourne, *J. Amer. Chem. Soc.* **89**, 760 (1967).
[12] S. Forsen and R. Hoffman, *J. Chem. Phys.* **40**, 1189 (1964).
[13] D. S. Kabakoff and E. Namanworth, *J. Amer. Chem. Soc.* **90**, 3234 (1970).
[14] I. C. Calder, P. J. Garratt, and F. Sondheimer, *Chem. Commun.* p. 41 (1967).
[15] I. C. Calder, P. J. Garratt, H. C. Longuet-Higgins, F. Sondheimer, and R. Wolowski, *J. Chem. Soc., C* p. 1041 (1967).
[16] I. C. Calder, Y. Gaoni, P. J. Garratt, and F. Sondheimer, *J. Amer. Chem. Soc.* **90**, 4954 (1968).
[17] J. K. Becconsall and S. O'Brien, *J. Organometal. Chem.* **9**, P27 (1967).
[18] J. Feeney and A. Heinrich, *Chem. Commun.* p. 295 (1966).
[19] B. M. Fung, *J. Chem. Phys.* **47**, 1409 (1967).
[20] S. Forsen and R. A. Hoffman, *J. Chem. Phys.* **45**, 2059 (1966).
[21] J. I. Brauman, D. F. McMillen, and Y. Kanazawa, *J. Amer. Chem. Soc.* **89**, 1728 (1967).
[22] S. H. Smallcombe, R. J. Holland, R. H. Fish, and M. C. Casserio, *Tetrahedron Lett.* p. 5987 (1968).
[23] J. W. Faller and M. J. Incorvia, *Inorg. Chem.* **7**, 840 (1968).
[24] J. W. Faller and M. J. Incorvia, *J. Organometal. Chem.* **19**, P13 (1969).

[25] J. W. Faller, M. E. Thomsen, and M. J. Incorvia, *J. Amer. Chem. Soc.* **93**, 2642 (1971).
[26] M. Saunders, *in* "Magnetic Resonance in Biological Systems" (A. Ehrenberg, B. G. Malmstroem, and T. Vaenngerd, eds.), p. 90. Pergamon, Oxford, 1967.
[27] B. G. Derendyaev, V. I. Mamatyuk, and V. A. Koptyug, *Tetrahedron Lett.* p. 5 (1968).
[28] H. Y. Carr and E. M. Purcell, *Phys. Rev.* **94**, 630 (1954).
[29] A. Abragam, "The Principles of Nuclear Magnetism." Oxford Univ. Press, London and New York, 1961.
[30] R. L. Conger and P. W. Selwood, *J. Chem. Phys.* **20**, 383 (1952).
[31] A. L. Van Geet and D. N. Hume, *Anal. Chem.* **37**, 979 and 983 (1965).
[32] B. M. Fung, *J. Amer. Chem. Soc.* **90**, 219 (1968).
[33] E. A. Guggenheim, *Phil. Mag.* [7] **2**, 538 (1926).
[34] A. Allerhand, H. S. Gutowsky, J. Jones, and R. A. Meinzer, *J. Amer. Chem. Soc.* **88**, 3185 (1966).
[35] C. S. Johnson, Jr., *Advan. Magn. Resonance* **1**, 33 (1965).
[36] L. W. Reeves, *Advan. Phys. Org. Chem.* **3**, 187 (1965).
[37] G. Binsch, *Top. Stereochem.* **3**, 79 (1968).
[38] J. W. Faller and M. E. Thomsen, *J. Amer. Chem. Soc.* **91**, 6871 (1969).
[39] R. E. Schirmer, J. H. Noggle, J. P. Davis, and P. A. Hart, *J. Amer. Chem. Soc.* **92**, 3266 (1970).
[40] J. W. Faller and A. S. Anderson, *J. Amer. Chem. Soc.* **92**, 5852 (1970).
[41] R. K. Harris and N. Sheppard, *J. Mol. Spectrosc.* **23**, 231 (1967).
[42] A. Jaeschke, H. Muensch, H. G. Schmid, H. Friebolin, and A. Mannschreck, *J. Mol. Spectrosc.* **31**, 14 (1969).
[43] J. H. Noggle and R. E. Shirmer, "The Nuclear Overhauser Effect." Academic Press, 1971.

Chemically and Electromagnetically Induced Dynamic Nuclear Polarization

3

RONALD G. LAWLER AND HAROLD R. WARD

I.	Introduction	99
	A. General	99
	B. Definitions of Some Physical Quantities	101
II.	Overhauser and Related Effects	103
	A. Coupled Relaxation of One Electron and One $I = \tfrac{1}{2}$ Nucleus	103
	B. Electromagnetically Pumped DNP	104
	C. Applications of Electromagnetically Pumped DNP	105
	D. Time-dependent Intramolecular Electron-nuclear Couplings in Organic Radicals	106
III.	Chemically Induced Dynamic Nuclear Polarization (CIDNP)	109
	A. Types of Spectra	109
	B. The Overhauser Model	111
	C. The Radical Pair Model	114
	D. Examples of CIDNP Spectra	127
	E. Applications of CIDNP	135
	F. Predicting the Future of CIDNP	143
	References	144

I. INTRODUCTION

A. General

A high-resolution NMR spectrum (indeed, any spectrum with well-defined lines) yields three principal types of information: (a) line *positions*, points of maximum intensity of individual lines or closely overlapping groups of lines, (b) line *widths* and *shapes* which describe the distribution of absorption or emission about these maxima, and (c) line *intensities* which are proportional to the integrated areas of envelopes described by positions, widths, and shapes and representing the rate of absorption of energy over the region of the line.

In the field of high-resolution NMR line position information has been most widely exploited since it is the source of the chemical shift and spin–spin splitting information of immense importance for structure determination.[1] Similarly, information gained from line widths and shapes is being increasingly applied to dynamic problems such as chemical rate processes.[2] It is only the line intensities which have been, until very recently, largely, and usually explicitly, ignored as a source of additional information. In structural applications of NMR it is usual to assume that the rate of absorption of photons is proportional to the number of nuclei with resonance frequencies under a given line and that the constant of proportionality is *the same* for all nuclei of the same general type in the sample. Thus the use of intensities has been limited almost exclusively to determining the relative numbers of magnetically equivalent nuclei absorbing at various line positions. It is the purpose of this chapter to describe conditions under which the intensities of NMR lines may depart dramatically, often by two or three orders of magnitude, from those normally encountered and to indicate the kinds of structural and dynamic information which have been or potentially may be obtained from examining these phenomena.

The effects discussed here will be restricted to those occurring in liquid samples containing $I = \frac{1}{2}$ nuclei which interact with unpaired electrons.[3] The production of altered NMR intensities via interaction of nuclei with a perturbed electron spin system has come to be known as dynamic nuclear polarization (DNP). An effect of this type was first predicted in 1953 by A. W. Overhauser[4] (and is therefore also often referred to as the Overhauser effect) and shortly thereafter was detected using the lithium nuclei in metallic lithium.[5] The technique was quickly taken up by both high-energy and solid-state physicists as a means of producing oriented nuclear targets[6] and of studying paramagnetic impurities and nuclear spin relaxation mechanisms in solids.[7] Despite the fact that some of the first systems studied were also the (by now chemically familiar) paramagnetic solutions of sodium naphthalene in 1,2-dimethoxyethane[8] and of sodium in liquid ammonia,[9] it was not until the early and mid-1960's that rather extensive application of the Overhauser effect to liquids was made by chemical physicists. The details of these experiments are now available in comprehensive reviews[10,10a,10b] from two of the groups which participated in these applications. Consequently the Overhauser effect will be dealt with here in highly abbreviated form.

Interest in DNP took a striking change of direction in 1967 when it was found[11,11a] that enhanced NMR absorption and emission may be observed with conventional analytical NMR instruments in samples which are undergoing rapid free radical reactions. By analogy with the electromagnetically pumped experiments the effects have come to be known as chemically induced dynamic nuclear polarization (CIDNP). Although the initial explanations[12,12a] of these effects were also analogous to those invoked for the Overhauser effect, it soon became apparent that the majority of the chemical effects were fundamentally different and required other explanations.[13-15a] It is with the physical origins and chemical applications of CIDNP that the bulk of this chapter will deal.

B. Definitions of Some Physical Quantities

The observable detected by a conventional high-resolution NMR spectrometer under nonsaturating, slow passage conditions[16] is the net rate at which energy in the form of monochromatic and coherent photons of frequency $v = \omega/2\pi$ is absorbed by the nuclear spin system. This rate, $P(\omega)$, is given[17] by

$$P(\omega) = (\hbar\omega) \sum_{i<j} W_{ij}(\omega)(n_i - n_j) \tag{1}$$

where n_i and n_j are the populations of the corresponding energy levels. The probability per unit time, $W_{ij}(\omega)$, of a stimulated transition occurring from energy level i to level j, where E_j is greater than E_i, is[17]

$$W_{ij}(\omega) = \tfrac{1}{2}\gamma_I^2 H_1^2 |\langle i|I_+|j\rangle|^2 g[\gamma_I H_0 - \hbar^{-1}(E_j - E_i)] \tag{2}$$

In Eq. (2), γ_I is the magnetogyric ratio of the nucleus, H_1 is half the amplitude of the linearly polarized applied radiofrequency magnetic field, H_0 is the static field, taken to be in the z direction, I_+ is the angular momentum raising operator which allows only transitions in which states i and j differ by one unit of angular momentum in the z direction, and $g[\gamma_I H_0 - \hbar^{-1}(E_j - E_i)]$ is a shape function with dimensions of frequency which allows for energy levels of a finite width but is large only when the Larmor frequency, $\gamma_I H_0$, is near the frequency of the transition between i and j.

In the most common kind of experiment the frequency v is held constant and H_0 is swept. Energy will then be *absorbed* by inducing a transition between i and j only if three conditions are satisfied simultaneously: (a) $\gamma_I H_0 = \hbar^{-1}(E_j - E_i)$, (b) $\langle i|I_+|j\rangle \neq 0$ and (c) $n_i > n_j$. (a) and (b) are simply the usual selection rules for spectroscopic transitions while (c) states the dependence of the absorption intensity on *populations*.

The relative populations of two levels which differ by one unit in the angular momentum quantum number m serve to define the *polarization* p_{ij} of the transition

$$p_{ij} = (n_i - n_j)/(n_i + n_j) \tag{3}$$

If the populations are described by a Boltzmann distribution, the equilibrium, or thermal, polarization p_{ij}° is

$$p_{ij}^\circ = \frac{1 - \exp\left(-\frac{\gamma_I \hbar H_0}{kT}\right)}{1 + \exp\left(-\frac{\gamma_I \hbar H_0}{kT}\right)} \simeq \frac{\frac{1}{2}\gamma_I \hbar H_0}{kT} \qquad (4)$$

For protons in samples at room temperature in fields of a few kilogauss, p_{ij}° is approximately 10^{-5}.

It has often proven useful to use the observed polarization via Eq. (4) as a measure of a *spin temperature* T_s. The magnitude of T_s may be much different from the actual temperature of the sample, or *lattice*, T_L, because the exchange of energy between a system of nuclear spins and the surrounding heat sink is often quite slow.[18] Thus if $p_{ij} \rangle \langle p_{ij}^\circ$, $T_s \langle \rangle T_L$; if $p_{ij} = 0$, $T_s = \infty$; and if p is negative, T_s will be negative!

When the populations deviate from their equilibrium values the NMR signal usually (but not always, as we shall see) deviates from its equilibrium intensity, $P^\circ(\omega)$. The deviation is described by the *enhancement factor*, V:

$$V = \frac{P(\omega) - P^\circ(\omega)}{P^\circ(\omega)} = \frac{\sum_{i<j} W_{ij}[(n_i - n_i^\circ) - (n_j - n_j^\circ)]}{\sum_{i<j} W_{ij}(n_i^\circ - n_j^\circ)} \qquad (5)$$

For the special case where absorption is due to two levels only, the enhancement factor reflects directly the deviations of populations of the levels from equilibrium (provided, of course, that the process giving rise to the population changes did not affect the lineshape function for the transition). In most of what follows we will consider the production of nonequilibrium populations of only two levels at a time, recognizing that the generation of enhancement factors for multiple level systems from populations via Eq. (5) is straightforward.

Equations (1)–(5) describe NMR transitions in microscopic terms. NMR absorption may, however, be described by the phenomenological Bloch equations[16] which are the equations of motion of the macroscopic nuclear magnetization, $\mathbf{M}_I$. Under the experimental conditions described above, the absorption intensity is also proportional to the component M_{Iz} of the magnetization in the direction of the applied static field. M_{Iz} is given by

$$M_{Iz} = \sum_{\text{all nuclei}} \mu_{zi} = \gamma_I \hbar \sum m_i \qquad (6)$$

where μ_{zi} is the z component of the magnetic moment and m_i is the magnetic quantum number of the *i*th nuclear spin. For a collection of identical nuclei with spin I the polarization p is defined as

$$p = I^{-1}\left(\sum_m mn_m\right)\bigg/\left(\sum_m n_m\right) \tag{7}$$

where n_m is the number of nuclei with quantum number m. For the special case where $I = \frac{1}{2}$, $m = \pm\frac{1}{2}$, and Eq. (7) reduces to the two-level polarization in Eq. (3). It is, in fact, p which is the correct definition of "polarization" as an indication of nuclear orientation.[6] We will, however, also use the term for p_{ij} as it applies via Eq. (3) to intensities of NMR transitions.

II. OVERHAUSER AND RELATED EFFECTS

A. Coupled Relaxation of One Electron and One $I = \frac{1}{2}$ Nucleus

The Overhauser effect is of the "cross-relaxation" type in which an electron spin system which has been driven away from equilibrium, that is, $T_s \neq T_L$, returns to thermal equilibrium or *relaxes*, at least partly, by exchange of energy with the nuclear spin system. The actual transitions responsible for relaxation are shown in Fig. 1 for the four energy levels arising from one

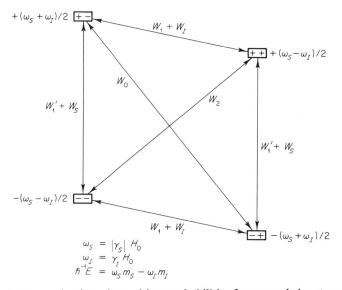

FIG. 1. Energy levels and transition probabilities for a coupled system of one $I = \frac{1}{2}$ nucleus and one electron, S. The first plus or minus sign in the energy level label refers to the corresponding sign of m_s. The energy, in radians per second, is shown alongside each level.

electron and one $I = \frac{1}{2}$ nucleus. The W's are time-independent transition probabilities (*wahrscheinlichkeiten*) for first-order decay of the deviations of the populations, $N_{m_S m_I}$, to their equilibrium values, $N^\circ_{m_S m_I}$. The transition probabilities with numerical subscripts arise from interaction of the two spins. Those with I and S subscripts refer to contributions to relaxation which would also be present when the spins are isolated from each other.

By recognizing that the z components of $\mathbf{M}_I$ and $\mathbf{M}_S$ (the electron spin magnetization) are related to the populations of the four levels by

$$M_{Iz} = \tfrac{1}{2}\gamma_I \hbar [N_{-+} + N_{++}) - (N_{+-} + N_{--})]$$
$$M_{Sz} = \tfrac{1}{2}\gamma_S \hbar [N_{++} + N_{+-}) - (N_{--} + N_{-+})] \tag{8}$$

it may be shown that[19]

$$d(M_{Iz} - M^\circ_{Iz})/dt = -T_{1I}^{-1}[(M_{Iz} - M^\circ_{Iz}) + (\gamma_I/\gamma_S)\S_I f_I (M_{Sz} - M^\circ_{Sz})] \tag{9a}$$

$$d(M_{Sz} - M^\circ_{Sz})/dt = -T_{1S}^{-1}[(\gamma_S/\gamma_I)\S_S f_S (M_{Iz} - M^\circ_{Iz}) + (M_{Sz} - M^\circ_{Sz})] \tag{9b}$$

where

$(W_0 + 2W_1 + W_2) + 2W_I = T_{1I}^{-1}$ = nuclear spin–lattice relaxation time^{-1}
$(W_0 + 2W_1' + W_2) + 2W_S = T_{1S}^{-1}$ = electron spin–lattice relaxation time^{-1}
$2W_I = T_{1I,0}^{-1}$ = nuclear relaxation time^{-1} in the absence of electron
$2W_S = T_{1S,0}^{-1}$ = electron relaxation time^{-1} in the absence of nucleus
$(W_2 - W_0)/(T_{1I}^{-1} - T_{1I,0}^{-1}) = \S_I$ = nuclear coupling parameter
$(W_2 - W_0)/(T_{1S}^{-1} - T_{1S,0}^{-1}) = \S_S$ = electron coupling parameter
$(T_{1I}^{-1} - T_{1I,0}^{-1})/T_{1I}^{-1} = f_I$ = nuclear leakage factor
$(T_{1S}^{-1} - T_{1S,0}^{-1})/T_{1S}^{-1} = f_S$ = electron leakage factor

The coupling parameters and leakage factors thus defined are dimensionless quantities with absolute values between zero and one. In the absence of coupling, the nuclear and electron magnetizations decay exponentially to their equilibrium values. With coupling, however, the nuclear magnetization is driven by the deviation of the *electron* magnetization via the second term in Eq. (9a).

B. Electromagnetically Pumped DNP

In electromagnetically pumped DNP experiments the electron spin transitions $(--) - (+-)$ and $(-+) - (++)$ are strongly irradiated (pumped) while the NMR transitions $(++) - (+-)$ and $(-+) - (--)$ are observed, most commonly under conditions of lower resolution than in analytical NMR.[10,10a,10b] Under steady state irradiation it follows from Eqs. (9) that

$$V = (M_{Iz} - M_{Iz}^\circ)/M_{Iz}^\circ = (\gamma_s/\gamma_I)\S_I f_I S_s \qquad (10)$$

where

$$S_s = (M_{Sz} - M_{Sz}^\circ)/M_{Sz}^\circ = \text{saturation factor} \qquad (11)$$

For protons $\gamma_s/\gamma_H = -658$, which accounts for the large enhancements which are available in systems with unpaired electrons. It should also be pointed out that this ratio is *smaller* for protons than for any other known nucleus except ^{3}H. For example, the corresponding ratios for ^{13}C, ^{15}N, ^{19}F, and ^{31}P are -2617, $+6493$, -699, and -1625, respectively. Complete saturation of the electron transitions ($S_s = 1$ or infinite electron spin temperature) is usually not achieved; however, measurement of V at several different pumping power levels makes it possible to extrapolate to the value of V for complete saturation, V_∞. The experimentally determined value of V_∞ thus depends on both $\S_I$ and f_I which depend in turn on the details of relaxation mechanisms and, therefore, on the structures and environments of the molecules containing the nuclei and unpaired electrons.

C. Applications of Electromagnetically Pumped DNP

All but one[20] of the studies of DNP in liquids so far reported have employed systems in which the unpaired electrons and nuclei are on *different* molecules. Consequently, the interactions responsible for coupled relaxation are fairly weak and usually vary rapidly with a frequency spectrum governed by Brownian motion. Furthermore, these applications have been limited to systems of stable free radicals and the enhanced nuclei have often been those in the solvent molecules. The principal application, therefore, has been in studying the weak interactions between structurally complex, stable free radicals and diamagnetic compounds. The relationship between structure and the observed enhancement, however, depends directly on incompletely developed and documented time-dependent models used for describing the interactions. The models so far used to describe these weak intermolecular interactions are discussed in the reviews[10,10a,10b] previously mentioned and will not be dealt with here.

We concentrate instead on intramolecular interactions which give rise to coupled electron–nuclear relaxation in the free radicals themselves. This is also an area, however, in which there are virtually no experimental data available for the reactive free radicals of most interest in organic chemistry. We therefore must confine ourselves to a description of the structural parameters on which relaxation rates and enhancement factors might be expected to depend in these species. Information on the rather extensive studies of nuclear relaxation in chemically stable paramagnetic species may be found in recent reviews of the subject.[21,21a–c]

D. Time-Dependent Intramolecular Electron–Nuclear Couplings in Organic Radicals

Two principal types of electron–nuclear coupling are responsible for coupled relaxation of nuclei and electron spins in the same free radical.

1. Dipole–Dipole Interaction

This interaction arises for the classical energy of coupling of two magnetic dipoles, $\gamma_I \hbar \mathbf{I}$ and $\gamma_s \hbar \mathbf{S}$. The Hamiltonian for this interaction is

$$\mathcal{H}^D = -\gamma_s \gamma_I \hbar^2 r^{-3}[\mathbf{I} \cdot \mathbf{S} - 3(\mathbf{S} \cdot \mathbf{r})(\mathbf{r} \cdot \mathbf{I}) r^{-2}] \tag{12}$$

where $\mathbf{r}$ is the vector joining the two dipoles. For molecules tumbling in solution $\mathcal{H}^D$ will be time dependent because of the time dependence of the direction and, when $\mathbf{I}$ and $\mathbf{S}$ are not on the same molecule, also will depend on changes in the magnitude of $\mathbf{r}$. For the case of interest here, where the interaction is intramolecular, the transition probabilities are[17]

$$W_0 = \tfrac{1}{10} K[1 + (\omega_s + \omega_I)^2 \tau_c^2]^{-1}$$
$$W_1 = \tfrac{3}{20} K[1 + \omega_I^2 \tau_c^2]^{-1}$$
$$W_1' = \tfrac{3}{20} K[1 + \omega_s^2 \tau_c^2]^{-1}$$
$$W_2 = \tfrac{6}{10} K[1 + (\omega_s - \omega_I)^2 \tau_c^2]^{-1}$$
$$K(\text{seconds}^{-1}) = \gamma_I^2 \gamma_s^2 \hbar^2 (\tau_c / b^6)$$
$$= 4.22 \times 10^{24} (\gamma_I / \gamma_s)^2 \tau_c (\text{seconds}) b^{-6} \text{ (angstroms)} \tag{13}$$

In Eqs. (13), b is the distance between the nucleus and the unpaired electron (assumed to be at a point). The correlation time τ_c is a time characteristic of the tumbling motion of molecules in solution. Roughly speaking, it is the reciprocal of the "tumbling rate" of the radical. τ_c is often approximated[17] by the Debye formula for a spherical molecule of radius a in a liquid of viscosity η at temperature T:

$$\tau_c = 4\pi \eta a^3 / 3kT \tag{14}$$

Consider the application of Eqs. (13) and (14) to the estimation of the (so far hypothetical) enhancement factor obtained by complete saturation of the unpaired electron in the phenyl radical. We assume the structure in Fig. 2 and imagine that it is dissolved in CCl_4 at room temperature (viscosity 0.91 cP) in a magnetic field of 14,000 G (60 MHz proton NMR). The following

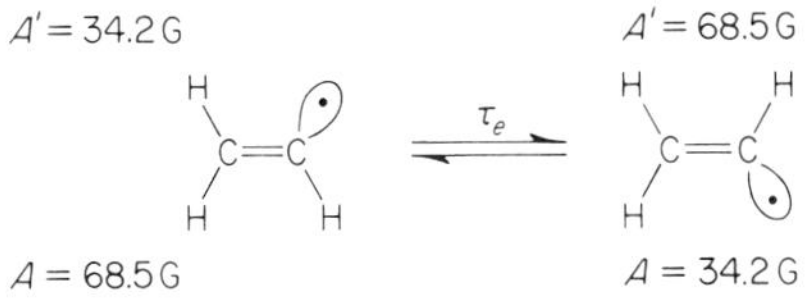

$A' = 34.2\,G$ $A' = 68.5\,G$

$$\begin{array}{c} H \\ \diagdown \\ C=C \\ \diagup \quad \diagdown \\ H \quad H \end{array} \underset{\tau_e}{\rightleftarrows} \begin{array}{c} H \quad H \\ \diagdown \quad \diagup \\ C=C \\ \diagup \quad \diagdown \\ H \end{array}$$

$A = 68.5\,G$ $A = 34.2\,G$

FIG. 2. Structural parameters used in the text to compute the enhancement factor and T_1 for the *ortho* protons in phenyl radical.

values are then obtained for the interaction parameters and characteristic times

$W_0 = 0.90 \times 10^4$ seconds^{-1} $\tau_c = 1.4 \times 10^{-11}$ second
$W_1 = 18.55 \times 10^4$ Assume $f_I = 1$
$W_1' = 1.35 \times 10^4$ Then $T_{1I} = 2.3 \times 10^{-6}$ second
$W_2 = 5.42 \times 10^4$ $\mathcal{S}_I = 0.10$
     $V_\infty = -66$

It is clear that estimates such as these are extremely sensitive to both a and b, or conversely, that experimental determination of τ_c and K should provide good values for these two structural parameters (within the framework of point dipole and spherical molecule approximations). It is, in fact, possible to determine separately the correlation time by taking advantage of the field dependence (through ω_s and ω_I) of the denominators in Eq. (13). At fields sufficiently low that $\omega_s^2 \tau_c^2 < 1$, for example, the enhancement in the present pure dipole case approaches $\frac{1}{2}(\gamma_s/\gamma_I) = -329$ and relaxation of the nuclear magnetization occurs predominantly via the cross-relaxation process, W_2. This is the maximum enhancement allowed by a purely dipolar interaction. By fitting observed and calculated field dependences of the enhancement factor and determining T_{1I}^{-1} when the dipole interaction dominates ($f_I = 1$) one can, in principle, determine both K and τ_c and therefore extract both b and a in the above model. This has so far never been carried out for the intramolecular interactions in a free radical, but the field dependence of V_∞ has played a primary role in analyzing the enhancement resulting from intermolecular cross-relaxation processes.[10a]

2. Time-Dependent Scalar Interaction

In addition to the direct interaction in Eq. (12), nuclei and electrons in free radicals are coupled by a scalar interaction

$$\mathcal{H}^s = A(t)\mathbf{I}\cdot\mathbf{S} \qquad (15)$$

with a coupling constant, or hyperfine splitting, A, which does not vary as the molecule tumbles in solution but may be time dependent because of conformational or chemical exchange processes involving the free radical. It was, in fact, a time-dependent A for nuclei interacting with conduction electrons in metals which formed the basis for Overhauser's original proposal.[4] These effects are now well known in ESR spectra and give rise to selective line broadening.[22]

Random modulation of the value of A contributes only to the cross-relaxation W_0 in Fig. 1. In particular, if A varies randomly with a mean square amplitude $\langle A^2 \rangle$ (in seconds^{-2}) with a frequency τ_e^{-1} which is greater than the electron relaxation rate, $T_{1S,0}^{-1}$, the cross-relaxation transition probability is[23]

$$W_0 = \langle A^2 \rangle \tau_e [1 + (\omega_s + \omega_I)^2 \tau_e^2]^{-1} \qquad (16)$$

As an example we consider the conformation change in the vinyl radical, shown in Fig. 3, which interchanges the hyperfine splitting of the two β protons.[24] From the averaging effects on the ESR line positions, but the absence of significant linewidth effects, it has been estimated that $3 \times 10^{-8} < \tau_e < 3 \times 10^{-10}$ second. Since $\langle A^2 \rangle = \frac{1}{4}(A - A')^2$ it follows that $7 \times 10^5 < W_0 < 7 \times 10^7$ second^{-1}. Furthermore, spin–lattice relaxation of both the electron and nuclei would probably be dominated by W_0 in this case. Consequently, $\S_I = +1$ and $V = +658$, that is, *the sign of the enhancement factor for protons is positive when the scalar interaction predominates and negative for dominant dipolar cross relaxation.*

The last observation points out the principal advantage which DNP has over conventional measurements of relaxation times. DNP's specificity for cross-relaxation processes makes it generally more sensitive to the type(s) of mechanism(s) responsible for[6,10,25] relaxation and justifies the necessity for more complex instrumentation if one is interested primarily in understanding the nature of thermal spin relaxation in systems with both nuclei and unpaired electrons.

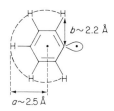

FIG. 3. Effect of conformation change on the hyperfine splittings of the β protons of vinyl radical.

III. CHEMICALLY INDUCED DYNAMIC NUCLEAR POLARIZATION (CIDNP)

A. Types of Spectra

Developments in the field of CIDNP during the four years of its existence have run a course familiar to practitioners of observational sciences such as astronomy but foreign and unsettling to those who have grown up with the idea that chemistry must be an experimental science. The initial developments in the field were for the most part purely observational, each new measurement giving an unanticipated result. An empirical accumulation of spectra gradually became available but was put to use initially only as an indicator of the intermediacy of free radicals in chemical reactions. It has only been quite

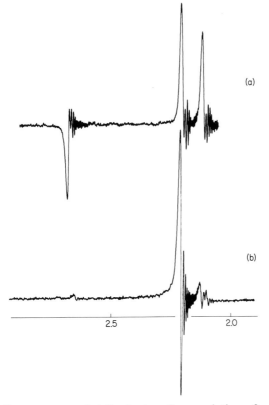

FIG. 4. (a) Spectrum recorded 5 minutes after a solution of acetyltrichloroacetyl peroxide and iodine in carbon tetrachloride was warmed to 50°. Assignments of lines are: CH_3I (2.1 δ), peroxide (2.2 δ), CH_3CCl_3 (2.7 δ). (b) Spectrum taken at 0° immediately following scan (a).

recently that a satisfactory theory has been developed which makes possible predictions, stringent tests, and detailed interpretation of the experimental results.

The experimentally observed CIDNP phenomena fall into two categories.

1. Net Polarization

Spectra of this type are illustrated in Fig. 4 by the methyl protons in CH_3CCl_3 (I) and CH_3I (II) formed in the reaction

$$CH_3CO_2CO_2CCl_3 + I_2 \longrightarrow CH_3I + CH_3CCl_3 + ICCl_3 + 2CO_2 \quad (17)$$
$$\quad\quad\quad\quad\quad\quad\quad\quad\quad\quad\quad (II) \quad\quad (I)$$

The absorption line from protons in each product has an integrated area which is more than an order of magnitude greater than it should be (compare spectra A and B). In the case of CH_3CCl_3 the enhancement is negative, i.e., emission (E), while CH_3I exhibits an enhanced absorption (A) line of nearly the same absolute intensity. Polarization of this type, in which the integrated intensity of a line or set of lines arising from a single set of magnetically equivalent nuclei differs from its equilibrium value, has become known as *net polarization*.

2. Multiplet Effect

An example of the second kind of enhancement commonly encountered is shown in Fig. 5 in the protons of styrene (III), isobutylene (IV), and ethylbenzene (V) formed in the reaction of *t*-butyllithium with (2-bromoethyl)-benzene.

$$C_6H_5CH_2CH_2Br + LiC(CH_3)_3 \longrightarrow$$
$$\quad\quad\quad\quad\quad\quad C_6H_5CH=CH_2 + CH_2=C(CH_3)_2 + C_6H_5CH_2CH_3 \quad (18)$$
$$\quad\quad\quad\quad\quad\quad\quad (III) \quad\quad\quad\quad (IV) \quad\quad\quad\quad (V)$$

In each of these products the integrated area of a multiplet arising from a single set of protons is the same as it would be in the unenhanced product although individual lines within the multiplet are strongly enhanced. This occurs because the emission of one side of the multiplet (low field in this example) is canceled by the enhanced absorption on the other side. These types of spectra are commonly said to exhibit a *multiplet effect*. Each of the multiplets shown in Fig. 5 has a *phase* of enhancement which would be labeled E/A, signifying emission downfield of absorption.

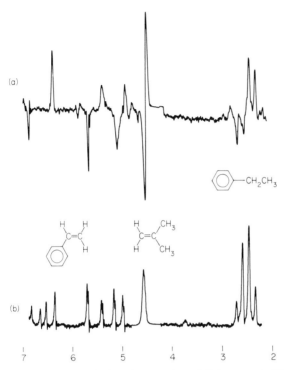

FIG. 5. (a) Spectrum taken during the reaction of (2-bromoethyl)benzene with *tert*-butyllithium in pentane showing CIDNP from protons in the three products indicated. (b) Spectrum of a synthetic mixture showing the normal intensities of the lines from the reaction products.

B. The Overhauser Model

The earliest explanations of CIDNP invoked an analogy with electromagnetically induced DNP discussed in the previous section. Although this explanation has now been superseded by a model conceptually completely different, the effects which it predicts are still possible, and it should only be a matter of time until they are unambiguously documented. Therefore we present briefly the essential features of this model and point out the failures which necessitated a search for a new explanation for most CIDNP effects.

1. Effect of Chemical Reaction on Nuclear and Electron Magnetization in a Free Radical

One of the first examples of CIDNP[11] was the observation of an emission line from the protons of benzene produced as a product S of scavenging by solvent proton abstraction of phenyl radicals, F, produced by thermal decomposition of benzoyl peroxide, R.

$$\text{Ph-}(CO_2)_2 \xrightarrow{k_R} 2\, \text{Ph}\cdot \xrightarrow[HR']{k_S} \text{Ph-H} \xrightarrow{T_{1,0}^{-1}} \text{Ph-H}$$

$$\quad R \qquad\qquad F \qquad\qquad S \qquad\qquad S_0$$

The rate constant $T_{1,0}^{-1}$ is for spin–lattice relaxation of the protons in the benzene product.

To show how the combination of chemical reaction and electron–nuclear cross relaxation may induce enhancement, we now make the following assumptions: (a) The phenyl radicals produced by decomposition of R are formed with $M_{Sz}^F = 0$, that is, the electron spin levels are formed with an infinite spin temperature by virtue of unpaired electron spins having been formed from a singlet state reactant with no preference for $m_S = \pm\frac{1}{2}$; (b) The nuclear magnetization of the reactant M_{Iz}^R remains at its equilibrium value throughout the reaction; (c) The nuclear magnetization of the free radical M_{Iz}^F is preserved in the scavenging step; and (d) the elementary steps obey pseudo-first-order kinetics. If these assumptions hold, Eqs. (9) are modified by the reaction, and we may write the following expressions for the rates of change of magnetization in the free radical and product:

$$d(M_{Iz}^S - M_{Iz}^{S,0})/dt = k_s(M_{Iz}^F - M_{Iz}^{F,0}) - T_{1,0}^{-1}(M_{Iz}^S - M_{Iz}^{S,0}) \qquad (19a)$$

$$d(M_{Iz}^F - M_{Iz}^{F,0})/dt = -T_{1I}^{-1}[1 + k_s/T_{1I}^{-1})(M_{Iz}^F - M_{Iz}^{F,0})$$
$$+ (\gamma_I/\gamma_S)\S_I f_I(M_{Sz}^F - M_{Sz}^{F,0})] \qquad (19b)$$

$$d(M_{Sz}^F - M_{Sz}^{F,0})/dt = -T_{1S}^{-1}[(\gamma_S/\gamma_I)\S_S f_S(M_{Iz}^F - M_{Iz}^{F,0})$$
$$+ (1 + k_s/T_{1S}^{-1})(M_{Sz}^F - M_{Sz}^{F,0})] - 2k_R(\gamma_S/\gamma_I)M_{Iz}^{R,0} \qquad (19c)$$

If we make the steady-state approximation for the electron and nuclear magnetizations in the free radical we obtain for the product nuclear magnetization

$$d[M_{Iz}^S - M_{Iz}^{S,0}]/dt = 2k_R V_F M_{Iz}^{R,0} - T_{1,0}^{-1}(M_{Iz}^S - M_{Iz}^{S,0}) \qquad (20)$$

where V_F, the enhancement factor for nuclear polarization in the free radical, is given by

$$V_F = (\gamma_S/\gamma_I)\S_I f_I[T_{1I}^{-1}/(T_{1I}^{-1} + k_s)]S_F \qquad (21)$$

and the saturation factor S_F for the electron magnetization in the free radical is

$$S_F = (M_{Sz}^{F,0} - M_{Sz}^F)/M_{Sz}^{F,0}$$
$$= k_s\{k_s + T_{1S}^{-1}[1 - \S_S f_S \S_I f_I T_{1I}^{-1}/(T_{1I}^{-1} + k_s)]\}^{-1} \qquad (22)$$

2. Value of the Enhancement Factor

Note the similarity between the expression for the enhancement factor in Eq. (21) and that given in Eq. (10) for electromagnetically pumped DNP. The main difference is that both the saturation factor and V_F itself now depend on the competition between chemical reaction of the radical via k_S and relaxation by way of T_{1S}^{-1} and T_{1I}^{-1}. We also see that V_F vanishes for *both* long and short values of k_S. If $k_S \gg T_{1S}^{-1}$, then $S_F = 1$, but V_F is decreased by the ratio of T_{1I}^{-1} to k_S which will be small because the nuclear relaxation rate in the free radical will never be more than a few times faster than the electron relaxation rate. Conversely, if $k_S \ll T_{1I}^{-1}$ the full effect of S_F is felt by V_F, but S_F is now decreased by the ratio of k_S to T_{1S}^{-1}.

We can obtain the maximum value of V_F for pure dipolar or pure scalar coupled relaxation by assuming that $f_S = f_I = 1$ and that the field is sufficiently low that the denominators in Eqs. (13) are unity. By substituting the appropriate quantities in Eq. (21) and finding the value of k_S which maximizes V_F, we find for dipolar relaxation a maximum value of -88 and for pure scalar relaxation a maximum of $+329$ for protons. By contrast, ordinary DNP yields maximum enhancements of -329 and $+658$, respectively, for the two types of interaction. This reflects primarily the less effective "pumping" by the chemical reaction.

The value of V_F is further reduced in the product S because of competing spin–lattice relaxation via the second term in Eq. (20). If $k_R < T_{1,0}^{-1}$, the enhancement in this example would be reduced by the ratio $2k_R/T_{1,0}^{-1}$.

3. Failures of the Overhauser Model

In Fischer's first report of nuclear polarization from the above reaction he found a maximum enhancement of -20 for the protons in benzene at 40 MHz and was able to reconcile this value with reasonable values for the transition probabilities (as in our example in Section II,D,1) and chemical rate constants, assuming a dominant dipolar interaction. Subsequently, however, reports appeared of enhancement factors of several hundred[25a] during some reactions, which clearly could not be accommodated by the above model.

A second difficulty arose in the other initial report[11a] of CIDNP in the protons of 1-butene formed in the reaction

$$CH_3CH_2CH_2CH_2Br + CH_3CH_2CH_2CH_2Li \longrightarrow$$
$$2CH_3CH_2CH_2CH_2 \cdot \longrightarrow CH_3CH_2CH^* = CH_2^* + \text{butane} + \text{octane} \quad (23)$$
$$4321$$

It was observed that the multiplets arising from the asterisked protons exhibited both some emission and some absorption lines of nearly the same intensity (multiplet effect). Although one could qualitatively explain enhanced absorption as arising from modulation of the hyperfine coupling of the 2 protons by rotation about the C—C bond,[12a] this could not explain the appearance of absorption lines for the 1 protons as well. Another difficulty is that the ESR spectra of alkyl radicals taken under somewhat similar conditions[24] do not seem to exhibit the effects expected[22] for significant coupled relaxation of electron and nuclear spins.

The problem, in short, with the Overhauser model is that its effects would be smaller than often observed, and it could not give rise to both enhanced absorption and emission lines in multiplets arising from the same set of magnetically equivalent nuclei. Consequently, it was necessary to look elsewhere for the correct explanation for the CIDNP phenomena.

C. The Radical Pair Model

1. Qualitative Description

As we have seen in Section I.A, in a slow passage, high-resolution NMR experiment the intensity I_{nm} of a transition between levels n and m is given by

$$I_{nm} = K_{nm}(N^n - N^m) \qquad (24)$$

where N is the population of a level and K_{nm} depends on instrumental parameters and a transition probability determined by the chemical shifts and spin–spin splittings. The NMR spectrometer then becomes simply a device for measuring the *relative concentrations* of two molecular species which differ in the magnitude of their nuclear spin angular momenta.

The original explanations (Section III.B) of CIDNP assumed that these concentrations were driven away from equilibrium by selective exchange of energy with the surroundings *via* simultaneous relaxation of electron and nuclear spins. There is, however, a conceptually simpler explanation for the observed enhancements: Since the species of interest are the products of chemical reactions, such effects also will arise if the *yield* of product is greater for some spin states of the nuclei than for others, that is, *if the rate of formation of product is dependent on the nuclear spin state of the molecules undergoing reaction*. Since it is the electrostatic interaction of the electrons with the nuclei and with each other which is primarily responsible for determining reaction rate constants, a suitable mechanism for such nuclear spin selective reactions must be one in which, at a critical step in the reaction pathway, the nuclear magnetic moments can exert a force on the electrons which is comparable to the electrostatic forces which the electrons experience. The starting point for the alternative model recently proposed[13] to account for CIDNP is the supposition that these conditions are satisfied in the limit of weak coupling of electron spins during free radical encounters at the distances characteristic of radical pair separation in a solvent cage. It then becomes possible for nuclear magnetic moments to induce intersystem crossing between singlet and triplet electronic states of radical pairs and thereby control the rates of reactions which depend on electron spin multiplicity.[26]

The present model for nuclear spin selective reactions assumes that

(a) Product formation occurs only from singlet electronic states (although it could be modified easily to include the unlikely possibility of products formed in triplet states)

(b) Intersystem crossing is induced by the magnetic fields generated by the nuclear spins *via* the hyperfine interaction

(c) Nuclear spin states do not change during the rapid formation or breaking of chemical bonds.

Justification for each of these assumptions is given below.

a. Product Formation. In the formation of a two-electron chemical bond between two atoms, the lowest lying singlet and triplet states differ in energy. This so-called "exchange energy" is usually such that the triplet state is of higher energy than the singlet state for all internuclear distances. Thus, a pair of atoms approaching each other in the singlet state will be attracted to form a stable bond, while a pair in the triplet state will repel each other. The repulsive interaction can be equated with a low probability for radical–radical reaction.[26]

b. Intersystem Crossing. The Hamiltonian representing the exchange coupling of electron spins[27] S_1 and S_2 is

$$H_{\text{exchange}} = -J(\tfrac{1}{2} + 2S_1 \cdot S_2) \tag{25}$$

In Eq. (25), the exchange integral J is usually negative and decreases rapidly with increasing internuclear distance. Calculations for the hydrogen molecule[29] have shown that J falls exponentially from approximately 5 eV at the equilibrium bond distance to a value smaller than the electron–nuclear hyperfine interaction at distances greater than 5 Å. Although little is known about the variation of J with distance for more complex systems,[30] the picture presented here is likely to be at least qualitatively correct. For caged radical pairs in solution, of course, the modulation of interradical separation by Brownian motion gives rise to a time dependence of J related to rotational and translational motions of molecules in liquids.[31]

A qualitative idea of the requirements[33,34,34a] for the mixing of singlet and triplet states of weakly coupled radical pairs in the presence of internal and external magnetic fields may be obtained by referring to Fig. 6, which presents

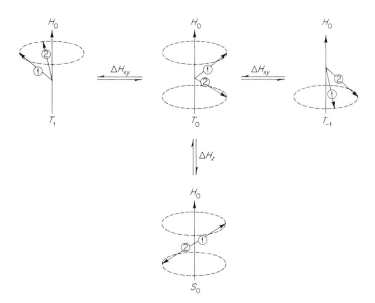

FIG. 6. Schematic representation of the motion of two weakly coupled electron spins in a magnetic field. The z axis is the direction of the applied field, H_0.

a schematic representation of the motion of two electron spin vectors which are weakly coupled to each other but strongly coupled to a magnetic field. The three magnetic substates of the triplet state T differ in the Z components of the net spin $(0, \pm 1)$, while in the singlet state S_0, all components of the two electron spins cancel. For the special case of mixing of S_0 and T_0 it can be seen that all that is required is a rotation of one of the electron spin vectors relative to the other about the Z axis. The magnetic field in the Z direction experienced by an electron includes both the applied field and the internal fields arising from nearby nuclear spins and electron orbital motion. Thus, in a large external field and with weak exchange coupling the electrons S_1 and S_2 precess about the Z axis with Larmor frequencies given in Eq. (26a).

$$\omega_1 = \beta \hbar^{-1} \left[g_1 H_0 + \sum A_1 m \right]$$
$$\omega_2 = \beta \hbar^{-1} \left[g_2 H_0 + \sum A_2 m \right] \quad (26a)$$

$$(\omega_1 - \omega_2) \equiv 2\delta_n = \beta \hbar^{-1} \left[(g_1 - g_2)H_0 + \left(\sum A_1 m - \sum A_2 m \right) \right] \quad (26b)$$

The first term arises from the Zeeman interaction with (possibly) different g factors for the two electrons and the second term from precession in the "hyperfine field" arising from nuclei with hyperfine splittings A (in gauss) and magnetic quantum numbers m.

If the two unpaired electrons in the radical partners have the projections of their spins in the X–Y plane separated by an angle $\theta(0)$ at $t = 0$, after a time τ, the angle between the spins will be

$$\theta(\tau) = \theta(0) + 2\delta_n \tau \tag{27}$$

Since $\theta = 0°$ and $180°$ correspond to triplet and singlet states, respectively, if $\delta_n \neq 0$, a pair which is born in the singlet state will develop some triplet character at later times and *vice versa*. Furthermore, the rate of mixing of states will depend on differences of both the g factors and hyperfine fields characterizing S_1 and S_2. The former term connotes a field dependence of the intersystem crossing rate, the latter term a nuclear spin dependence. The more rigorous extension of this classical picture presented in the following section shows that both terms must be included to account for the observed effects of spin selective reactions, since dephasing induced by g factor differences may be canceled or reinforced by the hyperfine fields arising from nuclei with opposite spins. It also should be pointed out that in magnetic fields such as those typically employed for high-resolution NMR (which are much larger than the internal fields, i.e., greater than a few hundred gauss), the hyperfine field components in the X–Y plane will be essentially zero, thus preventing mixing of the states $T_{\pm 1}$.[34a]

Typical values of Zeeman and hyperfine field differences for organic free radicals are a few gauss,[35] which amounts to a frequency of dephasing of electron spins ("intersystem crossing" frequency) of 10^7–10^8 rad second^{-1}. During the lifetime of a typical radical encounter pair in solution[36] (10^{-9}–10^{-10} second) one would, therefore, expect a percent or so of mixing of singlet and triplet states with $\delta_n \neq 0$ in the limit of very weak exchange coupling, but less mixing if the exchange interaction is larger than δ_n. This small admixture gains in significance if we reflect that 50% mixing of an unreactive triplet for some states ($\delta_n \tau > 1$) and no mixing for others ($\delta_n = 0$) would give rise to an NMR enhancement of approximately 10^5, or about 10^2 larger than has so far been observed!

c. Nuclear Spin States. The assumption that nuclear spin states do not change during the steps of bond making or breaking is based on the rapidity with which such processes are expected to occur. In order for the nuclear-spin angular momentum to change direction during a molecular vibration ($\approx 10^{-13}$ second) an interaction much larger than the hyperfine coupling would be required. Furthermore, in high fields the nuclear spin orientation will not change even during the cage lifetime[37] because of the absence of the necessary internal fields in the X and Y directions. In this picture of the cage reaction at high fields, the nuclei serve as unchanging labels for the molecular fragments which are carried from the reactant, through radical pair, to product. The first opportunity for scrambling of the labels then comes only after the radicals have escaped from the cage and takes place *via* thermal relaxation at a rate several orders of magnitude slower than the intersystem crossing rates of concern here (see Section II,D).

2. Quantitative Formulation

Although the preceding classical model serves as an argument for the plausibility of nuclear spin effects on rates of radical recombination reactions, its qualitative nature makes it of little value as a device for quantitative predictions. We therefore present a more rigorous treatment of this mechanism of singlet–triplet mixing—an extension of that presented elsewhere [13-15a,34a-c] which makes possible, in principle, quantitative predictions of enhancement factors.

The solution of the problem has three parts:

(a) Obtaining an expression for the intersystem crossing rate which depends on nuclear spin quantum numbers

(b) Incorporating this rate into a scheme for the overall dynamic behavior of the system to obtain relative populations of nuclear spin states in the products of the reaction

(c) Bridging the gap between populations of nuclear spin states and the observable NMR spectrum.

Operations (a) and (c) are straightforward and require only a knowledge of the parameters describing phenomenological spin Hamiltonians for the radical pair and products. In fact, the present knowledge of magnetic resonance parameters for many free radicals and diamagnetic compounds [35,38,38a,38b] is sufficiently good that the only imponderable in (a) and (c) is the average strength of the electron exchange coupling in a radical pair in solution. The procedure (b), however, requires a knowledge of such mechanistically important, but poorly understood, quantities as the mean lifetime of a caged radical pair, the relative rates of geminate recombination and escape from a cage encounter ("cage effect"), the rates of nuclear relaxation in free radicals and the rate constants for atom transfer reactions, and diffusive encounters of free radicals. Fortunately, (a) and (c) often are sufficient for an understanding of the *relative* intensities of NMR lines arising from a single product, although (b) also must be considered for a proper accounting of the *absolute* intensities of lines from reaction products.

a. Rate of Singlet–Triplet Mixing. A fundamental assumption in the preceding section was that the rate of reaction of a radical pair is proportional to the probability of finding the pair in the singlet state at some time after its birth. This probability $\rho_s^n(t)$ is the electronic singlet state diagonal element of the density matrix [39] for the radical pair in the nuclear spin state n. It may be calculated for the present problem [13–15a,34a] by solving the equation [39]

$$i\dot{\rho} = [H, \rho] \tag{28a}$$

using the truncated spin Hamiltonian [13–15a,34a]

$$H = \left[\tfrac{1}{2}(\omega_1 + \omega_2) + \tfrac{1}{2}\left(\sum A_1 I_z + \sum A_2 I_z\right)\right] S_z - J(S^2 - 1)$$
$$+ \left[\tfrac{1}{2}(\omega_1 - \omega_2) + \tfrac{1}{2}\left(\sum A_1 I_z - \sum A_2 I_z\right)\right](S_{1z} - S_{2z}) \tag{28b}$$

where $\omega_1 = \hbar^{-1} g_1 \beta H_0$, $S = S_1 + S_2$, and A, J, and H are in rads/second. After an initial sudden change during the radical pair forming step, J is assumed to be time independent and the magnetic field is assumed to be high enough that the electron Zeeman interaction is sufficiently large [13–15a,34a–c] to justify neglect of off-diagonal terms in the hyperfine interaction which mix $T_{\pm 1}$ with S_0 and T_0. We thus use a basis set $\chi_i = |S, m_s\rangle |m_I\rangle$ for the representation of ρ, which is a product of electron singlet or triplet functions, $|S, m_s\rangle$ and the nuclear spin functions $|m_I\rangle$. Solving for $\rho_s^n(t)$ for the nth state of the nuclear spin system, we obtain

$$\rho_s^n(t) = \rho_s^n(0) - \tfrac{1}{2}[\rho_s^n(0) - \rho_t^n(0)](\delta_n^2/\omega_n^2)[1 - \cos 2\omega_n t] \tag{29}$$

where $\omega_n = (J^2 + \delta_n^2)^{1/2}$; $\rho_s^n(0)$ and $\rho_t^n(0)$ are the probabilities that the radical pair is born in the S_0 or T_0 states, respectively; and δ_n is defined in Eq. (26b). For the special case of a pair formed initially in the singlet state $[\rho_s^n(0) = 1, \rho_t^n(0) = 0]$, Eq. (29) becomes identical to the expressions previously given.[13–15a] Furthermore, we see that $\rho_s^n(t)$ oscillates in time, as in our qualitative picture. The frequency depends on the values of J and δ_n, while the amplitude of oscillation depends both on the ratio δ_n^2/ω_n^2 and on the initial probabilities. For example, as $[\rho_s^n(0) - \rho_t^n(0)]$ approaches zero (equal probabilities of S_0 and T_0 state birth) the amplitude of oscillation drops to zero. In general, both the amplitude and frequency of oscillation will depend on the nuclear spin quantum numbers except for the special (and probably unrealistic) case corresponding to the classical model, $\delta_n^2 \gg J^2$, in which only the frequency varies with nuclear spin state.

b. Classifications of Reactions and Products. To correlate $\rho_s^n(t)$ with an observable reaction rate it is useful to trace the history of the molecular fragment which contains a given set of nuclei as it travels through the reaction sequence. The chemical species, or host molecules, in which the nuclei may reside, and the reactions they undergo are summarized below.

Host molecules:

R = diamagnetic reactant
C_g = geminate combination product
F = free radical
C_d = combination product from "diffusive" or "random" encounters of free radicals
S = product formed by trapping of F with scavenger

The concentrations of these species may be changed by the following set of reactions, where $\phi^n_{s(t)}$ and ϕ^n_r are nuclear spin dependent probabilities for product formation in radical–radical reaction steps (a) and (c), and the k's are, for simplicity, pseudo-first-order rate constants.

Reaction steps:

(a) $R^n \xrightarrow{k_r \phi^n_{s(t)}} C_g^n$ Geminate combination via radical pair born in singlet (s) or triplet (t) state

(b) $R^n \xrightarrow{k_r(1-\phi^n_{s(t)})} F^n$ Escape from geminate encounter

(c) $F^n \xrightarrow{k_d \phi^n_r} C_d^n$ Random diffuse combination

(d) $F^n \xrightarrow{k_s} S^n$ Scavenging of free radicals

(e) $X^n \xrightarrow{T_x^{-1}} X_0^n$ Nuclear spin–lattice relaxation in R, F, C, and S

For a complex set of reactants there will in general be several different species which fall into the categories C_g, C_d, and S. For example, C_g and C_d include both coupling and disproportionation products which may be formed in different proportions or with different radical partners in geminate and diffusive combinations. Typical values (second^{-1}) of the rate constants in this set of reactions under the conditions where CIDNP is observed will be $k_r \approx 10^{-2}$; T_R^{-1}, T_C^{-1}, $T_S^{-1} \approx 0.1$–1; $k_d \approx 10^3$; $T_F^{-1} \approx 10^4$; $k_S \approx 10^3$–10^8.

Although the above set of coupled first-order reactions can be solved straightforwardly for the functional dependence of level populations on k's and ϕ's, the general solution leaves much to be desired in the way of physical insight. Therefore, the evaluations of ϕ^n_s, ϕ^n_t, and ϕ^n_r are first considered, and then some simple specific examples are given of the enhancement expected and observed for products of the type C_g, C_d, and S.

c. Nuclear Spin Dependent Radical Pair Reaction Probabilities. A radical pair P formed in spin state n either from a reactant or by a diffusive encounter of free radicals will disappear by one of the two processes shown in (30).

$$C^n \xleftarrow{k_c \rho_s^n(t) + k'_c} P^n \xrightarrow{\tau^{-1}} F^n \qquad (30)$$

(a) The fragments may diffuse apart at a rate which we will take to be described by a mean diffusion lifetime,[40] τ, or (b) they may react to form a

3. Dynamic Nuclear Polarization 121

combination product with a rate constant which we assume will be $k_c \rho_s^n(t) + k'_c$, where k_c is the rate constant for reaction of a singlet state pair and k'_c is the rate constant for possible reaction pathways which are independent of intersystem crossing induced by the mechanism considered here. The three rate constants may be related by two nuclear spin *independent* quantities: $a_s = (k_c + k'_c)(k_c + k'_c + \tau^{-1})^{-1} =$ cage effect for singlet pairs, and $r = k_c(k_c + k'_c)^{-1} =$ spin selectivity. The probability $P^n(t)$ that P will exist at time t if it was born at $t = 0$ is given by

$$P^n(t) = \exp\left[-t(\tau^{-1} + k'_c) - k_c \int_0^t \rho_s^n(t')\,dt'\right] \qquad (31)$$

If it is also assumed that there is negligible probability that P will be reformed from the original fragments once they have diffused apart (or reacted), the probability ϕ^n that the pair will eventually react to give a combination product becomes

$$\phi^n = 1 - \tau^{-1} \int_0^\infty P^n(t)\,dt \qquad (32)$$

Because of the dependence of $\rho_s^n(t)$ on the electron spin state at $t = 0$, Eq. (32) must in general be averaged over a distribution of values of $\rho_s^n(0)$ and $\rho_t^n(0)$. However, $\rho_s^n(0)$ and $\rho_t^n(0)$ are not completely independent of each other, being related by the condition that $\rho_s^n(0) + \rho_t^n(0) + \rho_+^n(0) + \rho_-^n(0) = 1$, where $\rho_\pm^n(0)$ are the probabilities of the pair being formed in the other two triplet levels.

The observable average value of ϕ^n will then be

$$\langle \phi^n \rangle = 1 - \tau^{-1}\left[\oint\int_0^\infty \phi^n(t)g(\rho_i)\,dt\,d\rho_i \Big/ \oint g(\rho_i)\,d\rho_i\right] \qquad (33a)$$

where $\oint d\rho_i$ denotes integration over the space spanned by the four initial probabilities, subject to the restriction that they sum to unity and have positive values, and $g(\rho_i)$ is a distribution function within this space. Although Eq. (33a) may in principle be evaluated for arbitrary distributions of initial conditions, the evaluation becomes especially simple if one expands $\rho^n(t)$ as a polynomial in $\rho_s^n(t)$ and discards terms higher than second order. This amounts to the assumption that $k_c \ll (k'_c + \tau^{-1})$, i.e., the nuclear spin dependent reaction pathway is minor, as has been implicitly assumed in previous treatments of this model.[13-15a] Ignoring nuclear spin independent terms higher than first order in k_c and assuming that $\delta_n \tau < 1$, one obtains

$$\langle \phi^n \rangle = a_s \{1 - r(1 - a_s)[1 - \langle \rho_s(0) \rangle]\}$$
$$- 2a_s r W_n (1 - a_s)^3 \delta_n^2 \tau^2 \{[\langle \rho_s(0) \rangle - \langle \rho_t(0) \rangle](1 + 2a_s r W_n)$$
$$- 2a_s r[\langle \rho_s^2(0) \rangle - \langle \rho_s(0) \rho_t(0) \rangle](1 + W_n)\} \quad (33b)$$

where

$$W_n = [1 + 4\omega_n^2 \tau^2 (1 - a_s)^2]^{-1}$$

Three types of average values for the initial probabilities in brackets are expected to be most commonly encountered[34a]:

(s) singlet precursor:

$\langle \rho_s(0) \rangle = 1;$ $\quad \langle \rho_t(0) \rangle 0;$ $\quad \langle \rho_s^2(0) \rangle = 1;$ $\quad \langle \rho_s(0) \rho_t(0) \rangle = 0$

(t) triplet precursor:

$\langle \rho_s(0) \rangle = 0;$ $\quad \langle \rho_t(0) \rangle = \frac{1}{3};$ $\quad \langle \rho_s^2(0) \rangle = 0;$ $\quad \langle \rho_s(0) \rho_t(0) \rangle = 0$

(r) random precursor:

$\langle \rho_s(0) \rangle = \frac{1}{4};$ $\quad \langle \rho_t(0) \rangle = \frac{1}{4};$ $\quad \langle \rho_s^2(0) \rangle = (\frac{1}{4})^2;$ $\quad \langle \rho_s(0) \rho_t(0) \rangle = 0$

Substitution of these initial conditions in Eq. (33b) yields:

Singlet precursor:

$$\phi_s^n = a_s - 2\delta_n^2 \tau^2 a_s r (1 - a_s)^3 W_n (1 - 2a_s r) \quad (34a)$$

Triplet precursor:

$$\phi_t^n = a_s[1 - r(1 - a_s)] + \tfrac{2}{3}\delta_n^2 \tau^2 a_s r (1 - a_s)^3 W_n (1 + 2a_s r W_n) \quad (34b)$$

Random precursor:

$$\phi_r^n = a_s[1 - \tfrac{3}{4}r(1 - a_s)] + \tfrac{1}{4}\delta_n^2 \tau^2 a_s^2 r^2 (1 - a_s)^3 W_n (1 + W_n) \quad (34c)$$

In the limit of small singlet cage effect, Eq. (34a) reduces to Eq. (34a')

$$\phi_s^n = a_s[1 - 2r\delta_n^2 \tau^2/(1 + 4\omega_n^2 \tau^2)] \quad (34a')$$

which is also obtained[13-15a] by averaging $\rho_s^n(t)$ over a distribution of electron–spin independent lifetimes for a pair born in the singlet state.

Inspection of Eqs. (34a)–(34c) gives rise to the following qualitative observations regarding the relative magnitudes of the nuclear spin dependent parts of $\langle \phi^n \rangle$ for the three types of initially formed pairs with the same values of δ_n, ω_n, and τ.

1. Large values of δ_n^2 increase ϕ_r^n and ϕ_t^n and decrease ϕ_s^n
2. The nuclear spin dependent parts of ϕ_t^n and ϕ_s^n are always of opposite sign
3. The nuclear spin dependence of all three probabilities tends to zero for both large and small singlet cage effects
4. The nuclear spin dependent part of ϕ_r^n decreases much more rapidly with decreasing singlet cage effect and spin selectivity than does that of either ϕ_t^n or ϕ_s^n, that is, *random encounters behave like triplet born encounters with reduced effectiveness*. This is in accord with qualitative predictions[41,41a] based on the assumption that random encounters should give rise to product preferentially from those pairs which undergo intersystem crossing most rapidly.

d. Geminate Combination Products. The majority of examples of CIDNP so far reported are for the case of products of reactions where only "geminate" encounters of radicals are considered. Strictly, this case applies only to reactions run with sufficiently high scavenger concentration to prevent diffusive random encounters of free radicals. The kinetic scheme for this case is

$$R^n \xrightarrow{k_r \phi_{ns(t)}} C_g^n \xrightarrow{T_c^{-1}} C_{g0}^n$$
$$\downarrow k_r(1-\phi_{ns(t)})$$
$$F_0^n \xleftarrow{T_F^{-1}} F^n \xrightarrow{k_s} S^n \xrightarrow{T_s^{-1}} S_0^n \tag{35}$$

The first-order kinetic equations for (35) may be readily constructed for the difference in concentration of two nuclear spin states n and m. If for simplicity the steady-state approximation is made for all species except R, which is assumed to be constant and in thermal equilibrium and if one ignores the contributions of unpolarized products, Eq. (36) is obtained:

$$(C_g^n - C_g^m) = \left(\frac{k_r}{T_C^{-1}}\right)(R_0^n + R_0^m)\tfrac{1}{2}(\phi_{s(t)}^n - \phi_{s(t)}^m) \tag{36a}$$

$$(S^n - S^m) = -\left(\frac{k_r}{T_S^{-1}}\right)\left(\frac{k_S}{k_S + T_F^{-1}}\right)(R_0^n + R_0^m)\tfrac{1}{2}(\phi_{s(t)}^n - \phi_{s(t)}^m) \tag{36b}$$

If we recall that the intensity of the NMR transitions between n and m will be proportional to the population differences given above, the following features of the spectra obtained from products of this simplified reaction sequence are apparent:

(a) The intensity decreases in direct proportion to the ratio of the rate constant for the slow reaction step to the reciprocal of the spin–lattice relaxation time in the product. Thus, variations of T_C^{-1} and T_S^{-1} from line to line in the spectrum will also show up as a change in intensity.

(b) The *sign* of the enhancement is *opposite* for combination products and products of scavenging.

(c) The absolute intensity in S is decreased by spin–lattice relaxation in the free radical F. It should be pointed out, however, that what has been simply written here as T_F^{-1} includes the interactions which were first suggested[12,12a] to be responsible for CIDNP and which could also be modified by electron polarization induced by the radical pair mechanism.[15,42] If the Overhauser mechanism were operating, one could then potentially have a situation in which T_F^{-1} is apparently *negative* and the absolute enhancement of S could be larger than in C_g! As yet, however, there is no experimental evidence for such effects.

If it is now assumed that the radical pair in this example is born in the singlet state and for simplicity[13–15a,34a] that a_s is small and $\omega_n \tau \ll 1$, Eq. (34a) may be used to show that

$$\tfrac{1}{2}(\phi_s^n - \phi_s^m) = a_s r \tau^2 (\delta_m^2 - \delta_n^2) \tag{36c}$$

Therefore

$$(C_g^n - C_g^m) = (k_r/T_c^{-1}) a_s r \tau^2 (\delta_m^2 - \delta_n^2)(R_0^n + R_0^m) = \Lambda_{c_g}(\delta_m^2 - \delta_n^2) \tag{37}$$

Equation (37) says qualitatively that a combination product from radical pairs born in the singlet state will be formed preferentially from those pairs with small values of δ_n^2. Reference to Eqs. (34b) and (34c) shows that just the opposite is true for pairs formed in the triplet state or by random encounters, while (36b) shows that the opposite will also hold for scavenging products of pairs born in the singlet state. Equation (37) also shows that in the limit of small cage effect and $\omega \tau \ll 1$, the term Λ_{c_g} will be the same for all NMR transitions of C_g. That is, *relative* intensities of lines from a single product will be determined only by δ_n, which in turn is a function only of g factors and hyperfine splittings. In general, of course, the relative intensities may be expected to deviate somewhat from those predicted by Eq. (37).

e. NMR Intensities. To illustrate how Eq. (37) may be applied to predict the intensity of a given enhanced NMR transition, we consider a first-order transition involving the flip of only a single spin one-half nucleus i, and let n and m in Eq. (24) correspond to $m_i = +\tfrac{1}{2}$ and $-\tfrac{1}{2}$, respectively, with the quantum numbers of all other nuclei the same for both states. For a nucleus, such as the proton, with a positive magnetogyric ratio this transition would be an absorption line in the normal spectrum. Let nucleus i also reside on radical fragment 1 and have a nonzero hyperfine coupling constant A_i only with electron S_1. Combining this information and Eqs. (24), (26b), and (37) we obtain for a C_g product:

$$I_i^j = -K_i^j \Lambda_{c_g}(A_i/2)\left[\left(\hbar^{-1}\beta H_0\right)(g_1 - g_2) + \left(\sum_{i \neq j}(A_1 - A_2)m\right)\right] \quad (38)$$

The sums are over all of the other nuclei on both fragments, the states of which are collectively designated by a superscript j. There will, in general, be as many different transitions of i as there are states j, and the observed spectrum will be a superposition of all these transitions for all of the nuclei. Only two features of these spectra are considered here: (a) the *net* intensity averaged over all j and (b) the relative intensities of lines in a multiplet arising from spin–spin splitting of transitions of i by a set of magnetically equivalent nuclei, k.

f. Net Enhancement. If Eq. (38) is summed over all j and divided by the total number of states j, since the values of m are symmetric about zero, the hyperfine term averages to zero, yielding

$$\langle I_i \rangle_{\text{net}} = -\langle K_i \rangle \Lambda_{c_g}(A_i/2)\hbar^{-1}\beta H_0(g_1 - g_2) \quad (39)$$

Thus, the net enhancement of nucleus i depends on both the sign and magnitude of its hyperfine splitting and the g-factor difference between the two radical fragments. A change in sign of either quantity changes the sign of the enhancement. The sign of enhancements for protons with various combinations of signs for A_i and Δg are tabulated in Table I. It also should be pointed out that in this first-order approximation the constant K_i is proportional to the number of magnetically equivalent i nuclei.

g. Relative Intensities of Multiplet Lines. In the case where $\Delta g = 0$, Eq. (39) shows that the net enhancement disappears. Reference to Eq. (38), however, shows that the second term remains and can give rise to enhancement for individual states j. Consider the case where i becomes coupled in the product C_g to a set of magnetically equivalent spin-$\frac{1}{2}$ nuclei with quantum number m_k. Since i is coupled only to one set of nuclei, we may, as before, average to zero the contributions of all other nuclei in Eq. (38). We are thus left with

$$I_i^k = -K_i^k \Lambda_{c_g}(A_i/2)[\hbar^{-1}\beta H_0(g_1 - g_2) \pm A_k m_k] \quad (40)$$

where $+$ and $-$ pertain to nuclei on the fragments 1 and 2, respectively. For the particular case where $\Delta g = 0$ the ratio of intensity of two lines, k and l, in the multiplet becomes

$$(I_i^k/I_i^l) = (I_{i,0}^k m_k)/(I_{i,0}^l m_l) \quad (41)$$

where $I_{i,0}^k$ is the unenhanced intensity for the kth line in the multiplet, which will be proportional to a binomial coefficient. Equation (41) reproduces

TABLE I

Enhanced Spectra of Geminate Cage Combination Products[a]

			A. Protons on both fragments				B. Protons on one fragment		
			$H_AR_1\cdot + \cdot R_2H_B \rightarrow H_AR_1R_2H_B$				$R_1\cdot + \cdot R_2 \underset{H_B}{\overset{H_A}{\diagup\!\!\diagdown}} \rightarrow R_1R_2 \underset{H_B}{\overset{H_A}{\diagup\!\!\diagdown}}$		
	A_A	A_B	H_A		H_B		H_A		H_B
(1)	−	−	A	E/A	E	E/A	E	A/E	E	A/E
(2)	−	+	A	A/E	A	A/E	E	E/A	A	E/A
(3)	+	−	E	A/E	E	A/E	A	E/A	E	E/A
(4)	+	+	E	E/A	A	E/A	A	A/E	A	A/E

[a] These predictions are valid if $J_{AB} > 0$ and $g_1 > g_2$. If $g_1 < g_2$ the net polarization changes sign, but the multiplet effect is unchanged. If $J_{AB} < 0$, the multiplet effect changes sign but the net polarization is unchanged. If the product forms from a triplet geminate encounter or in a diffusive encounter, both the phase of the multiplet effect and the sign of net enhancement are changed.

exactly the empirically derived expression for intensity ratios for multiplets in organohalides enhanced during free radical reactions with organolithium compounds.[43] One interesting consequence of Eq. (41) is the prediction that the center line ($m_k = 0$) of a multiplet arising from an even number of protons will be unenhanced in first-order spectra. Whether the multiplet will exhibit high field emission (A/E) or low field emission (E/A) depends on the signs of A_i, A_k, and m_k. The proper assignment of m_k to the lines of a multiplet requires, in turn, a knowledge of the sign of the nuclear spin–spin coupling constant. A useful rule to remember is that if the nuclear spin–spin coupling J_{NN} is *positive*, the *low* field lines of a multiplet will have *negative* values of m_k, provided that the magnetogyric ratio of i is positive and that the NMR experiment is carried out by varying the magnetic field rather than the frequency. The predicted "phase" of the multiplet effect under the above set of conditions is summarized in Table I for various signs of A_i and A_k.

We have found the summary of signs of enhancement given in Table I to be quite useful for rapid analysis of enhanced spectra to give a qualitative notion of the origin (singlet or triplet geminate cage, diffusive cage, etc.) of a reaction product.[43a] Interpretation of spectra exhibiting combined net polarization and multiplet effects must be carried out with some caution, however.[43b] The apparent "phase" of a multiplet effect superimposed on a large net polarization may in some cases be reversed from those given in the tables when g-factor differences are large. Likewise, even small degrees of intensity borrowing in multiplet spectra can give the appearance of net polarization or a reversal of the phase of the multiplet effect.

For computation of relative intensities of lines in spectra exhibiting non-first-order effects, it is necessary to apply Eq. (37) to each transition and substitute the population into Eq. (24). This may be done rapidly by hand[41] for two- and three-spin systems, or for more complex systems by suitable modification[44] of available spectrum simulation computer programs. Two such simulated spectra (to be discussed in more detail below) based essentially on computer evaluation of Eqs. (24) and (37) for all of the transitions, are shown in Figs. 7 and 8. The agreement between computed and observed spectra, while not perfect, is sufficiently encouraging that no major revision of this model for relative intensities seems to be necessary within the accuracy of currently available intensity determinations.

We now discuss qualitatively some experimental examples of spin selection. A complete treatment of these examples would require a more detailed consideration and, indeed, knowledge of reaction kinetics than is either warranted here or presently possible for most of the reactions to be discussed.

D. Examples of CIDNP Spectra

1. Spin Selection in Geminate Encounters

The reaction between α-chlorotoluene and ethyllithium[45] (familiar from its use in the Gilman titration of organolithium reagents) provides a suitable example of the spectrum obtained from products formed during a geminate encounter. The free radical reaction is

$$C_6H_5CH_2Cl + LiCH_2CH_3 \xrightarrow{-LiCl} [C_6H_5CH_2\cdot \ \cdot CH_2CH_3] \longrightarrow C_6H_5CH_2CH_3CH_3 \quad (42)$$

Although a full discussion of the mechanism of reaction of alkyllithium reagents with alkyl halides is not appropriate here, it is necessary to comment on the free radical character of the reaction which will be assumed in this discussion. Free radical formation by electron transfer was suggested long ago,[46] but only recently has it received acceptance. The reports of CIDNP in products of such reactions[11a,47-50] vindicated this proposal and were supported subsequently by direct ESR observations of the radical intermediates.[51,52] Even more recently, closely related coupling reactions have been shown to give racemized products,[53] consistent with radical intermediates in alkyl–lithium–secondary alkyl halide reactions. Stereospecificity (inversion) has been reported only in reactions of resonance-stabilized organolithium reagents (allyl and benzyl[54]) and in these cases a nucleophilic displacement

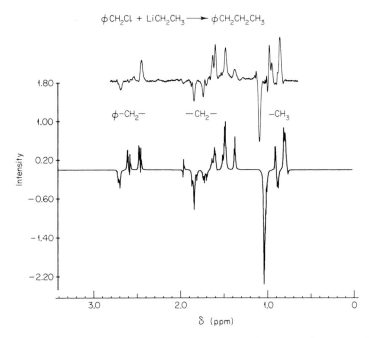

FIG. 7. Experimental (upper) and simulated (lower) enhanced spectra of the propyl groups of n-propylbenzene formed during the reaction of α-chlorotoluene with ethyllithium.

appears most likely. Halogen–metal exchange reactions between alkyllithium reagents and alkyl iodides can be observed by NMR in ethereal solvents at low temperatures, but no CIDNP has as yet been detected in the exchange products under these conditions.

NMR enhancement for the 7-proton sidechain of the n-propylbenzene formed as a C_g product of the above reaction is shown in Fig. 7. A simulation of the spectrum is given and agrees rather well with the experimental spectrum. It is useful also to consider the qualitative agreement of Fig. 7 with what is anticipated from a knowledge of signs of hyperfine splittings, g factors, and spin–spin splittings. This information is summarized below:

1. The signs of both vicinal spin–spin splittings in the propyl group are positive.[38b]

2. The hyperfine splittings for the protons of interest are benzyl,[55] $a_\alpha = -16.35$ G; ethyl,[24] $a_\alpha = -22.48$ G, $a_\beta = +26.99$ G.

3. The g factors of unsubstituted hydrocarbon radicals[35] are sufficiently similar that this contribution to the enhancement should be negligible except

in an extremely high external field, and therefore no net polarization is expected from this radical pair.

If the relative intensities of lines in the triplets from the benzylic and methyl protons of the products are considered, it can be seen that the benzylic protons and the α protons of ethyl form a pair of the type given in row 1 of Table I, Part A. The α- and β-protons of ethyl are of the type in row 2 of Table I, Part B. The prediction for both triplets is an E/A phase with relative intensities from Eq. (41) of $-2:0:2$ for the benzylic protons and $-3:0:3$ for the methyl group if all hyperfine splittings were of equal magnitude. The observed deviations from these intensities most likely arise from non-first-order effects and differences in the magnitudes of the hyperfine coupling constants, both of which have been taken into account in the computer simulated spectra. The methylene protons of the ethyl group are a combination of the types from row 1 of Table I, Part A and row 2 of Table I, Part B, both of which predict the observed E/A phase.

The reaction of α,α-dichlorotoluene and ethyllithium[56] provides an opportunity to explore the effect of a g-factor difference in the radicals of the primary encounter pair. Electron transfer should lead to α-chlorobenzyl and ethyl radicals, which on coupling give α-chloropropylbenzene. The α-chlorobenzyl radical should have a higher g-factor

$$C_6H_5CHCl_2 + LiCH_2CH_3 \xrightarrow{-LiCl} [C_6H_5CHCl\cdot \ \cdot CH_2CH_3] \longrightarrow$$
$$C_6H_5CHClCH_2CH_3 \quad (43)$$

(approximately 2.004 estimated from the simulated spectrum) than the ethyl ($g = 2.003$),[24] and as a result, the benzylic protons should show net enhanced absorption (Table I, Part A, row 1). The β protons, which were α to the electron in the ethyl radical (and therefore coupled to it with a negative hyperfine splitting), should show emission, and the methyl protons (coupled to the electron in the ethyl radical by a positive hyperfine splitting) should show enhanced absorption. Inspection of Fig. 8 shows a clear confirmation of these expectations for the α and β protons; the methyl proton absorption is obscured by the signal from butane which also shows signal enhancement. A multiplet effect of the type observed in n-propylbenzene also is expected and can be observed by comparison of relative intensities of the lines within the multiplets from the α and β protons during the reaction. The enhanced absorption observed during the reaction of gem-dichloropropanes with alkyllithium reagents[57] also results from a chlorocyclopropyl radical and an alkyl radical.

During the past two years, several reports of CIDNP in thermal rearrangements have appeared[58-67] and most are explicable by spin selection in a singlet geminate encounter. For example, the rearrangement of VI to VII

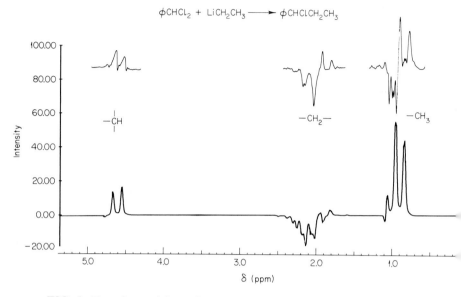

FIG. 8. Experimental (upper) and simulated (lower) enhanced spectra of the propyl group of α-chloropropyl benzene formed during the reaction of α,α-dichlorotoluene with ethyllithium. The experimental spectrum is a composite of scans taken during three separate reactions, a *modus operandi* necessitated by the rapidity of the reaction.

reported by Schöllkopf[60] shows emission for the benzylic protons of VII and enhanced absorption for the proton α to the sulfur atom, as predicted from the anticipated g-factor differences in the radical pair.

$$\underset{(VI)}{C_6H_5CH_2 \overset{+}{\underset{\underset{CH_3}{|}}{S}} \overset{-}{\overset{\overset{O}{\|}}{C}}HCC_6H_5} \longrightarrow [C_6H_5CH_2\cdot\ CH_3\overset{\cdot}{S}\overset{\overset{O}{\|}}{C}HCC_6H_5] \longrightarrow$$

$$\underset{(VII)}{\overset{E\quad A}{C_6H_5CH_2CH\overset{\overset{O}{\|}}{C}C_6H_5}} + (C_6H_5{-}CH_2)_2 \quad (44)$$
$$\underset{}{\underset{SCH_3}{|}} \qquad\qquad \text{minor}$$

2. Spin Selection in Products of Radicals Escaping Geminate Encounters

From Eq. (36) it can be seen that if those radicals escaping a geminate encounter are trapped by an atom transfer in a time short compared to the nuclear relaxation time in the radical, a nonequilibrium spin distribution should be observed in the product. A clear example of such an effect can be seen during the thermal decomposition of propionyl peroxide in the presence of molecular iodine. The first step of such a decomposition is that of oxygen–oxygen bond cleavage, either coincident with or shortly followed by decarboxylation of the acyloxy radicals,[68] thus generating two ethyl radicals in a geminate cage.

$$CH_3CH_2CO_2)_2 \rightarrow [2CH_3CH_2CO_2\cdot] \xrightarrow{-CO_2}]CH_3CH_2\cdot \ \cdot O_2CCH_2CH_3] \xrightarrow{-CO_2}$$
$$CH_3CH_2I \xleftarrow{I_2} [CH_3CH_2\cdot \ \cdot CH_2CH_3] \quad (45)$$

If an intermediate geminate pair consisting of a propionoxy and an ethyl radical were to exist long enough to allow nuclear spin induced singlet–triplet mixing, the characteristic net polarization of spin selection in a radical pair of differing g factors would be expected in products of cage reactions and cage escape. The failure to observe net polarization in these products[68a] is quite consistent with the difficulty in trapping free alkylacyloxy radicals,[68] and the conclusion is that decarboxylation proceeds to completion in the geminate encounter cage. Spin selection in the geminate pair of ethyl radicals will have the same effect on a single ethyl radical as it will in the previously considered case where the partner radical was benzyl. We therefore expect scavenged ethyl radicals to exhibit a phase opposite to that observed in the *n*-propylbenzene. Figure 9(a) clearly shows the A/E phase for both the methyl and methylene protons of ethyl iodide. Furthermore, Eq. (41) applied to both sets of protons predicts that relative intensities of the lines, from low field to high, will be $+3$, $+3$, -3, -3, and $+6$, 0, -6, in qualitative agreement with the experimental intensities. Enhanced spectra arising from the products of radicals escaping a geminate cage were first published by Kaptein,[69] who has shown that alkyl peroxide decomposition in the presence of hexachloroacetone leads to trapping (by chlorine transfer) of the radicals escaping the geminate pair and that the resulting chlorides clearly show the expected A/E multiplet structure.

Few unambiguous examples are available of the spectra of the trapping product of radicals which escape a geminate encounter with a radical of a significantly different g factor. Fischer has recently suggested[70] that his first reported CIDNP system may in fact be such a case. The thermal decomposition of benzoyl peroxide gives two benzoyloxy radicals predominantly, but also, in much smaller amount, a benzoyloxy and a phenyl radical.

$$(C_6H_5CO_2)_2 \xrightarrow{\text{major}} [2C_6H_5CO_2\cdot]$$
$$\searrow \text{minor}$$
$$[C_6H_5CO_2\cdot \ \cdot C_6H_5] \longrightarrow C_6H_5\cdot \xrightarrow{SH} C_6H_6 \quad (46)$$

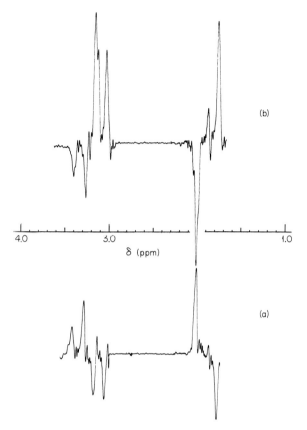

FIG. 9. Enhanced spectrum of ethyl iodide formed (a) during the decomposition of an o-dichlorobenzene solution of propionyl peroxide and iodine and (b) during the decomposition of an o-dichlorobenzene solution of benzoyl peroxide and ethyl iodide.

Spin selection arising from the g-factor difference (the g factor is higher for benzoyloxy) predicts a net emission in the benzene formed by hydrogen abstraction from the solvent, SH, by the phenyl radical (which has a positive hyperfine splitting) escaping the geminate pair (see Table 1, Part A, row 2 or 4).

3. Spin Selection in Diffusive Encounters of Free Radicals

The spin selection model applied to reactions occurring between two free radicals which diffuse into a solvent cage predicts that those radical pairs which have large values of δ_n will be more likely to attain the singlet character necessary for reaction. This is exactly the opposite of the prediction for selection in the geminate pair, and leads, for example, to expected A/E spectra for combination products of alkyl radicals with both α and β protons and E/A for products formed by scavenging of free radicals. Simple examples of the latter effect are provided by some of the first experiments illustrating CIDNP: the free radical reactions of iodoalkanes. Observation of enhanced spectra from the reactant iodide was explained[71] by invoking a rapid iodine transfer reaction between an alkyl radical and an alkyl iodide. The rates of such iodine transfers between alkyl radicals have been measured in the gas phase[72] and (for allyl radical and allyl iodide) in solution[73] and have been shown to proceed at a rate significantly greater than that expected for nuclear relaxation in the radical. This provides a mechanism by which alkyl radicals formed initially from an alkyl iodide can find their way back to reactant and carry with them the population changes arising from nuclear spin selection. The following reaction provides an illustration of both the enhancement of a reactant and spin selection occurring in a diffusive encounter:

$$(C_6H_5CO_2)_2 \longrightarrow 2C_6H_5\cdot + 2CO_2 \quad (47a)$$

$$C_6H_5\cdot + ICH_2CH_3 \longrightarrow C_6H_5I + \cdot CH_2CH_3 \quad (47b)$$

$$CH_3CH_2I + \cdot CH_2CH_3 \longrightarrow CH_3CH_2\cdot + ICH_2CH_3 \quad (47c)$$

The corresponding reaction scheme is shown in Eq. (48):

$$\begin{array}{c} R_0^n \xleftarrow{T_R^{-1}} R^n \\ \uparrow \downarrow \scriptstyle{k'_{-r}\,\|\,k'_r} \\ F_0^n \xleftarrow{T_F^{-1}} F^n \xrightarrow{k_d\phi_r^n} C_d^n \end{array} \quad (48)$$

The constant k'_r is for reaction (47b) and contains the concentration of phenyl radicals. Reaction (47c) is indicated by k'_{-r}. The only nuclear spin dependent step is assumed to be the formation of combination product, C_d. The suggestion has been made[41a] that spin selection may occur during the abstraction steps (47b) and (47c), but such a process is not required to explain the data and involves consideration of quartet states and interactions which are poorly understood.

The qualitative features of scheme (48) are:

a. The nonequilibrium distribution of nuclear spins in F will be opposite of that in C_d and the same as that in a C_g product of singlet origin.

b. The population distribution in F will be efficiently transferred to R if k'_{-r} competes successfully with T_F^{-1}.

c. k_d also must compete with T_F^{-1} if enhancement is to be observed in R since it is the diffusive combination step which produces spin selection.

The spectrum of ethyl iodide reacting in a thermally decomposing solution of benzoyl peroxide is shown in Fig. 9(b), and is in good accord with the prediction just given. The negative enhancement is weaker than in Fig. 9(a) since a significant contribution to the intensity is made by unenhanced reactant molecules.

An illustration of the effect of a g-factor difference in a secondary encounter is provided by the decomposition of trichloroacetyl peroxide in a solution of tetramethylethylene (TME).[74]

$$(CCl_3CO_2)_2 \longrightarrow [2CCl_3^{\cdot}] \longrightarrow CCl_3^{\cdot} + TME \longrightarrow \underset{(VIII)}{\cdot\!\!\!\!>\!\!\!\!-\!\!\!\!<\!\!\!\!-CCl_3}$$

$$\underset{(IX)}{>\!\!\!\!-\!\!\!\!<\!\!\!\!-CCl_3} + HCCl_3 \longleftarrow [(VIII) + \cdot CCl_3] \underset{\text{encounter}}{\overset{\text{diffusive}}{\longleftarrow}} \quad (49)$$

In this system there is no possibility of spin selection in a geminate encounter, nor is there any reason to expect spin selection on addition of the trichloromethyl radical to TME. Diffusive encounter of $\cdot CCl_3$ and (VIII), however, should lead to spin selection with protons in (VIII), β to the unpaired electron, leading to more rapid reaction if their spins are of minus spin. The spectra of the vinyl and allyl protons of (IX) and the chloroform proton all show strong emission lines (Fig. 10) with enhancement factors of ≈ -200 at 60 MHz. The observation of the same enhancement factors, after correcting for differences in T_1, for the proton transferred to form chloroform as for those protons remaining behind in (XI) suggests strongly that no spin selection occurs in the hydrogen atom transfer process. Trozzolo[75] has reached similar conclusions concerning the decompositions of deuterated peroxides.

No spectral enhancements are observed for the nonallylic methyls in (IX), which allows an alternative reaction path *via* the intermediacy of (X) to be dismissed since all protons in (X) should be

$$Cl_3C\cdot + TME \longrightarrow \underset{(X)}{>\!\!\!\!-\!\!\!\!<} + CHCl_3 \overset{\cdot CCl_3}{\longrightarrow} \underset{(IX)}{>\!\!\!\!-\!\!\!\!<\!\!\!\!-CCl_3} \quad (50)$$

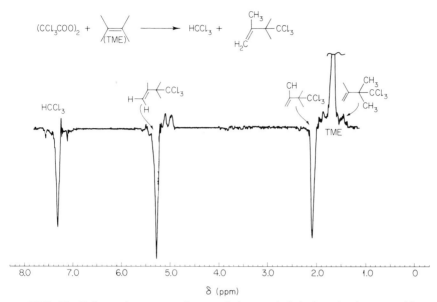

FIG. 10. Enhanced spectrum (composite) recorded during the decomposition of trichloroacetyl peroxide in tetramethylethylene/carbon tetrachloride.

coupled to the electron and should influence reactivity. Enhancement factors of up to -1100 at 19.6 MHz have been observed in the chloroform formed during the decomposition of trichloroacetyl peroxide in the presence of 1,7-octadiyne.[74] This number is larger than that allowed by the simple Overhauser mechanism (limit of approximately -88 for dipolar coupling) and is a strong argument in favor of the spin selection model. High enhancement factors for chloroform reported by Koenig[76] probably have a similar origin.

The predictions of the spin selection model for the random diffusive encounter cage apply qualitatively to geminate encounter cages formed from triplet precursors, although the relative magnitudes of the effects depend in different ways on the parameters describing cage encounters [Eqs. (34b) and (34c)]. Closs was first to apply this model,[25a,77] originally to radical pairs generated by hydrogen transfer to a triplet state ketone. Recently,[15a,78] by systematically varying the g-factor differences he has elegantly demonstrated the spin selection leading to net absorption or emission in both geminate and diffusive encounters.

E. Applications of CIDNP

The occurrence of CIDNP is still sufficiently unpredictable that it is more of a "phenomenon" than a "technique." In spite of this, however, when conditions are suitable for its observation a surprising amount of mechanistic and structural information can be obtained, much of it easily obtainable by no other method. We summarize here and give examples of some realized and potential applications of CIDNP. Many of these follow immediately by properly applying Table I to predict the phase of enhancement from a particular radical pair, some examples of which were presented in the preceding section. It has been our experience that proficiency in making such predictions and checking their consistency with experiment can be achieved with very little practice.

1. Proof of Radical Mechanism for Product Formation

One of the first reports[11a] of CIDNP also marked its first application to a reaction mechanism. The large enhancements observed during coupling reactions of organolithiums with organohalides [Eq. (42), for example] was used to infer the intermediacy of species involving unpaired electrons. That is, the mechanisms

$$\text{LiR} + \text{R'X} \xrightarrow{S_N 2} \text{R—R'} + \text{LiX} \tag{51a}$$

$$\text{LiR} + \text{R'X} \longrightarrow \text{LiX} + \overline{\text{R}\cdot \ \cdot \text{R'}} \longrightarrow \text{R—R'} \tag{51b}$$

were distinguishable.

While some of the later examples of CIDNP have been from reactions such as decompositions of peroxides[37a,69,79-84] and azocompounds[77,84] which are well recognized as free radical, many of the examples[58-67] have involved reactions for which reasonable ionic mechanisms may be written. The sulfonium ylid rearrangement in Eq. (44), for example, might equally well be written as a concerted 1,2 shift or proceeding via the intermediacy of an ion pair, (XI).

$$[\text{PhCH}_2{}^+ \ \text{CH}_3\overset{-}{\text{S}}\text{CHCOPh}]$$
(XI)

One of the principal uses of CIDNP has, therefore, been to distinguish ionic and free radical pathways. The development of the radical pair model has not changed the categorical statement that *CIDNP in a product proves that at least a portion of the product was formed via the intermediacy of a free radical*. Two qualifications are necessary, however. First, the magnitude of enhancement depends on some very poorly understood and unpredictable parameters such as "cage lifetimes" and electron coupling strength in radical pairs. As a consequence, one *cannot* say that the *absence* of CIDNP proves an ionic or concerted mechanism. Secondly, the spectre of parallel ionic and radical pathways which continues to haunt the mansion of mechanistic organic chemistry cannot be put to rest by CIDNP. It is unlikely that in the near future one will be able to calculate for the vast majority of reactions *a priori* enhancement factors with sufficient accuracy to rule out even a 10 to 1 excess of ionic over free radical pathways. CIDNP only says that *some* of the product is formed via free radicals. Whether one chooses to believe in parallel reaction paths with different degrees of electron pairing but similar rates must for the time being remain a matter of personal choice.

2. Multiplicity of Radical Pair Precursors

In photochemical reactions and reactions involving divalent carbon intermediates it is frequently of interest to know (a) the electron spin multiplicity of the primary reactive intermediate, i.e., the excited state or carbene, and (b) whether the subsequent bimolecular reactions of this intermediate are rapid and concerted or proceed via a radical pair intermediate.

It is known, for example, that UV-irradiated benzophenone undergoes a facile reductive coupling reaction with hydrogen donating hydrocarbons such as toluene.[25a] Three possible mechanisms for this process are shown below, and the predicted type of enhancement of the methylene protons of the 1,1,2-triphenylethanol product is indicated for each. (N) refers to *no enhancement* and the other symbols are as previously used.

Concerted insertion:

$$Ph_2C=O \xrightarrow{h\nu} {}^{1,3}[Ph_2C=O] \xrightarrow{PhCH_2-H} Ph_2C(OH)CH_2Ph \quad (N) \quad (52a)$$

Hydrogen abstraction by excited singlet:

$$Ph_2C=O \xrightarrow{h\nu} {}^{1}[Ph_2C=O] \xrightarrow{PhCH_3} \overline{Ph_2\dot{C}OH \cdot \dot{C}H_2Ph}^S \longrightarrow$$
$$Ph_2C(OH)CH_2Ph \quad (E) \quad (52b)$$

Hydrogen abstraction by excited triplet:

$$Ph_2C=O \xrightarrow{h\nu} {}^{3}[Ph_2C=O] \xrightarrow{PhCH_3} \overline{Ph_2\dot{C}OH \cdot \dot{C}H_2Ph}^t \longrightarrow$$
$$Ph_2C(OH)CH_2Ph \quad (A) \quad (52c)$$
$$\text{(observed)}$$

The experimentally observed [25a] result is enhanced absorption (A), confirming that this reaction proceeds via the coupling of a radical pair generated by abstraction of a hydrogen atom from toluene by a triplet state of benzophenone.

It is also possible to separate the effects of direct photolysis from triplet or singlet sensitization. An example of this is the light induced decomposition of propionyl peroxide recently reported by Kaptein et al.[85]

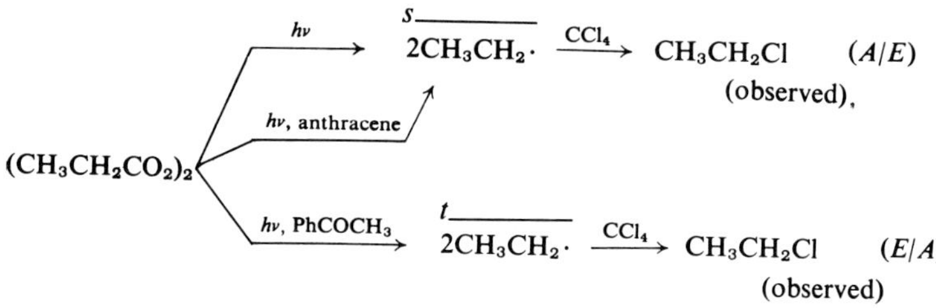

$$(CH_3CH_2CO_2)_2 \begin{cases} \xrightarrow{h\nu} \overset{S}{\overline{2CH_3CH_2 \cdot}} \xrightarrow{CCl_4} CH_3CH_2Cl \quad (A/E) \quad (53a) \\ \quad\quad\quad\quad\quad\quad\quad\quad\quad\quad\quad\quad\quad\quad (\text{observed}), \\ \xrightarrow{h\nu,\ \text{anthracene}} \\ \xrightarrow{h\nu,\ PhCOCH_3} \overset{t}{\overline{2CH_3CH_2 \cdot}} \xrightarrow{CCl_4} CH_3CH_2Cl \quad (E/A) \quad (53b) \\ \quad\quad\quad\quad\quad\quad\quad\quad\quad\quad\quad\quad\quad\quad (\text{observed}) \end{cases}$$

Sensitization of the decomposition by energy transfer from an excited *singlet* state of anthracene was suggested because of the increased enhancement (characteristic of a more efficient or faster reaction) observed when anthracene was added to the peroxide solution.

The multiplicity of two different carbene intermediates can also be inferred from the reactions shown below:[77,86]

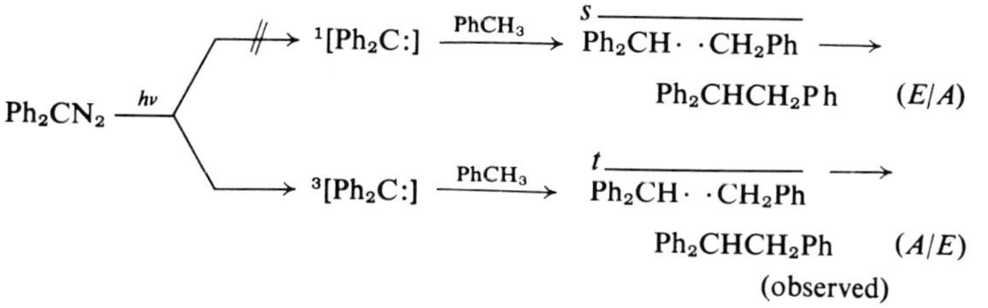

$$Ph_2CN_2 \xrightarrow{h\nu} \begin{cases} \not\rightarrow {}^1[Ph_2C:] \xrightarrow{PhCH_3} \overset{S}{\overline{Ph_2CH \cdot \cdot CH_2Ph}} \longrightarrow \\ \quad\quad\quad\quad\quad\quad\quad\quad Ph_2CHCH_2Ph \quad (E/A) \quad (54a) \\ \rightarrow {}^3[Ph_2C:] \xrightarrow{PhCH_3} \overset{t}{\overline{Ph_2CH \cdot \cdot CH_2Ph}} \longrightarrow \\ \quad\quad\quad\quad\quad\quad\quad\quad Ph_2CHCH_2Ph \quad (A/E) \quad (54b) \\ \quad\quad\quad\quad\quad\quad\quad\quad (\text{observed}) \end{cases}$$

$$CH_3O_2CCHN_2 \xrightarrow{h\nu} \begin{cases} \rightarrow {}^1[CH_3O_2CCH:] \xrightarrow{CCl_4} \overset{S}{\overline{CH_3O_2CCHCl \cdot \cdot CCl_3}} \longrightarrow \\ \quad\quad\quad\quad\quad\quad\quad\quad\quad\quad g_1 < g_2 \\ \quad\quad\quad\quad\quad\quad\quad\quad CH_3O_2CCHClCCl_3 \quad (E) \quad (55a) \\ \quad\quad\quad\quad\quad\quad\quad\quad (\text{observed}) \\ \not\rightarrow {}^3[CH_3O_2CCH:] \xrightarrow{CCl_4} \overset{t}{\overline{CH_3O_2CCHCl \cdot \cdot CCl_3}} \longrightarrow \\ \quad\quad\quad\quad\quad\quad\quad\quad CH_3O_2CCHClCCl_3 \quad (A) \quad (55b) \end{cases}$$

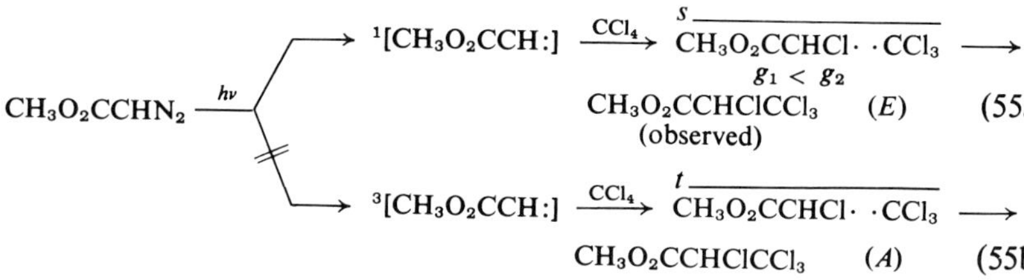

There are, of course, other ways of obtaining information of this type, such as direct ESR observation of triplet intermediates or the use of stereochemical probes of the reactivity of a radical pair. However, these are often either not applicable under the conditions where reactions are usually run or are more difficult and tedious to apply than is CIDNP, provided, of course, that one can observe it.

3. Identity and Life History of Radical Pairs in Which Polarization Occurs

When benzoyl peroxide is decomposed in the presence of methyl iodide a small amount of methyl benzoate is formed which exhibits an intense NMR emission line from the methyl group.[80b,87] Two alternative pathways for formation of this product are shown below.

Diffusive recombination of free radicals:

$$(PhCO_2)_2 \longrightarrow Ph\cdot + PhCO_2\cdot$$
$$Ph\cdot + ICH_3 \longrightarrow PhI + \cdot CH_3$$
$$PhCO_2\cdot + \cdot CH_3 \longrightarrow \overline{PhCO_2\cdot \,\,\cdot CH_3}^{\,r} \longrightarrow PhCO_2CH_3 \quad (A) \quad (56a)$$
$$g_2 > g_1$$

Pair substitution:

$$(PhCO_2)_2 \longrightarrow \overline{PhCO_2\cdot \,\,\cdot Ph}^{\,s} \xrightarrow{ICH_3} \overline{PhCO_2\cdot \,\,\cdot CH_3}^{\,s} \longrightarrow$$
$$PhCO_2CH_3 \quad (E) \quad (56b)$$
$$\text{(observed)}$$

The result is clearly at odds with a diffusive coupling of methyl and benzoyloxy radicals and provides good evidence for the occurrence of a pair substitution process in which *paired* radicals are scavenged. The presence of such a scavenging reaction could most likely have been detected only by laboriously studying the effect of concentration on product yields.

An even more dramatic example is that of the origin of methyl acetate formed from the thermal decomposition of acetyl peroxide. *Six* different possible mechanisms immediately come to mind to someone who is not an expert in acyl peroxide chemistry.

Concerted, intramolecular rearrangement:

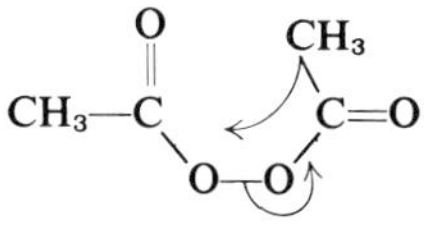

$$CH_3-C\underset{O-O}{\overset{O}{\diagdown}}\!\!\diagup\!\!\underset{}{\overset{CH_3}{\diagup}}C=O \longrightarrow CH_3CO_2CH_3 + CO_2 \quad (N) \text{ non radical} \quad (57a)$$

Radical displacement of CO_2:

$$(CH_3CO_2)_2 \longrightarrow \overline{CH_3CO_2\cdot \,\, CH_3-CO_2\cdot} \longrightarrow CH_3CO_2CH_3 + CO_2 \quad (N)$$

nuclei too weakly coupled to electrons, identical radicals, no net polarization $\quad$ (57b)

Induced decomposition by acetoxy radical:

$$CH_3CO_2\cdot + CH_3\text{—}CO_2\text{—}O_2\text{—}CCH_3 \longrightarrow CH_3CO_2CH_3 + CO_2 + \cdot O_2CCH_3 \quad (N)$$

acetyl weakly coupled; methoxy never in radical   (57c)

Induced decomposition by methyl radical:

$$(CH_3CO_2)_2 \xrightarrow{\quad S \quad} \overline{CH_3CO_2\cdot \;\cdot CH_3} \longrightarrow CH_3\cdot \quad (57d)$$

$$CH_3\cdot + O\text{=}C(CH_3)O\text{—}O_2CCH_3 \longrightarrow$$
$$CH_3O_2CCH_3 + \cdot O_2CCH_3 \quad (A)\text{ methoxy, } (N)\text{ acetyl}$$

Diffusive recombination of methyl and acetoxy:

$$(CH_3CO_2)_2 \longrightarrow CH_3CO_2\cdot + \cdot CH_3$$
$$CH_3CO_2\cdot + \cdot CH_3 \longrightarrow \overline{CH_3CO_2\cdot \;\cdot CH_3}^{\,r} \longrightarrow$$
$$CH_3CO_2CH_3 \quad (A)\text{ methoxy, } (N)\text{ acetyl} \quad (57e)$$

Geminate cage recombination of methyl and acetoxy:

$$(CH_3CO_2)_2 \longrightarrow \overline{CH_3CO_2\cdot \;\cdot CH_3}^{\,s} \longrightarrow$$
$$CH_3CO_2CH_3 \quad (E)\text{ methoxy, } (N)\text{ acetyl (observed)} \quad (57f)$$

Only the last mechanism agrees with the observed enhancement.[69] Some of the other mechanisms could, of course, be ruled out on other grounds, and some could also be operating simultaneously with the last one. The sign of the enhancement says, however, that this is the process giving rise to the enhancement.

4. Distance of Separation of Caged Radical Pairs

The exchange interaction J serves as the probe of distance since it varies rapidly with the separation of the two radical fragments. Very little information is yet available on quantitative dependence of J on distance.[29,30] Studies of rigid intramolecular biradicals which are now underway in several laboratories may help to clarify the situation. Unfortunately, the models of high-field polarization are not sensitive to J when it is either much larger or much smaller than the hyperfine couplings. Low-field enhancement, however, should be a much more sensitive function of the exchange interaction.[37b]

5. Lifetimes of Radical Pairs

Qualitatively, one can say (Section III,C) that unless the pair "exists" for a time which is not too much less than the reciprocal of the hyperfine frequency (10^{-8}–10^{-9} second), detectable effects will not arise. More quantitative estimates of lifetimes[40] require specific information about the dynamics of motion of radical pairs, the strength of electron–electron coupling, and the availability of electron spin independent reaction pathways. A lower limit to an effective lifetime may, however, be obtained by assuming only spin-dependent reaction paths and negligible exchange interaction. Values of τ estimated in this way are between 10^{-9} and 10^{-10} seconds. Additional refinements of the theoretical models will probably make it possible for more accurate values to be obtained.

6. Enhancement of Unstable Diamagnetic Reaction Intermediates

The very large enhancements frequently observed make possible the detection of NMR signals from very low concentrations of molecules produced as intermediates in chemical reactions. Since enhancements of 10^3 are now fairly common, it is possible to easily detect 0.1% of a product or intermediate. The intermediate must, however, not be so reactive that its lifetime is less than a second or so; otherwise, lifetime broadening of the NMR lines will occur and make detection more difficult. Small amounts of broadening, however, could in principle provide the lifetime itself. This aspect of CIDNP has not yet been exploited, but almost certainly has been operating in numerous spectra where weak, transient, unidentified lines frequently appear.

7. Relative Magnitudes and Signs of Free Radical g Factors and Hyperfine Splittings

Since the singlet–triplet mixing parameter δ_n depends on hyperfine splittings and g factors, it is possible in favorable cases to determine one or more of these parameters from experimentally observed CIDNP spectra.

For example, Closs[78] has carried out an elegant study of the reaction

$$p\text{-XC}_6\text{H}_4\text{CHO} + \text{CH}_2(\text{C}_6\text{H}_4\text{-}p\text{Y})_2 \xrightarrow{h\nu} \overset{t}{\overline{p\text{-XC}_6\text{H}_4}\underset{g_2}{\text{CH(OH)}}\cdot\cdot\underset{g_1}{\text{CH(C}_6\text{H}_4 p\text{Y})_2}} \quad (58)$$
$$\downarrow$$
$$p\text{-XC}_6\text{H}_4\text{CH(OH)CH(C}_6\text{H}_4)\text{-}p\text{Y})_2$$

By assuming that the hyperfine splittings of the benzylic protons are insensitive to substitution in the ring, he was able to estimate the following g-factor differences:

X	Y	$(g_1 - g_2) \times 10^3$
Br	H	2.7
Cl	H	1.5
H	H	0.47
H	Cl	−0.33
H	Br	−2.7

Similarly, as already mentioned, the spectrum in Fig. 9 is simulated best with a g-factor difference of 1×10^{-3} between ethyl and α-chlorobenzyl radicals.

By employing radicals with known hyperfine splittings and g factors as radical pair probes, Fischer[88] has been able to calculate both the g factor (2.0080 ± 0.0003) and proton hyperfine splitting $(-17.0 \pm 0.1$ G$)$ for the previously uncharacterized dichloromethyl radical. He employed the reaction scheme

$$(PhCO_2)_2 \xrightarrow{h\nu} Ph\cdot$$

$$Ph\cdot + CH_2Cl_2 \longrightarrow PhH + \cdot CHCl_2$$

$$Ph\cdot + XR \longrightarrow PhX + \cdot R$$

$$R\cdot + \cdot CHCl_2 \xrightarrow{\quad r \quad} \overline{R\cdot \cdot CHCl_2} \longrightarrow R-CHCl_2$$

$$XR = ICH_3, CH_3COCH_3, CH_2ClCOOH, CCl_4 \tag{59}$$

It is therefore clear that CIDNP can serve as an extremely useful supplement to ESR in determining g factors and hyperfine splittings of highly reactive free radicals. Furthermore, the relative *signs* of hyperfine splittings follow immediately from CIDNP spectra but are usually impossible to obtain by conventional ESR on highly reactive free radicals. (For a rather special exception, see Fessenden and Schuler.[89]) The applications to nuclei other than protons appear particularly promising in view of the recent observations of CIDNP from ^{13}C nuclei in unenriched samples reported by Lippmaa et al.[90]

8. Determination of Rate Constants and Relaxation Times

Since the intensities of CIDNP lines depend in general on a number of kinetic constants and relaxation times (see Section III,D), careful study of the effects on intensities of varying concentration and temperature should make it possible to extract some of these parameters. This has so far been little exploited, a notable exception being the determination by Closs[91] of the ratio of the spin–lattice relaxation rate in a free radical to the rate of its reaction in a scavenging step.

The use of chemical pumping of lines in diamagnetic products promises to be of considerable use as a supplement to pulsed NMR and other methods of determining relaxation times. The determination of the spin–lattice relaxation time in a product of a photochemical reaction is especially simple: One just shuts off the light and measures the time constant for decay of the line. (For a possible complication due to slow dark reactions, see Cocivera and Roth.[86])

F. Predicting the Future of CIDNP

It is clear that the chemical and physical applications of CIDNP are still in their infancy. Beyond the obvious goal of finding even more reactions which exhibit the effects, however, what can one predict about the future developments in this field?

First, the surface has barely been scratched in examining nuclei other than protons. The ubiquitous ^{13}C nucleus shows special promise since Lippmaa's demonstration[90] that enhancement factors with ^{13}C are often an order of magnitude larger than those for protons in the same molecule. This follows both from the longer T_1's for ^{13}C and from the somewhat larger hyperfine splittings for this nucleus in organic free radicals. ^{15}N and ^{31}P are of obvious biological interest, and ^{19}F has already served as a useful probe to simplify the problem of having available in a complex spectrum a window through which to view enhancement.[92]

Secondly, the chemical world is presently on the threshold of a great increase in the availability of pulsed NMR spectrometers and associated computing facilities for performing multiple channel sensitivity enhancement by the Fourier transform technique.[93] The same equipment, however, may and certainly will be applied to recording rapidly changing spectra exhibiting CIDNP. Thus the problem of comparing intensities of lines at the same reaction time will be considerably alleviated.

Thirdly, the experimental and theoretical results discussed here apply only to reactions run and observed in magnetic fields substantially higher than the internal fields present in reacting radical pairs. It has already been shown,[37] however, that observable, and often quite different, effects can be obtained when reactions are run at low fields and observed with high fields. Preliminary theoretical extension[37b] of the radical pair model to lower fields suggests that these experiments will give valuable additional information about cage effects, lifetimes, and electron interactions in radical pairs. Proper understanding of field effects is especially important in view of the fact that the magnetic field is essentially the only external variable which can be changed in CIDNP experiments which would not be expected to change the course of the chemical reaction.

Fourth, additional refinements in the methods used to generate free radicals can be expected. Better probe designs are sure to increase the efficiency with which light can be fed to an NMR sample. The recent development of high power lasers with output in the near UV, which have recently been used to generate radicals for ESR,[94] is also sure to be exploited as a means of inducing photo-CIDNP. Similarly, development of a convenient, workable liquid flow system for use in high resolution NMR will make possible much better control of the conditions under which radicals can be generated and rapid reactions studied.

Fifth, the relationships between CIDNP and the less extensively studied and documented phenomenon of *electron* polarization observed when rapid radical reactions are studied by ESR[94,94a] need to be vigorously explored. Especially important will be finding reactions in which both types of polarization may be observed under the same experimental conditions.

Finally, the fact that an NMR emission line serves as a source of radio-frequency radiation is sure to prove an irresistible temptation for someone with a talent for engineering. Although such a chemical radiofrequency maser is sure to have an embarrassingly low thermodynamic efficiency and, therefore, probably possess little commercial attractiveness, it would provide a dramatic illustration of spin-correlation effects in chemical reactions.

It should be obvious from the above that CIDNP is here to stay. Future developments will be limited only by the inherent properties of chemical reactions and the ability of human ingenuity to take advantage of them.

Acknowledgments

Our thanks go to a considerable number of investigators of CIDNP in other laboratories, whose names appear in the references, for their communication of results prior to publication either as preprints or in conversation. Both authors gratefully acknowledge financial support from the National Science Foundation and the Alfred P. Sloan Foundation. Portions of the manuscript were completed while R.G.L. was a guest of Dr. M. P. Klein of the Chemical Biodynamics Laboratory, University of California, Berkeley, and H.R.W. was a visitor at the laboratory of Professor H. Fischer, University of Zürich. The hospitality of these institutions and individuals was greatly appreciated.

References

[1] See Chapter 7, Vol. 3 this series.
[2] For a recent review, see C. S. Johnson, Jr., *Advan. Magn. Resonance* **1**, 33 (1965).

3. Dynamic Nuclear Polarization   145

[3] For effects due to double resonance in nuclear-spin only systems, see Chap. 3, vol. 4.
[4] A. W. Overhauser, *Phys. Rev.* **92**, 411 (1953).
[5] T. R. Carver and C. P. Slichter, *Phys. Rev.* **92**, 212 (1953).
[6] C. D. Jeffries, "Dynamic Nuclear Orientation." Wiley (Interscience), New York, 1963; J. M. Daniels, "Oriented Nuclei: Polarized Targets and Beams." Academic Press, New York, 1965; M. E. Rose, "Nuclear Orientation." Gordon & Breach, New York, 1963.
[7] A. Abragam, "The Principles of Nuclear Magnetism," Chapter IX. Oxford Univ. Press, London and New York, 1961.
[8] L. H. Bennett and H. C. Torrey, *Phys. Rev.* **108**, 499 (1957).
[9] T. R. Carver and C. P. Slichter, *Phys. Rev.* **102**, 975 (1956).
[10] K. H. Hausser and D. Stehlik, *Advan. Magn. Resonance* 3, 79 (1968).
[10a] R. A. Dwek, R. E. Richards, and D. Taylor, *Annu. Rev. NMR Spectrosc.* **2**, 293 (1969).
[10b] R. E. Richards, *in* "Magnetic Resonance in Biological Systems" (A. Ehrenberg, B. G. Malmström, and T. Vänngård, eds.), p. 53. Pergamon, Oxford, 1967.
[11] J. Bargon, H. Fischer, and U. Johnsen, *Z. Naturforsch. A* **22**, 1551 (1967).
[11a] H. R. Ward and R. G. Lawler, *J. Amer. Chem. Soc.* **89**, 5518 (1967).
[12] J. Bargon and H. Fischer, *Z. Naturforsch. A* **22**, 1556 (1967).
[12a] R. G. Lawler, *J. Amer. Chem. Soc.* **89**, 5518 (1967).
[13] R. Kaptein and L. J. Oosterhoff, *Chem. Phys. Lett.* **4**, 214 (1969).
[14] G. L. Closs, *J. Amer. Chem. Soc.* **91**, 4552 (1969).
[15] H. Fischer, *Chem. Phys. Lett.* **4**, 611 (1970).
[15a] G. L. Closs and A. D. Trifunac, *J. Amer. Chem. Soc.* **92**, 2184 (1970).
[16] F. Bloch, *Phys. Rev.* **70**, 460 (1946).
[17] N. Bloembergen, E. M. Purcell, and R. V. Pound, *Phys. Rev.* **73**, 679 (1948).
[18] A. Abragam and W. G. Proctor, *Phys. Rev.* **109**, 1441 (1958); A. G. Redfield, *Science* **164**, 1015 (1969).
[19] I. Solomon, *Phys. Rev.* **99**, 559 (1958).
[20] D. Stehlik and K. H. Hausser, *Z. Naturforsch. A* **22**, 914 (1967).
[21] D. R. Eaton and W. D. Phillips, *Advan. Magn. Resonance* **1**, 103 (1965).
[21a] E. deBoer and H. van Willigen, *Progr. NMR Spectrosc.* **2**, 111 (1967).
[21b] A. L. Buchachenko and N. N. Sysoeva, *Russ. Chem. Rev.* **37**, 798 (1968).
[21c] G. A. Webb, *Annu. Rep. NMR Spectrosc.* **3**, 211 (1970).
[22] J. R. Bolton and P. D. Sullivan, *Advan. Magn. Resonance* **4**, 39 (1970); A. Hudson and G. R. Luckhurst, *Chem. Rev.* **69**, 191 (1969).
[23] A. Abragam, "The Principles of Nuclear Magnetism," p. 308. Oxford Univ. Press, London and New York, 1961.
[24] R. W. Fessenden and R. H. Schuler, *J. Chem. Phys.* **39**, 2147 (1963).
[25] Measurements of both spin-lattice (T_1) and spin-spin (T_2) relaxation times under a range of conditions may also yield information regarding the mechanism(s) responsible for relaxation [H. S. Gutowsky and J. C. Tai, *J. Chem. Phys.* **39**, 208 (1963)]. Precise sets of T_1 and T_2 measurements are, however, often more difficult and time consuming than DNP, provided, of course, that the necessary apparatus for the latter is available.
[25a] G. L. Closs and L. E. Closs, *J. Amer. Chem. Soc.* **91**, 4549 and 4550 (1969).
[26] The role of electron spin in chemical reactions has received renewed attention

recently [see, for example, F. A. Matsen and D. J. Klein, *Advan. Photochem.* **7**, 1 (1969); R. D. Burkhart and J. C. Merrill, *J. Phys. Chem.* **73**, 2699 (1969); P. S. Engel and P. D. Bartlett, *J. Amer. Chem. Soc.* **92**, 5883 (1970)].

[27] J. H. Van Vleck, "The Theory of Electric and Magnetic Susceptibilities," p. 318. Oxford Univ. Press, London and New York, 1932. Note that this interaction appears as a scalar product of the type often encountered in angular momentum problems and interpreted as a magnetic interaction. It is really of electrostatic origin, however, and results from the combined effects of electron repulsion and the Pauli Exclusion Principle. The true magnetic dipole interactions between electrons and nuclei have been ignored in this and other treatments for the sake of simplicity but at the expense of rigor. The time scale for the events considered here is sufficiently short that the usual models[28] for the time dependence of such dipole couplings may not be applicable without modification.

[28] A. Abragam, "The Principles of Nuclear Magnetism," Chapter VIII. Oxford Univ. Press, London and New York, 1961.

[29] J. E. Harriman, M. Twerdochlib, M. B. Milleur, and J. O. Hirschfelder, *Proc. Nat. Acad. Sci. U.S.* **57**, 1558 (1967); W. Kolos and L. Wolniewicz, *J. Chem. Phys.* **43**, 2429 (1965); J. N. Murrell and J. J. C. Teixeira-Dias, *Mol. Phys.* **19**, 521 (1970).

[30] Electron spin resonance measurements on stable biradicals [S. H. Glarum and J. H. Marshall, *J. Chem. Phys.* **47**, 1374 (1967); G. R. Luckhurst, *Mol. Phys.* **10**, 543 (1966)], radical pairs trapped in crystalline solids [K. Itoh, H. Hayashi, and S. Nagakura, *ibid.* **17**, 561 (1969)], and glasses [S. I. Weissman and N. Hirota, *J. Amer. Chem. Soc.* **86**, 2548 (1964)] and measurements of spin exchange rates between radicals in solution [W. Plachy and D. Kivelson, *J. Chem. Phys.* **47**, 3312 (1967)] indicate that the exchange interaction may be comparable to the hyperfine interaction at distances of the order of a molecular diameter.

[31] The time dependence of J plays an explicit and central role in the modification of the radical pair model put forward by H. Fischer[15,32] and S. H. Glarum [Abstracts of papers presented at 159th American Chemical Society Meeting, paper ORGN 40, February 1970]. They assume large changes in J during the lifetime of an encounter pair while the theory discussed here assumes that after an initial large change these changes are small and distributed about a time independent mean value. A treatment of the effects of diffusion on radical pair reactivity [F. J. Adrian, *J. Chem. Phys.* **53**, 3374 (1970)], on the other hand, assumes that $J = 0$. Since J is a conveniently ambiguous parameter it is not surprising that its value can be manipulated to make any of several models produce predicted enhancements of the right order of magnitude.

[32] H. Fischer, *Z. Naturforsch. A* **25**, 1957 (1970).

[33] The physical concepts underlying the idea of nuclear spin selection via intersystem crossing seem to have been anticipated in a discussion of the formation of excited states during recombination of radiolysis intermediates [B. Brocklehurst, *Nature (London)* **221**, 921 (1969)]. The possibility of observing effects of a magnetic field on the rate of electron exchange [H. Shimizu, *J. Chem. Phys.* **42**, 3599 (1965); C. S. Johnson, Jr., *Mol. Phys.* **12**, 25 (1967)] and triplet-triplet annihilation [R. E. Merrifield, *J. Chem. Phys.* **48**, 4318 (1968)] has also been discussed. The picture presented here of a phase shift between electron spins induced by nuclear spins is analogous to one treatment of transverse

relaxation arising from nuclear spin phase shifts induced by chemical exchange processes (deBoer and van Willigen,[21a] p. 140).

[34] H. R. Ward, *Accounts Chem. Res.* **5**, 18 (1972).

[34a] R. G. Lawler, *Accounts Chem. Res.* **5**, 25 (1972).

[34b] H. Fischer, *Topics in Current Chemistry*, **24**, 1 (1971).

[34c] G. L. Closs, *Spec. Lect. XXIIIrd. Int. Congr. Pure Appl. Chem.*, **4**, 19 (1971).

[35] For a comprehensive listing of g-factors and hyperfine splittings covering the literature through March 1964, see H. Fischer, *in* "Magnetic Properties of Free Radicals," Landolt-Bornstein, New Series, Vol. 1. Springer Publ., New York, 1965.

[36] Discussion of the current status of theories of diffusive behavior of reactive molecules in solution may be found in comprehensive reviews [R. M. Noyes, *Progr. Chem. Kinet.* **1**, 129 (1961); A. M. North, "Collision Theory of Chemical Reactions in Liquids." Methuen, London, 1964; A. M. North, *Quart. Rev. Chem. Soc.* **20**, 421 (1966)].

[37] For reactions run at low fields, however, effects have been observed which require simultaneous electron and nuclear spin flips and are not in accord with the model presented here [H. R. Ward, R. G. Lawler, H. Y. Loken, and R. A. Cooper, *J. Amer. Chem. Soc.* **91**, 4928 (1969)]. (Also see references 37a–c.)

[37a] M. Lehnig and H. Fischer, *Z. Naturforsch A* **24**, 1771 (1969).

[37b] J. F. Garst, R. H. Cox, J. T. Barbas, R. D. Roberts, J. I. Morris, and R. C. Morrison, *J. Amer. Chem. Soc.* **92**, 5761 (1970).

[37c] J. F. Garst and R. H. Cox, *J. Amer. Chem. Soc.* **92**, 6389 (1970); J. L. Charlton and J. Bargon, *Chem. Phys. Lett.* **8**, 442 (1971).

[38] J. D. Memory, "Quantum Theory of Magnetic Resonance Parameters." McGraw-Hill, New York, 1968.

[38a] P. L. Corio, "Structure of High-Resolution NMR Spectra." Academic Press, New York, 1966.

[38b] A. A. Bothner-By, *Advan. Magn. Resonance* **1**, 195 (1965).

[39] R. C. Tolman, "The Principles of Statistical Mechanics," Chapter 9. Oxford Univ. Press, London and New York, 1938.

[40] What we have called τ is actually a rather poorly defined quantity. It incorporates all processes which destroy the encounter pair and do not lead to a combination product or depend on the singlet character of a radical pair. These processes may include (a) a reaction which competes with combination and is spin independent, e.g. fragmentation of one of the partner radicals (b) reaction of one partner with a scavenger molecule or (c) diffusive separation of the pair to a distance where the probability of encounters with partners of other pairs is high. The former two processes are likely to be capable of at least approximate description by a mean lifetime, implying first-order loss of the pair. Alternative (c), however, is not a true first-order process and its time dependence falls off more nearly as $t^{-3/2}$ than exponentially [R. M. Noyes, *J. Chem. Phys.* **22**, 1349 (1954)]. The latter model has recently been explored by F. J. Adrian [*ibid.* **53**, 3374 (1970)]. The crucial point is that unless an *effective* τ, of whatever origin, of at least the order of 10^{-9} seconds duration is operative, the predicted effects are smaller than usually observed, simply because the singlet-triplet mixing frequency (ca. 10^8 rad/sec.) is turned on for an insufficient length of time. Recent observations of "pair substitution" reactions (cf. Section 3.5c) promise to provide a link between the dynamics of radical pair behavior and the rates of ordinary bimolecular reactions.

[41] G. L. Closs and A. D. Trifunac, *J. Amer. Chem. Soc.* **92**, 2186 (1970).
[41a] F. Gerhart and G. Ostermann, *Tetrahedron Lett.* p. 4705 (1969).
[42] R. Kaptein and L. J. Oosterhoff, *Chem. Phys. Lett.* **4**, 195 (1969).
[43] A. R. Lepley, *J. Amer. Chem. Soc.* **91**, 749 (1969).
[43a] The same information is also obtained by applying a short set of rules. R. Kaptein, *Chem. Commun.*, p. 732 (1971).
[43b] Relaxation effects in multiple spin systems may also cause deviations from predicted intensities. K. Müller and G. L. Closs, *J. Amer. Chem. Soc.* **94**, 1002 (1972).
[44] A modification by G. T. Evans of the program NMRIT is suitable for systems of up to 8 spins. Larger spin systems with magnetic equivalence have been treated using the method of magnetic equivalence factoring [Abstracts of papers presented at 159th American Chemical Society Meeting, paper ORGN 70, February 1970]. Closs has also reported a modification of the program LAOCOON II [G. L. Closs and A. D. Trifunac, *J. Amer. Chem. Soc.* **92**, 7227 (1970)].
[45] J. S. Wishnok, M. Halfon, and R. A. Cooper, unpublished results (1968–70).
[46] C. S. Marvel, F. D. Hager, and D. D. Coffman, *J. Amer. Chem. Soc.* **49**, 2323 (1927).
[47] A. R. Lepley, *J. Amer. Chem. Soc.* **90**, 2710 (1968).
[48] H. R. Ward, R. G. Lawler, and R. A. Cooper, *J. Amer. Chem. Soc.* **91**, 746 (1969).
[49] A. R. Lepley and R. L. Landau, *J. Amer. Chem. Soc.* **91**, 748 (1969).
[50] A. R. Lepley, *Chem. Commun.* p. 64 (1969).
[51] G. A. Russell and D. W. Lamson, *J. Amer. Chem. Soc.* **91**, 3967 (1969).
[52] H. Fischer, *J. Phys. Chem.* **73**, 3834 (1969).
[53] J. Sauer and W. Braig, *Tetrahedron Lett.* p. 4275 (1969).
[54] W. D. Korte, L. Kinner, and W. Kaska, *Tetrahedron Lett.* p. 603 (1970); L. H. Sommer and W. D. Korte, *J. Org. Chem.* **35**, 22 (1970).
[55] A. Carrington and A. D. McLachlan, "Introduction to Magnetic Resonance," p. 91. Harper, New York, 1967.
[56] H. R. Ward, R. G. Lawler, H. Y. Loken, and R. A. Cooper, *J. Amer. Chem. Soc.* **91**, 4928 (1969).
[57] H. R. Ward, R. G. Lawler, and H. Y. Loken, *J. Amer. Chem. Soc.* **90**, 7359 (1968).
[58] A. R. Lepley, *J. Amer. Chem. Soc.* **91**, 1237 (1968).
[59] J. E. Baldwin and J. E. Brown, *J. Amer. Chem. Soc.* **91**, 3647 (1969).
[60] U. Schöllkopf, G. Ostermann, and J. Schossig, *Tetrahedron Lett.* p. 2619 (1969); U. Schöllkopf, J. Schossig, and G. Osterman, *Justus Liebiss Ann. Chem.* **737**, 158 (1970); H. Iwamura, M. Iwamura, T. Nishida, M. Yoshida, and J. Nakayama, *Tetrahedron Lett.* p. 63 (1971).
[61] R. W. Jemison and D. G. Morris, *Chem. Commun.* p. 1226 (1969).
[62] D. G. Morris, *Chem. Commun.* p. 1345 (1969); p. 221 (1971).
[63] U. Schöllkopf, U. Ludwig, G. Ostermann, and M. Patsch, *Tetrahedron Lett.* p. 3415 (1969); J. E. Baldwin, J. E. Brown, and G. Höfle, *J. Amer. Chem. Soc.* **93**, 788 (1971).
[64] A. R. Lepley, *Chem. Commun.* p. 1460 (1969).
[65] F. Gerhart, *Tetrahedron Lett.* p. 5061 (1969).
[66] J. E. Baldwin, J. E. Brown, and R. W. Cordell, *Chem. Commun.* p. 31 (1970); J. E. Baldwin, J. E. Brown, and G. Höfle, *J. Amer. Chem. Soc.* **93**, 788 (1971).

[67] J. E. Baldwin, W. F. Erickson, R. E. Hackler, and R. M. Scott, *Chem. Commun.* p. 576 (1970).
[68] J. C. Martin, J. W. Taylor, and E. H. Drew, *J. Amer. Chem. Soc.* **89**, 129 (1967).
[68a] R. A. Cooper, R. G. Lawler and H. R. Ward, *J. Amer. Chem. Soc.* **94**, 545 (1972).
[69] R. Kaptein, *Chem. Phys. Lett.* **2**, 261 (1968).
[70] H. Fischer and B. Blank, *Helv. Chim. Acta.* **54**, 905 (1971).
[71] H. R. Ward, R. G. Lawler, and R. A. Cooper, *Tetrahedron Lett.* p. 527 (1969).
[72] D. M. Tomkinson and H. O. Pritchard, *J. Phys. Chem.* **70**, 1579 (1966).
[73] R. G. Lawler, H. R. Ward, R. Allen, and P. Ellenbogen, *J. Amer. Chem. Soc.* **93**, 789 (1971).
[74] H. R. Ward, R. G. Lawler, and H. Y. Loken, unpublished results (1969).
[75] A. Trozzolo [Abstracts of papers presented at 159th American Chemical Society Meeting, paper ORGN 42, February 1970].
[76] T. Koenig and W. R. Mabey, *J. Amer. Chem. Soc.* **92**, 3804 (1970).
[77] G. L. Closs and A. D. Trifunac, *J. Amer. Chem. Soc.* **91**, 4554 (1969).
[78] G. L. Closs, C. E. Doubleday, and D. R. Paulson, *J. Amer. Chem. Soc.* **92**, 2185 (1970).
[79] J. Bargon and H. Fischer, *Z. Naturforsch. A* **23** 2109 (1968).
[80] S. V. Rykov and A. L. Buchachenko, *Dokl. Akad. Nauk SSSR* **185**, 870 (1969).
[80a] S. V. Rykov, A. L. Buchachenko, and V. I. Baldin, *Zh. Strukt. Khim.* **10**, 928 (1969).
[80b] A. L. Buchachenko, S. V. Rykov, A. V. Kessenikh, and G. C. Bylina, *Dokl. Akad. Nauk SSSR* **190**, 839 (1970).
[80c] S. V. Rykov, A. L. Buchachenko, V. A. Dodonov, A. V. Kessenikh, and G. A. Razuvaev, *Dokl. Akad. Nauk SSSR* **189**, 341 (1969).
[80d] S. V. Rykov, A. L. Buchachenko, and A. V. Kessenikh, *Spectrosc. Lett.* **3**, 55 (1970).
[81] S. R. Farenholtz and A. M. Trozzolo, *J. Amer. Chem. Soc.* **93**, 253 (1971).
[82] K. R. Darnall and J. N. Pitts, Jr., *Chem. Commun.* p. 1305 (1970).
[83] C. Walling and A. R. Lepley, *J. Amer. Chem. Soc.* **93**, 546 (1971).
[84] S. F. Nelsen, R. B. Metzler, and M. Iwamura, *J. Amer. Chem. Soc.* **91**, 5103 (1970); A. G. Lane, C. Rüchardt, and R. Werner, *Tetrahedron Lett.* p. 3213 (1969); A. Rieker, P. Niederer, and D. Leibfritz, *ibid.* p. 4287 (1969); J. Hollaender and W. P. Neumann, *Angew. Chem., Int. Ed. Engl.* **9**, 804 (1970).
[85] R. Kaptein, J. A. den Hollander, D. Antheunis, and L. J. Oosterhoff, *Chem. Commun.* p. 1687 (1970).
[86] M. Cocivera and H. D. Roth, *J. Amer. Chem. Soc.* **92**, 2573 (1970).
[87] R. A. Cooper, H. R. Ward, and R. G. Lawler, *J. Amer. Chem. Soc.*, **94**, 552 (1972).
[88] M. Lehnig and H. Fischer, *Z. Naturforsch. A* **25**, 1963 (1970).
[89] R. W. Fessenden and R. H. Schuler, *J. Chem. Phys.* **43**, 2704 (1965); R. W. Fessenden, *J. Magn. Resonance* **1**, 277 (1969).
[90] E. Lippmaa, T. Pehk, A. L. Buchachenko, and S. V. Rykov, *Chem. Phys. Lett.* **5**, 521 (1970).
[91] G. L. Closs and A. D. Trifunac, *J. Amer. Chem. Soc.* **92**, 7227 (1970); G. L. Closs and D. R. Paulson, *ibid.* p. 7229.
[92] J. W. Rakshys, Jr., *Chem. Commun.* 578 (1970); *Tetrahedron Lett.*, p. 4745 (1971).

[93] See, for example, R. Ernst, *Advan. Magn. Resonance* **2**, 1 (1967); T. C. Farrar and E. D. Becker, "Pulse and Fourier Transform NMR," Academic Press, New York, 1971.

[94] P. W. Atkins, I. C. Buchanan, R. C. Gurd, K. A. McLauchlan, and A. F. Simpson, *Chem. Commun.* p. 513 (1970).

[94a] B. Smaller, J. R. Remko, and E. C. Avery, *J. Chem. Phys.* **48**, 5174 (1968); S. H. Glarum and J. H. Marshall, *ibid.* **52**, 5555 (1970); H. Paul and H. Fischer, *Z. Naturforsch. A* **25**, 443 (1970); R. Livingston and H. Zeldes, *J. Chem. Phys.* **53**, 1406 (1970); P. W. Atkins, R. C. Gurd, K. A. McLauchlan, and A. F. Simpson, *Chem. Phys. Lett.* **8**, 55 (1971).

Ion Cyclotron Resonance Spectroscopy

4

JOHN I. BRAUMAN AND LARRY K. BLAIR

I.	Basic Principles of Ion Detection	152
II.	The Ion Cyclotron Double Resonance Technique for Studying Ion–Molecule Reactions	153
III.	Interpretation of the Double Resonance Experiment	154
IV.	Applications	155
V.	Thermochemical Quantities	156
VI.	Proton Transfer Reactions. Double Resonance Spectra	157
VII.	Acidities from Proton Transfer	161
VIII.	Reaction Pathways	162
IX.	Structures of Ions	163
X.	Concluding Remarks	165
	References	166

The study of ion–molecule reactions is of great importance in understanding the chemical reactions of species when they are removed from a solution environment. Thus, one can explore multiple reaction pathways and effects of structure on reactivity, without the incursion of solvation effects. When such gas-phase reactions are well understood, chemical behavior in solution can be analyzed in terms of both intrinsic (isolated molecule) and solvation effects. Consequently, we gain a better understanding of the important factors which contribute to chemical reactivity.

Ion cyclotron resonance (ICR) spectroscopy is finding increasing use as a tool in the study of ion–molecule reactions. The basic principles involved in ICR have been discussed extensively.[1-20] In this article, we briefly outline a description of the method and its application to a few selected problems. The purpose of this article is to provide an introduction to the ICR method, and consequently, extensive literature coverage has been avoided. Extensions to other problems and areas are left to the reader. Furthermore, in view of the intended readership, qualitative applications, rather than quantitative ones, will be stressed. More physical and quantitative applications are, however, becoming increasingly important, and will ultimately enhance the usefulness of the method.

I. BASIC PRINCIPLES OF ION DETECTION

Basically, the ICR spectrometer is a mass spectrometer whose design incorporates principles of the omegatron mass spectrometer[21] and the signal detection scheme used in magnetic resonance spectrometers. The ICR spectrometer is now commercially available (Dynaspec, Inc., Mount View, California).

Fundamental relationships which describe the motion of charged particles in magnetic and electric fields serve as a basis for understanding the detection of ions by the ICR spectrometer.[1] An ion of mass m and charge q moving in a uniform magnetic field of field strength H will be accelerated into a circular orbit whose plane is perpendicular to the magnetic field. The motion of the ion in this orbit will be characterized by a cyclotron frequency ω_c which is related to the above quantities by

$$\omega_c = qH/mc \tag{1}$$

where c is the speed of light. It is important to note that the cyclotron frequency ω_c is independent of the velocity of the ion. Thus, for a typical case which involves a number of ions with a distribution of velocities, all ions of a particular m/q value will be characterized by a unique cyclotron frequency independent of their velocities. The velocity distribution, however, will be reflected by a distribution of orbital radii since $\omega_c = v/r$.

The basic phenomenon which allows the detection of these ions is that an ion orbiting in a magnetic field with a cyclotron frequency characteristic of its mass:charge ratio can absorb power from an alternating electric field $E_1(t)$ of frequency ω_1 when $\omega_1 = \omega_c$. Experimentally, two parallel plates are situated between the pole faces of a magnet so that an alternating electric field $E_1(t)$ can be applied normal to H. The parallel plates are a part of the resonant circuit of a very sensitive marginal oscillator which serves as a source of $E_1(t)$. The absorption of energy by the ions of a particular mass: charge ratio whose ω_c is equal to ω_1 of $E_1(t)$ can be detected easily since the marginal oscillator level is extremely sensitive to small impedance changes in the resonant circuit. Thus, a mass spectrum can be obtained by sweeping the frequency ω_1 of the marginal oscillator at constant H through the range of cyclotron frequencies ω_c of the ions present while simultaneously monitoring the marginal oscillator level. Alternatively, a mass spectrum is obtained by

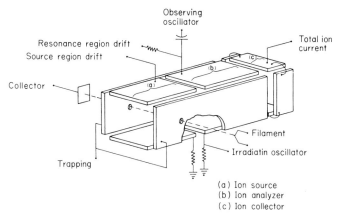

(a) Ion source
(b) Ion analyzer
(c) Ion collector

FIG. 1. ICR Cell. Ions are formed in the source region by an electron beam which is collinear with the magnetic field. Ions drift from the source region to the analyzer region where they are detected by the marginal oscillator. For this particular cell design, the double resonance experiment takes place in the analyzer region; reactant ions are irradiated by the auxiliary rf signal and the product ion intensities are monitored by the marginal oscillator, all in the analyzer region of the cell. The time an ion spends in the cell is determined by the source and analyzer drift voltages and to some extent, by trapping voltage which tends to keep the ions from escaping from the cell.

sweeping H at constant ω_1 while monitoring the marginal oscillator level. The latter technique has been used more extensively. A cutaway view of the spectrometer cell is shown in Fig. 1, along with a brief discussion of the various components.

II. THE ION CYCLOTRON DOUBLE RESONANCE TECHNIQUE FOR STUDYING ION–MOLECULE REACTIONS

The previous discussion of basic principles concerned the use of the ICR spectrometer as a mass spectrometer only. Its prime virtue lies not only in obtaining mass spectra, but also in its ease of adaptation for the study of ion–molecule reactions. For both positive and negative ions, reactions of the general type

$$A^+ + B \longrightarrow C^+ + D$$
$$E^- + F \longrightarrow G^- + H$$

are conveniently studied in the pressure range 10^{-6}–10^{-4} torr, where the frequency of bimolecular collisions is sufficient to be detectable. The lifetime of ions in the ICR can be of the order of milliseconds or longer; thus reactions can occur.

It has been observed that the rate constants of most ion–molecule reactions depend quite strongly on the kinetic energy of the reactant ions.[22] Consequently, changing the kinetic energy of the reactant ion A^+, for example, affects the rate constant and thus alters the (steady state) concentration of product ion C^+. The ICR double resonance technique is based on this phenomenon.

The double resonance experiment consists in observing the effect of perturbation of the translational energy of a particular ion on the intensity of another ion which may be coupled to it by an ion–molecule reaction. In the experiment, the intensity of the signal of a product ion C^+, for example, is monitored by the marginal oscillator at a frequency ω_c equal to the cyclotron frequency of C^+. Simultaneously, a reactant ion A^+ is irradiated with an auxillary oscillator at a frequency ω_A equal to the cyclotron frequency of A^+. From the previous relationship (1), it follows that the ratio of frequencies is related to the masses of the two ions by

$$\omega_A/\omega_c = m_c/m_A \qquad (2)$$

Usually, a double resonance spectrum is obtained by observing the intensity of the signal for a particular product ion at a fixed marginal oscillator frequency and the sweeping the frequency of the auxillary irradiating oscillator through the frequencies of the other ions present. Thus the resulting spectrum is an indication of the ions in the mixture which are coupled chemically to the observed ion.

III. INTERPRETATION OF THE DOUBLE RESONANCE EXPERIMENT

For most exothermic and thermoneutral ion–molecule reactions, it has been observed that the rate constant k decreases with increasing kinetic energy of the reactant ion, or in other words, the sign of dk/dE is $(-)$, although the simplest theory predicts no dependence if the interaction is an ion-induced dipole one. From the ICR double resonance experiment the sign of dk/dE can be obtained directly, since an increase or a decrease in product ion intensity observed when the reactant ion is irradiated indicates[4] a corresponding $(+)$ or $(-)$ dk/dE. Strictly, this assumes that in the experiment no reactant ions are removed from the cell when they are irradiated. It is, in principle, possible to obtain a double resonance signal for a reaction whose rate constant does not depend on reactant ion kinetic energy if the reactant ion is removed from the reaction chamber when irradiated. Thus, some indication that ions are not being removed from the cell by irradiation should be obtained in order to interpret an increase or decrease in product ion intensity as an indication of the sign of dk/dE. However, the interpretation of a double resonance experiment in terms of whether a reaction is occurring or not is a valid one. This has been the most general approach to the interpretation of the double resonance observations.[1,4]

If in a double resonance experiment a decrease in product ion concentration is observed when the reactant ion is irradiated, the reaction must be occurring in the absence of irradiation. Thus, in these cases, the double resonance experiment is merely a probe for detecting a reaction which is occurring in the ICR cell. Consequently, questions about the energy of a particular reactant ion need not consider the effect of irradiation, but rather the process by which the ion was formed and the relaxation the ion may have experienced before the reactive encounter.

The observation of an increase in the product ion signal when the reactant is irradiated indicates that the reaction is enhanced when the reactant ion is accelerated. It is quite likely that this is a reflection of a $(+)$ dk/dE. For such reactions it is not certain that the process was occurring when no irradiation was being applied. It is reasonable that an endothermic process, or a process requiring an activation energy, would show this behavior in a double resonance experiment.

The absence of any observable double resonance signal usually means that the reaction is not occurring at a detectable rate, although it is also possible that a reaction is occurring with $dk/dE = 0$. This later situation is, however, highly unlikely; it is more reasonable that a reaction is not occurring because of unfavorable energetics.

It should be noted that pressure-dependent studies of the more conventional type of studying ion–molecule reactions can also be done. These often provide confirmatory evidence for occurrence of reactions, although by their very nature, pressure dependent studies are less specific in the information they provide.

IV. APPLICATIONS

Applications of the ICR method can largely be divided into two main areas: (1) determination of absolute and relative thermochemical quantities,[4,17,19,23,24,25] and (2) studies of the occurrence and mechanisms of ion–molecule reactions[2,3,5-8,10-15] and measurement of rate constants.[9,16] In the first area are included studies of acidities, proton affinities, heats of formation, etc.; in the second are included systematic examination of reaction pathways and structures of ions. Both areas make extensive use of the double resonance method for probing the occurrence or nonoccurrence of reactions.

V. THERMOCHEMICAL QUANTITIES

The ICR method is of great utility in determining thermodynamic quantities. For example, Beauchamp and Buttrill[4] have determined the proton affinities of H_2O and H_2S by reference to known values for other species. In this section we describe in detail some of our work[23] on relative acidities of alcohols:

$$ROH + R'O^- \longrightarrow RO^- + R'OH \tag{3}$$

The purpose of this exposition is to demonstrate how the experiments are designed and interpreted

Alkoxide ions are not readily generated from alcohols. Therefore, water was added to the alcohols to serve as a primary source of negative ions which subsequently reacted with alcohols by proton transfer to yield alkoxide ion. The principle primary negative ion formed from electron interaction with water is H^- (at 6.5 eV) by the well-known[26] dissociative resonance capture reaction, Eq. (4). In addition, the rapid reaction (5) is known[27] to occur with the unusually large cross section of 160 $Å^2$. Since reaction (6) occurs,[28] it is probable that reaction (7) is also an important one for the production of RO^- in a mixture of alcohols and water.

$$H_2O + e^- \longrightarrow H^- + OH \cdot \tag{4}$$

$$H^- + H_2O \longrightarrow OH^- + H_2 \tag{5}$$

$$OH^- + ROH \longrightarrow RO^- + H_2O \tag{6}$$

$$H^- + ROH \longrightarrow RO^- + H_2 \tag{7}$$

Under ICR conditions, H^- was not observed directly; however, a plot (Fig. 2) of ion intensity for HO^-, CH_3O^-, and $C_2H_5O^-$ versus electron energy for a mixture of water, methanol, and ethanol is very similar to the plot obtained at low pressure for H^- from water,[26] and it suggests that H^- is the precursor, either directly or indirectly, of all the negative ions. However, one cannot easily determine which ion, H^- or OH^-, is more efficient for the production of alkoxide ions from alcohols. A reasonable explanation for the inability to observe H^- directly is that the H^- intensity is depleted by rapid reaction in the source region with water and/or alcohol.

A typical single resonance spectrum showing the negative ions present in a mixture of H_2O, CH_3OH, and C_2H_5OH at $\approx 10^{-5}$ torr and an electron

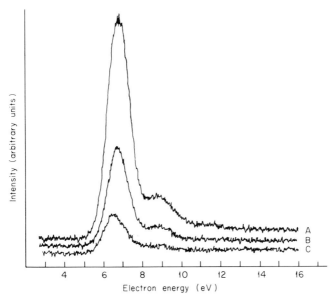

FIG. 2. Ion intensity of $C_2H_5O^-$ (curve A), CH_3O^- (curve B), and OH^- (curve C) versus electron energy in a mixture of ethanol, methanol, and water ($\approx 10^{-5}$ torr). These plots indicate that all the negative ions are coupled chemically to the H^- initially formed. (Electron energy scale, uncorrected) [Reprinted from *J. Amer. Chem. Soc.* **92**, 5986 (1970). Copyright 1970 by the American Chemical Society. Reprinted by permission of the copyright owner.]

energy of 6.5 eV is shown in Fig. 3. The major contributions occur at m/e 17, 31, and 45 for the $M - 1$ anions of water, methanol, and ethanol, respectively.

Deuterium labeling showed that it is the hydroxylic proton which is removed from the alcohols and that the negative ions are, in fact, alkoxide ions. For example, the single resonance spectrum of a mixture of D_2O and CH_3OD contains peaks only at m/e 18 and 31. In addition, single-resonance spectra of mixtures containing $(CD_3)_2CHOH$ show only an m/e 65 contribution indicating that the methyl protons are not being transferred.

VI. PROTON TRANSFER REACTIONS. DOUBLE RESONANCE SPECTRA

When ion cyclotron pulsed double resonance studies were made on a mixture of H_2O, CH_3OH, and C_2H_5OH, the following results were obtained: Irradiation of m/e 17 and 31 resulted in a decrease in the intensity of m/e 45.

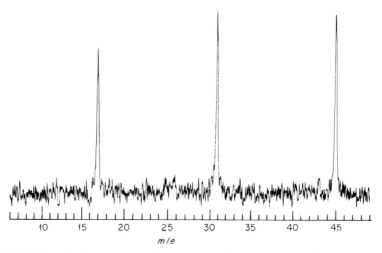

FIG. 3. Negative ions present in a mixture of water, methanol, and ethanol. Total pressure, $\approx 10^{-5}$ torr. This spectrum was obtained by using an electron energy modulation scheme. [Reprinted from *J. Amer. Chem. Soc.* **92**, 5986 (1970). Copyright 1970 by the American Chemical Society. Reprinted by permission of the copyright owner.]

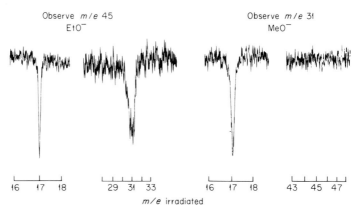

FIG. 4. Alkoxide ion double resonance results for a mixture of water, methanol, and ethanol (electron energy, 6.5 eV uncorrected; total pressure, 10^{-5} torr). The signal intensities of m/e 45 ($C_2H_5O^-$) and 31 (CH_3O^-) are displayed as a function of the frequency of the irradiating oscillator. Thus, irradiation of m/e 17 (OH^-) causes a decrease in the intensity of both $C_2H_5O^-$ and CH_3O^-. Irradiation of 31 causes a decrease in $C_2H_5O^-$; but irradiation of 45 has no effect on CH_3O^-. [Reprinted from *J. Amer. Chem. Soc.* **92**, 5986 (1970). Copyright 1970 by the American Chemical Society. Reprinted by permission of the copyright owner.]

Similarly, irradiation of m/e 17 caused a decrease in the intensity of m/e 31 (Fig. 4). In addition the intensity of m/e 17 was not changed when m/e 31 and 45 were accelerated. These double resonance results provide evidence that the following reactions [Eqs. (8)–10)] are occurring in the ICR cell and that the corresponding reverse reactions are not occurring, at least within the limits of measurability.

$$HO^- + C_2H_5OH \longrightarrow C_2H_5O^- + H_2O \tag{8}$$

$$HO^- + CH_3OH \longrightarrow CH_3O^- + H_2O \tag{9}$$

$$CH_3O^- + C_2H_5OH \longrightarrow C_2H_5O^- + CH_3OH \tag{10}$$

This conclusion results from the argument that perturbation of one ion can cause a decrease in the concentration of another ion only if the two ions are coupled chemically in the absence of perturbation. That is, a concentration could only be decreased if the perturbation in some way inhibited a reaction which had previously been occurring. Thus, for example, perturbation of kinetic energy of HO^- reduces the abundance of CH_3O^-; consequently, HO^- must be reacting with CH_3OH to produce CH_3O^-. The observations taken together result in the following order of ability to transfer a proton: $C_2H_5OH > CH_3OH > H_2O$.

Results for ion cyclotron pulsed double resonance experiments similar to the H_2O, CH_3OH, C_2H_5OH example for a number of alcohol mixtures with water are compiled in Table I. By the same type of reasoning and interpretation applied to the example, the following order of ability to transfer a proton is obtained:

$$(CH_3)_3CCH_2OH > (CH_3)_3COH > (CH_3)_2CHOH > C_2H_5OH$$
$$> CH_3OH > H_2O$$

and

$$(CH_3)_3COH \approx n\text{-}C_5H_{11}OH \approx n\text{-}C_4H_9OH > n\text{-}C_3H_7OH > C_2H_5OH$$

In addition to the alcohols, the double resonance results for toluene have been included in Table I. Initially, it was observed that OH^- abstracted a proton from toluene in a mixture of toluene and water and that benzyl anion did not abstract a proton from water. By investigation of a number of mixtures of toluene and water with alcohols, it was found that toluene lies between methanol and ethanol in its ability to transfer a proton. Transfer of the benzylic hydrogen from toluene was inferred from the observation that in mixtures of water and toluene with excess benzene, only $C_7H_7^-$ and no $C_6H_5^-$ was observed.

In order to check the double resonance technique as a method for ordering the ability to transfer protons, a system was chosen for which the energetics

TABLE I

Double Resonance Results for Forward (F) and Reverse (R) Proton Transfer Reactions [a,b]

Reaction				F	R
OH^-	$+ CH_3OH$	$= H_2O$	$+ CH_3O^-$	−	0
OD^-	$+ CH_3OD$	$= D_2O$	$+ CH_3O^-$	−	0
OH^-	$+ C_2H_5OH$	$= H_2O$	$+ C_2H_5O^-$	−	0
OH^-	$+ n\text{-}C_3H_7OH$	$= H_2O$	$+ C_3H_7O^-$	−	*
OH^-	$+ i\text{-}C_3H_7OH$	$= H_2O$	$+ i\text{-}C_3H_7O^-$	−	0
OH^-	$+ n\text{-}C_4H_9OH$	$= H_2O$	$+ n\text{-}C_4H_9O^-$	−	0
OH^-	$+ t\text{-}C_4H_9OH$	$= H_2O$	$+ t\text{-}C_4H_9O^-$	−	*
OH^-	$+ n\text{-}C_5H_{11}OH$	$= H_2O$	$+ n\text{-}C_5H_{11}O^-$	−	*
OH^-	$+ (CH_3)_3CCH_2OH$	$= H_2O$	$+ (CH_3)_3CCH_2O^-$	−	*
CH_3O^-	$+ C_2H_5OH$	$= CH_3OH$	$+ C_2H_5O^-$	−	0
$C_2H_5O^-$	$+ n\text{-}C_3H_7OH$	$= C_2H_5OH$	$+ n\text{-}C_3H_7O^-$	−	0
CH_3O^-	$+ i\text{-}C_3H_7OH$	$= CH_3OH$	$+ i\text{-}C_3H_7O^-$	−	0
$C_2H_5O^-$	$+ i\text{-}C_3H_7OH$	$= C_2H_5OH$	$+ i\text{-}C_3H_7O^-$	−	0
$n\text{-}C_3H_7O^-$	$+ (CD_3)_2CHOH$	$= n\text{-}C_3H_7OH$	$+ (CD_3)_2CHO^-$	−	−
$n\text{-}C_3H_7O^-$	$+ n\text{-}C_4H_9OH$	$= n\text{-}C_3H_7OH$	$+ n\text{-}C_4H_9O^-$	−	0
$i\text{-}C_3H_7O^-$	$+ n\text{-}C_4H_9OH$	$= i\text{-}C_3H_7OH$	$+ n\text{-}C_4H_9O^-$	−	−
$i\text{-}C_3H_7O^-$	$+ t\text{-}C_4H_9OH$	$= i\text{-}C_3H_7OH$	$+ t\text{-}C_4H_9O^-$	−	0
$C_2H_5O^-$	$+ n\text{-}C_5H_{11}OH$	$= C_2H_5OH$	$+ n\text{-}C_5H_{11}O^-$	−	0
$i\text{-}C_3H_7O^-$	$+ n\text{-}C_5H_{11}OH$	$= i\text{-}C_3H_7OH$	$+ n\text{-}C_5H_{11}O^-$	−	−
$n\text{-}C_4H_9O^-$	$+ n\text{-}C_5H_{11}OH$	$= n\text{-}C_4H_9OH$	$+ n\text{-}C_5H_{11}O^-$	−	−
$t\text{-}C_4H_9O^-$	$+ n\text{-}C_5H_{11}OH$	$= t\text{-}C_4H_9OH$	$+ n\text{-}C_5H_{11}O^-$	−	−
$t\text{-}C_4H_9O^-$	$+ (CH_3)_3CCH_2OH$	$= t\text{-}C_4H_9OH$	$+ (CH_3)_3CCH_2O^-$	−	+
OH^-	$+ C_6H_5CH_3$	$= H_2O$	$+ C_6H_5CH_2^-$	−	*
CH_3O^-	$+ C_6H_5CH_3$	$= CH_3OH$	$+ C_6H_5CH_2^-$	−	0
$C_6H_5CH_2^-$	$+ C_2H_5OH$	$= C_6H_5CH_3$	$+ C_2H_5O^-$	*	+
$C_6H_5CH_2^-$	$+ i\text{-}C_3H_7OH$	$= C_6H_5CH_3$	$+ i\text{-}C_3H_7O^-$	−	0
$C_6H_5CH_2^-$	$+ t\text{-}C_4H_9OH$	$= C_6H_5CH_3$	$+ t\text{-}C_4H_9O^-$	−	0

[a] The sign of the double resonance signal is given. A negative sign (−) means that the product ion concentration diminished when the reactant was irradiated. A negative sign is generally associated with an exothermic or thermoneutral reaction. A positive sign (+) is generally associated with endothermic reactions.[4] A zero (0) indicates that the reaction was investigated, but no signal change was observed. In conjunction with a forward (−) this suggests that the reverse reaction was not proceeding measurably. An asterisk (*) indicates the reaction was not investigated.

[b] Reprinted from *J. Amer. Chem. Soc.* **92**, 5986 (1970). Copyright 1970 by the American Chemical Society. Reprinted by permission of the copyright owner.

for proton transfer are known. The following reaction was investigated for a mixture of dry HCl and HBr; the forward reaction was confirmed by pulsed double resonance, and no reverse reaction was observed.

$$\text{HBr} + \text{Cl}^- \longrightarrow \text{HCl} + \text{Br}^-$$

In addition, both of the isotopic proton transfer reactions $^{79}\text{Br}^- + \text{H}^{81}\text{Br} \rightleftarrows \text{H}^{79}\text{Br} + {}^{81}\text{Br}^-$ and $^{35}\text{Cl}^- + \text{H}^{37}\text{Cl} \rightleftarrows \text{H}^{35}\text{Cl} + {}^{37}\text{Cl}^-$ were also observed.

VII. ACIDITIES FROM PROTON TRANSFER

An ideal approach to the determination of gas-phase acidities would be to measure the equilibrium concentrations of ions and neutrals for a general system $\text{HA} + \text{B}^- = \text{HB} + \text{A}^-$. However, with present mass spectrometers, it is experimentally difficult to trap ions for sufficient time for them to relax both thermally and chemically. Nevertheless, considerable success has been met in obtaining "thermochemical" quantities such as proton affinities of neutrals and heats of formation of ions under nonequilibrium conditions where the general approach to interpretation has been to consider the energetics of the reaction of an isolated ion and neutral molecule.

The simple criterion of the occurrence or nonoccurrence of an ion–molecule reaction has often been employed for determining whether a reaction is exothermic or endothermic. This criterion is expected to be more reasonable for reactions involving simple atom transfer or proton transfer than for those reactions involving considerable structural rearrangements. For the proton transfer studies involving alkoxide ions and alcohols, we investigated both forward and reverse reactions, and we assumed that the occurrence or nonoccurrence of a reaction is a reflection of the energetics for the proton transfer, that is, we assume that for a reaction to occur the reactants must have sufficient energy to form the products in their ground states. A reaction might not occur either because it is endothermic or because it has an unusually large activation energy. However, when a reaction is observed to occur in one direction, but not the reverse, considerations of microscopic reversibility make it clear that provided the reactants are in their ground states, only thermodynamic quantities are important. For reactions of the type indicated in Eq. (3), $\Delta H^0 = \Delta E^0$, thus the predominant direction of reaction provides evidence regarding the sign of ΔH^0. Finally, it is certain that the intrinsic ΔS^0 must be approximately zero for these reactions since contributions to the entropy are essentially identical for both reactants and products, excluding symmetry and statistical effects. Consequently, the observed order of ability to transfer a proton is interpretable as an accurate reflection of the sign of ΔG^0, and thus reflects a true gas-phase acidity. Customarily, the proton affinity ($-\Delta H°$ for the reaction), which is the quantity obtained directly, is actually quoted. While we have, in fact, directly determined the relative proton affinities of the alkoxide ions, we relate these to acidity for "chemical" purposes, so they can be compared with solution behavior.

We emphasize the importance of examining reactions in both directions since it is possible that the presence of internally excited reactant ions can lead to incorrect results about ground-state energetics when only a single reaction is considered. The internal consistency of sets of paired experiments makes it highly unlikely that excited states are responsible for the results. As an example, the observation that $C_2H_5O^-$ abstracts a proton from propanol but not from methanol provides strong evidence that propanol is a stronger acid than methanol, regardless of the structure or energetics of $C_2H_5O^-$. (The observation of the reverse proton transfer from ethanol to CH_3O^- suggests that there is no unusual activation barrier for this reaction.) Although our observations of proton transfer in both directions for some of the larger alcohols can be explained by the presence of internally excited ions, they are more likely the result of these reactions being nearly thermoneutral for the ground-state ions. We have assumed that for those cases in which proton transfer is observed in both directions, the two alcohols are approximately equal in acidity.

The significance of these results has previously been discussed.[23] It should be clear, however, that determinations such as these are of substantial value in understanding the intrinsic ionic behavior and reactions of many species.[24,25]

VIII. REACTION PATHWAYS

The work of Gray[7] illustrates the use of ICR in elucidating reaction pathways. Consider the positive ions present in acetonitrile at $\approx 10^{-5}$ torr. After the molecular ion, the second most intense ion is m/e 42, CH_3CNH^+, and it is of interest to determine which species are involved in its formation. We may assume that the neutral reactant is in all cases CH_3CN since no other neutral species is present in sufficiently high concentration to react. By sweeping the irradiating oscillator, and observing m/e 42, Gray was able to show that species at m/e 41, 40, 39, 28, 27, 26, and 14 were reacting. Some of these nominal masses may correspond to more than one ion (e.g., H_2CN^+ and $C_2H_4^+$). However, by a combination of high-resolution mass spectra and isotopic labeling in CD_3CN and $CH_3C^{15}N$, Gray was able to deduce the formulas of these ions and arrive at the set of reactions shown in Table II. Reaction (1) was deduced from experiments with labeled compounds. Reaction (2) is a particularly interesting one, since it can occur formally either via proton transfer from CH_3CN^+ or via hydrogen atom transfer to

TABLE II

Reactions of Acetonitrile [a,b]

(1)	CH_3CNH^+	$+ CH_3CN \rightarrow CH_3CN$	$+ CH_3CNH^+$
(2)	CH_3CN^+	$+ CH_3CN \rightarrow CH_2CN$	$+ CH_3CNH^+$
(3)	CH_2CN^+	$+ CH_3CN \rightarrow CHCN$	$+ CH_3CNH^+$
(4)	$CHCN^+$	$+ CH_3CN \rightarrow CCN$	$+ CH_3CNH^+$
(5)	CH_2N^+	$+ CH_3CN \rightarrow HCN$	$+ CH_3CNH^+$
(6)	HCN^+	$+ CH_3CN \rightarrow CN$	$+ CH_3CNH^+$
(7)	$C_2H_2^+$	$+ CH_3CN \rightarrow C_2H$	$+ CH_3CNH^+$
(8)	CH_2^+	$+ CH_3CN \rightarrow CH$	$+ CH_3CNH^+$

[a] The formulas of neutral products are inferred from conditions of mass balance.
[b] Data from Gray.[7]

this ion. The experiments $CH_3CN^+ + CD_3CN \rightarrow CD_3CNH^+$; $CH_3CN^+ + CD_3CN \nrightarrow CH_3CND^+$; and $CD_3CN^+ + CH_3CN \rightarrow CH_3CND^+$; $CD_3CN^+ + CH_3CN \nrightarrow CD_3CNH^+$ demonstrate unequivocally that only proton transfer is important. Thus, protonated acetonitrile can be shown to be produced by a minimum of eight pathways.

IX. STRUCTURES OF IONS

One of the most interesting applications of ICR has been to the determination of structures of ions. We discuss here the essentials of a study by Diekman, MacLeod, Djerassi, and Baldeschwieler[10] on the structures of ions produced in the McLafferty rearrangement. It is well established[29] that ketones containing a γ-hydrogen undergo, following an electron impact, a reaction which produces an olefin and an enol ion, for example,

$$\underset{(I)}{CH_3\overset{O^{+\cdot}}{\underset{|}{C}}CH_2CH_2\overset{H}{\underset{|}{C}}HCH_3} \longrightarrow \underset{(IIa)}{CH_3\overset{OH^{+\cdot}}{\underset{|}{C}}=CH_2} + CH_2=CHCH_3 \qquad (11)$$

The structure of the ion (II) is believed to be enolic rather than ketonic. When a ketone contains two alkyl groups, each containing a γ-hydrogen, the possibility exists for further reaction, and another McLafferty rearrangement often ensues. There has been considerable controversy regarding the structure of the resulting ion.[10,30] Specifically [Eq. (12)] the ion may have the structure corresponding to (V) or (II), depending upon whether the hydrogen is transferred to oxygen or carbon. There is, furthermore, another possible ion of the same mass corresponding to the ketonic structure (VI), which is believed not to be important.

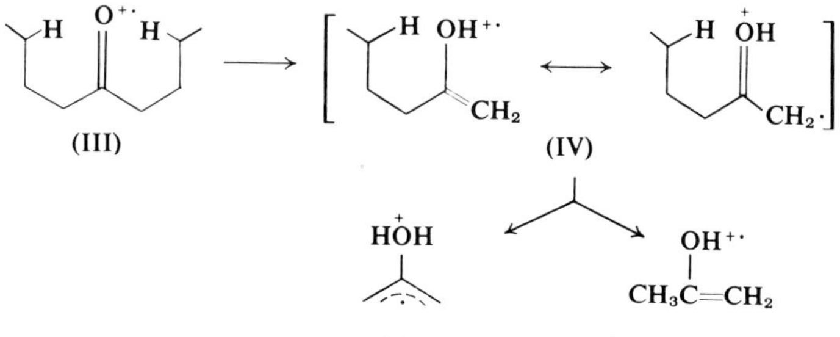

(12)

Ions of some of the various structures can be generated unequivocally by reactions (13) and (14)[29,31] in addition to (11) and (12):

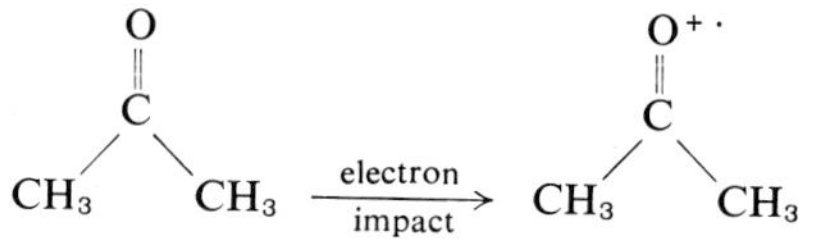

(13)

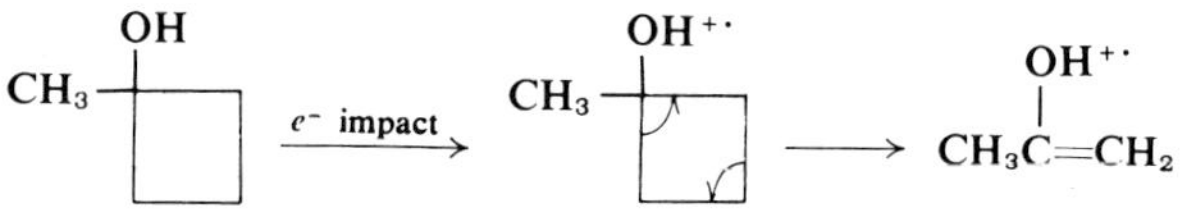

(14)

The ICR method offers a unique opportunity for solving this problem, since the ions (II), (V), and (VI), having different structures, would be expected to undergo different and distinguishable reactions with various neutral species. In fact, when the behavior of these ions was studied (often in mixtures with deuterium-labeled precursors to allow differentiation), some seven reactions were found which distinguished (VI) from (II). Furthermore, in all seven reactions, (IIa) (from a single McLafferty rearrangement) behaved identically with (IIc) (from methylcyclobutanol). Finally, in all reactions studied, (IIb) (from the double McLafferty rearrangement) was indistinguishable in its reactions from (IIa) and (IIc). Thus, the ion formed in the ICR experiments via Eq. (8) is undoubtedly (II), and all of the enol ions (IIa–c) have the same structure. Nevertheless, the possibility existed that this ion was only formed under the rather long lifetime conditions in the ICR.

This structure of the ion, however, was further substantiated by another series of experiments by Eadon, Diekman, and Djerassi.[14] Making use of the appropriately deuterium-labeled ketone, 4-nonanone-1,1,1-d_3, and the knowledge that secondary hydrogens participate 10 to 20 times as readily as primary hydrogens, it was possible to generate a monodeuterio double McLafferty ion whose structure must be either (VII) or (VIII) [Eq. (15)]:

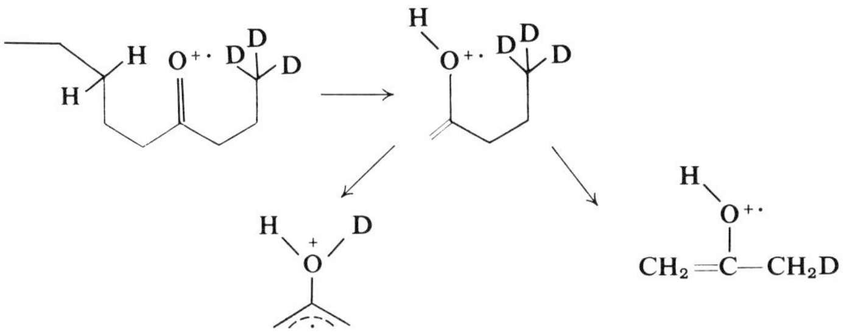

$$\text{(15)}$$

(VII) (VIII)

Since it is well established that proton transfer reactions of McLafferty ions involve only protons bound to oxygen, lack of D^+ transfer would rule out (VII) as a possible structure. Indeed, only H^+ transfer was observed. Furthermore, when 4-nonanone-7,7-d_2 was used, ion (IX) was formed and was found to transfer only D^+. The possibility of isomerization of (VII) to (VIII) or (IX) is excluded since the appropriate H^+ or D^+ transfer was unique and indicated no scrambling. Thus, one must conclude that the enol ion (II, VIII, IX) is produced directly and is the predominant structure formed in this reaction.

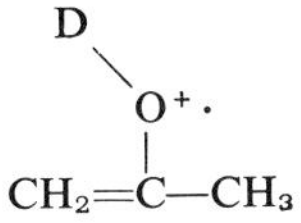

(IX)

X. CONCLUDING REMARKS

This very brief description of the ICR method has, of necessity, not included discussions of much interesting and important work. It has not dealt at all with some of the more quantitative aspects, many of which are discussed by Henis,[32] such as measurements of rate constants and line shapes or improvements in analysis and detection. Future applications to the technique will surely take advantage of many of these refinements. Also, use of the instrument in an analytical sense has great potential. It has, for example, been used[33] in studies of photodetachment [Eq. (16)] because of its great sensitivity and long ion lifetimes:

$$Z^- + h\nu \longrightarrow Z\cdot + e^- \tag{16}$$

The ICR method has already proved to be a valuable addition to the stock of chemical tools and instrumentation, and offers promise of many additional applications.

Added in proof

Since this chapter was completed, important, comprehensive reviews of ICR have been published by Baldeschwieler and Woodgate,[34] Gray,[35] Jennings,[36] and Beauchamp.[37] These articles should serve to bring the reader up to date in various applications of ICR and provide more detail in experiments. Finally, we call attention to recent utilization[38] of McIver's trapped ion cell,[20] which indicates that it may become possible to determine equilibrium constants quantitatively by trapping ions for long periods of time.

References

[1] For a review, see J. D. Baldeschwieler, *Science* **159**, 263 (1968). Also, references 2–20 for applications of the method and developments.
[2] L. R. Anders, J. L. Beauchamp, R. D. Dunbar, and J. D. Baldeschwieler, *J. Chem. Phys.* **45**, 1062 (1966).
[3] J. L. Beauchamp, L. R. Anders, and J. D. Baldeschwieler, *J. Amer. Chem. Soc.* **89**, 4569 (1967).
[4] J. L. Beauchamp and S. E. Buttrill, Jr., *J. Chem. Phys.* **48**, 1783 (1968).
[5] J. M. S. Henis, *J. Amer. Chem. Soc.* **90**, 844 (1968).
[6] F. Kaplan, *J. Amer. Chem. Soc.* **90**, 4483 (1968).
[7] G. A. Gray, *J. Amer. Chem. Soc.* **90**, 2177 and 6002 (1968).
[8] R. C. Dunbar, *J. Amer. Chem. Soc.* **90**, 5676 (1968).
[9] S. E. Buttrill, Jr., *J. Chem. Phys.* **50**, 4125 (1969).
[10] J. Diekman, J. K. MacLeod, C. Djerassi, and J. D. Baldeschwieler, *J. Amer. Chem. Soc.* **91**, 2069 (1969).
[11] R. M. O'Malley and K. R. Jennings, *Int. J. Mass Spectrom. Ion Phys.* **2**, 441 (1969).
[12] W. T. Huntress, Jr., J. D. Baldeschwieler, and C. Ponnamperuma, *Nature (London)* **223**, 468 (1969).
[13] M. Inoue and S. Wexler, *J. Amer. Chem. Soc.* **91**, 5730 (1969).
[14] G. Eadon, J. Diekman, and C. Djerassi, *J. Amer. Chem. Soc.* **91**, 3986 (1969)
[15] R. P. Clow and J. H. Futrell, *J. Chem. Phys.* **50**, 5041 (1969).
[16] M. T. Bowers, D. D. Elleman, and J. King, Jr., *J. Chem. Phys.* **50**, 1840 (1969).
[17] D. Holtz and J. L. Beauchamp, *J. Amer. Chem. Soc.* **91**, 5913 (1969).
[18] W. T. Huntress, Jr. and J. L. Beauchamp, *Int. J. Mass Spectrom. Ion Phys.* **3**, 149 (1969).
[19] P. Kriemler and S. E. Buttrill, Jr., *J. Amer. Chem. Soc.* **92**, 1123 (1970).
[20] R. T. McIver, Jr., *Rev. Sci. Instrum.* **41**, 555 (1970).
[21] R. W. Kiser, "Introduction to Mass Spectrometry and its Applications," p. 72. Prentice-Hall, Englewood Cliffs, New Jersey, 1965.
[22] For general discussions of ion-molecule reaction rate energy dependence, see *Advan. Chem. Ser.* **58** (1966); V. L. Tal'rose and G. V. Karachevtsev, *Advan.*

Mass Spectrom. **3**, 211 (1966); D. P. Stevenson, *in* "Mass Spectrometry" (C. A. McDowell, ed.), p. 65. McGraw-Hill, New York, 1963; C. E. Melton, *in* "Mass Spectrometry of Organic Ions" (F. W. McLafferty, ed.), p. 65. Academic Press, New York, 1963; F. W. Lampe, J. L. Franklin, and F. H. Field, *Progr. React. Kinet.* **1**, 69 (1961); K. R. Ryan and J. H. Futrell, *J. Chem. Phys.* **43**, 3009 (1965).

[23] J. I. Brauman and L. K. Blair, *J. Amer. Chem. Soc.* **90**, 6561 (1968); **92**, 5986 (1970).
[24] J. I. Brauman and L. K. Blair, *J. Amer. Chem. Soc.* **90**, 5636 (1968).
[25] J. I. Brauman and L. K. Blair, *J. Amer. Chem. Soc.* **91**, 2126 (1969).
[26] See R. N. Compton and L. G. Christophorou, *Phys. Rev.* **154**, 110 (1967), and references cited therein.
[27] J. A. D. Stockdale, R. N. Compton, and P. W. Reinhardt, *Phys. Rev.* **184**, 81 (1969).
[28] V. L. von Trepka, *Z. Naturforsch. A* **18**, 1122 (1968); D. Vogt and H. Neuert, *Z. Phys.* **199**, 32 (1967).
[29] See Diekman[10] for leading references: also, H. Budzikiewicz, C. Djerassi, and D. H. Williams, "Mass Spectrometry of Organic Compounds," pp. 155–162. Holden-Day, San Francisco, California, 1967; F. W. McLafferty, *Anal. Chem.* **31**, 82 (1959).
[30] F. W. McLafferty and W. T. Pike, *J. Amer. Chem. Soc.* **89**, 5953 (1967).
[31] P. Ausloos and R. E. Rebbert, *J. Amer. Chem. Soc.* **83**, 4897 (1961).
[32] J. M. S. Henis, manuscript in preparation (1972).
[33] J. I. Brauman and K. C. Smyth, *J. Amer. Chem. Soc.* **91**, 7778 (1969); K. C. Smyth and J. I. Brauman, *J. Chem. Phys.* **56**, 1132 (1972).
[34] J. D. Baldeschwieler and S. S. Woodgate, *Accts. Chem. Res.* **4**, 114 (1971).
[35] G. A. Gray, *Advan. Chem. Phys.* **19**, 141 (1971).
[36] G. C. Goode, R. M. O'Malley, A. J. Ferrer-Correia, and K. R. Jennings, *Nature* **227**, 1093 (1970).
[37] J. L. Beauchamp, *Ann. Rev. Phys. Chem.* **22**, 527 (1971).
[38] M. T. Bowers, D. H. Aue, H. M. Webb, and R. T. McIver, Jr., *J. Amer. Chem. Soc.* **93**, 4314 (1971); R. T. McIver, Jr., and J. R. Eyler, *ibid.* **93**, 6334 (1971).

Nuclear Quadrupole Resonance in Organic and Metalloorganic Chemistry

M. G. VORONKOV AND V. P. FESHIN

I. Introduction	169
II. Distribution and Assignment of Frequencies in the Spectral Bandwidth	174
III. The Nature of the Chemical Bond	176
IV. The Correlation of NQR Frequencies of the Type RX, RZX, and RR'R"MX with Substituent Inductive Constants and with Chemical and Physical Properties	177
V. Transmission of Electronic Effects across Saturated Carbon and Heteroatomic Systems	184
VI. Electronic Effects of Alkyl Groups	191
VII. Alicyclic Compounds	195
VIII. Aromatic Systems	199
IX. Araliphatic Compounds	204
X. Heterocyclic Compounds	205
XI. Phosphorus Compounds	209
XII. Sulfur Compounds	210
XIII. Intra- and Intermolecular Coordination in Metalloorganic Compounds	211
XIV. Molecular Complexes	216
XV. Polymeric Compounds	222
XVI. Some Other Areas of Application of the NQR Method	223
XVII. Conclusion	224
References	224

I. INTRODUCTION

Nuclear quadrupole resonance (NQR) is one of the newest physical methods for studying the structural features of crystalline substances. The nature of the NQR effect, which was first observed in 1949,[1] the interpretation of data obtained by this technique, and the possibilities for its application have been treated in detail.[2-31] However, much experimental data have been accumulated recently, and new areas for the use of NQR have opened. Fundamental surveys on NQR which have appeared in the last few years (e.g., O'Konski[12] and Lucken[31]) have been written, primarily, for specialists in the area of solid-state radiospectroscopy and contain only limited information on the possibilities of applying this method for the solution of chemical structure problems.

In the present work the authors have tried to survey, in a form maximally suited for a wide range of chemists, the available information on the application of the NQR method to organic and metalloorganic chemistry. In attempting to meet this goal, it will be necessary to present briefly the fundamentals of the NQR method which characterize the scope of its application to the solution of problems in theoretical chemistry as well as to the identification and structure determination of organic and metalloorganic compounds.

The NQR method is based on the use of radio-frequency irradiation (up to 1000 MHz) for the excitation of transitions between quadrupole energy levels in elliptical atomic nuclei which have quadrupole moments (nuclear spin quantum number $I \geq 1$). The quadrupolar nucleus is surrounded by an electric field which can be nonuniform as a result of asymmetry in the distribution of electrons in the molecule.

The quadrupole energy levels of a nucleus with spin $I \geq 1$, which is located in an electric field with an axially symmetric field gradient q_{zz} ($q_{zz} = \partial^2 V/\partial z^2$, where V is the potential at the center of the nucleus arising from all external charges, and z is the fixed axis directed along the chemical bond) relative to the direction of the chemical bond, are described by

$$E = eQq_{zz}[3m^2 - I(I + 1)]/4I(2I - 1)] \qquad (1)$$

where eQq_{zz} is the quadrupole interaction constant, I is the nuclear spin, m is the magnetic quantum number, e is the charge of an electron, and Q is the nuclear quadrupole moment, which is a measure of the deviation of the electric charge distribution in the nucleus from spherical symmetry.

If the nucleus has spherical symmetry ($Q = 0$) or its electronic environment is spherical ($q_{zz} = q_{xx} = q_{yy}$, where q_{xx}, q_{yy}, and q_{zz} are components of the electric field gradient q along the axes x, y, and z), the quadrupole energy levels are degenerate, and the NQR effect is not observed.

As the quantity m^2 enters into Eq. (1), it follows that in an axially symmetrical field the quadrupole energy levels will be doubly degenerate with respect to m. Thus for nuclei with spin $I = \frac{3}{2}$ (^{35}Cl, ^{37}Cl, ^{79}Br, ^{81}Br, ^{33}S, ^{201}Hg,

5. NQR in Organic and Metalloorganic Chemistry

etc.) there exists two doubly degenerate energy levels, and therefore, in the NQR spectrum, one transition frequency (v) is observed:

$$v = 3eQq_{zz}(2|m| - 1)/4hI(2I - 1) \qquad (2)$$

The NQR frequency for a nucleus with spin $I = \frac{3}{2}$ in an axially symmetrical electric field ($q_{zz} \neq q_{xx} = q_{yy}$) is related to the quadrupole interaction constant by the simple formula $eQq = 2hv$ (h is Planck's constant). Generally, the value for eQq is expressed in terms of frequency (in megahertz), leaving out the quantity h, and thus, $eQq = 2v$. In the case of an asymmetric field gradient ($q_{zz} \neq q_{xx} \neq q_{yy}$) this relationship does not hold since the NQR frequency

$$v = \frac{eQq_{zz}\sqrt{1 + \eta^2/3}}{2h} \qquad (3)$$

now depends on two parameters, q_{zz} and η [$\eta = (q_{xx} - q_{yy})/q_{zz}$], the asymmetry parameter of the electric field gradient which is equal to zero in an axially symmetric electric field. When $I = \frac{3}{2}$, the asymmetry parameter may be determined by NQR line splitting in a weak magnetic field (the Zeeman effect).

For the ^{14}N nucleus with spin $I = 1$, which is located in an asymmetrical environment

$$v = 3eQq_{zz}(1 \pm \eta/3)/4h \qquad (4)$$

that is, when $\eta \neq 0$, two transition frequencies are observed which permit the calculation of the quantities eQq_{zz} and η.

For nuclei with spin $I = \frac{5}{2}$ (^{127}I, ^{121}Sb, etc.), two NQR frequencies are also observed.

Since the electric quadrupole moment of a given nuclide is a constant, the NQR frequency measures the degree of anisotropy of the electronic environment of the resonating nucleus, that is, the distribution of electron density in its vicinity and thus, to an extent, of the entire molecule.

The major contribution to the electric field gradient is made by the valence p electrons. Filled electron shells and s electrons (in the absence of polarizing fields) do not affect the quantity q_{zz}, and thus have no effect on v, because they are spherically symmetrical. Electrons in the d and f shells have low density at the nucleus, and their contribution to the field gradient is therefore insignificant (less than 10% of the contribution of p electrons with the same principal quantum number). Thus, in first approximation, the NQR frequency is determined by the asymmetry in the distribution of p electrons in the vicinity of the nucleus under investigation.

A halogen atom, for example, has five valence np electrons. The electric field gradient (q_{zz}) or the NQR frequency of the atom is determined by the difference between the number of p electrons in one of the atomic orbitals, for example, N_z, and the average number of p electrons in the remaining two orbitals, $(N_x + N_y)/2$:

$$q_{zz} = [(N_x + N_y)/2 - N_z]q_0 \tag{5}$$

where q_0 is the electric field gradient arising from one p_z electron in a free halogen atom.

If the halogen atom takes part only in σ bonding, then $N_x = N_y = 2$, but N_z varies according to the nature of that bond (its hybridization and ionic character). The greater the magnitude of N_z (that is, the higher the sp hybridization and ionic character of the bond), the smaller q_{zz}, and thus, eQq_{zz} and ν. The participation of the unshared electron pairs of a halogen atom (N_x and N_y) in the formation of a bond, that is the π character of the bond, has an opposite effect on q_{zz} (eQq_{zz} and ν). This makes possible the use of NQR spectra for drawing conclusions about the nature of the chemical bond, the mutual interaction of atoms in molecules, the equivalence or nonequivalence of atoms with resonating nuclei and the location of these atoms in the molecule, crystal, etc.

There is as yet no exact theory which links NQR frequency or the electric field gradient with the nature of the chemical bond. From present concepts, it is possible to evaluate the hybridization type, the degree of ionic character of the chemical bond, and its π character from NQR data, but only in a very approximate way. Thus the interpretation of NQR spectra of given compounds from this point of view cannot be considered satisfactory. At present, it is most fruitful to search for empirical relationships by studying isostructural compounds in which the value for the asymmetry parameter remains approximately constant so that the contribution of only one of the three above-mentioned effects (the ionicity, π character, and hybridization) will change. Research in this direction has already attained considerable success.

The electric field gradient, which characterizes the distribution of electron density surrounding the resonating nucleus, is determined not only by the nature of the bonds to adjacent atoms, but also by the effect of more remote atoms and groups. Therefore, in considering the NQR spectra of a series of isostructural compounds of the type X—M—Y (Y is the substituent varied, X is the atom with the resonating nucleus, and M is the atom or atomic system which separates X and Y), it is possible to study the effect of substituent Y on the electron density of atom X, and thus, the electronic transmittance of system M.

In a series of compounds of the type Y—X or Y—M—X, where X is the reaction center and for which the quantitative characterization of the effect of substituent Y on the electron density and hence on the reactivity at X, the so-called σ constants (σ^* proposed by Taft, σ proposed by Hammett, etc.[32-34]) are widely employed. It would be expected that for a compound of the indicated type (X is the atom with the resonating nucleus) σ constants for the substituents should be related to the NQR frequency which also characterizes the electron density X. Indeed, between the quantities eQq or v^T (usually $T = 77°K$) and the σ constants, a correlation of the following type is observed:

$$v^T = v_0^T + \rho \Sigma \sigma \tag{6}$$

or

$$v^T = v_0^T + \rho_I \Sigma \sigma_I + \rho_c \Sigma \sigma_c \tag{7}$$

where $\Sigma \sigma$, $\Sigma \sigma_I$, and $\Sigma \sigma_c$ are terms which characterize the total electronic effect of the substituents on the atom with the resonating nucleus; v_0^T is the NQR frequency when $\Sigma \sigma = 0$, $\Sigma \sigma_I = 0$, and $\Sigma \sigma_c = 0$; and ρ, ρ_I, and ρ_c are the coefficients for the transmittance of the electronic effect of the substituents on the electron density of atom X. At present, such correlations are widely employed in studying electronic effects in organic and metalloorganic compounds, and substantial attention is given to them below.

The NQR frequency depends on the temperature at which it is measured (it usually falls with increasing temperature). The temperature coefficient of NQR frequency $(1/v)(\partial v/\partial T)$ is of the order -10^{-4} deg^{-1}. Therefore, comparative measurements should be preferably taken at a standard temperature of 77°K (liquid nitrogen). NQR frequencies determined at a temperature $T°K$ is indicated by v^T (v^{77}, v^{196}, v^{273}, etc.).

The use of NQR radiospectroscopy for the study of the mutual interaction of atoms in molecules has many advantages over chemical and other physical methods (the unusual sensitivity of the NQR method, the absence of solvent effects, etc.), and frequently it gives specific information which is unobtainable from any other source. Nevertheless, the method also has its drawbacks and limitations (the absence of an exact theory for NQR, some ambiguity observed in a number of instances resulting from the crystal lattice effects, etc.). Therefore, in making generalizations, results obtained by NQR radiospectroscopy (similar to other physical and chemical methods of investigation) must be checked and supported by data obtained in other ways.

In the present work we will not touch on instrumental questions. It is only necessary to mention that at present two fundamentally different methods are used for measuring the frequency of transitions between nuclear quadrupole energy levels—the stationary and impulse (or spin echo) methods. The latter allows in addition the easy measurement with high precision of the spin–lattice (T_1) and spin–spin (T_2) relaxation times.

II. DISTRIBUTION AND ASSIGNMENT OF FREQUENCIES IN THE SPECTRAL BANDWIDTH

Values for NQR frequencies of chemical compounds may vary from a few kilohertz to ≈ 1000 MHz,[29,35-37] depending on the value of the quadrupole moment of the resonating nucleus and its electronic environment, which is determined predominantly by the chemical bonds to adjacent atoms.[35-37]

The greatest interest for the chemist is in the chemical shifts in the NQR frequencies resulting from changes in the chemical bonds to the quadrupolar nucleus. Thus, for example, the NQR frequencies of ^{35}Cl in chlorine-containing compounds so far studied are found in the range 4–75 MHz and generally increase with an increase in the electronegativity of the bonded atom, M (Fig. 1). NQR frequencies for ^{35}Cl (ν^{77}) in organic and metalloorganic compounds which have a C—Cl bond are found in the range of ≈ 28–45 MHz; for an S—Cl bond the range is 31–40 MHz (Fig. 1). For molecules which have an Si—Cl, Ge—Cl, or Sn—Cl bond, the value for ν^{77} is in the low-frequency region (16–21 MHz, 17–26 MHz, and 15–25 MHz, respectively); when a N—Cl bond is present, the signal is observed in the high-frequency range (43–56 MHz).

While NQR frequencies of quadrupole nuclei of monovalent atoms depend in large measure on the valence state of the neighboring atom, a significant

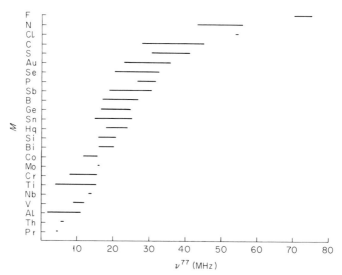

FIG. 1. Chlorine-35 NQR frequency ranges for compounds containing an M—Cl bond.

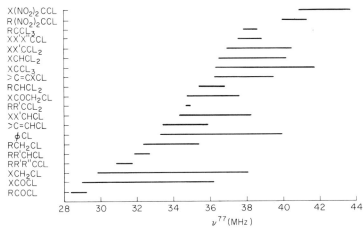

FIG. 2. Chlorine-35 NQR frequency ranges for organic and metalloorganic compounds containing a C—Cl bond.

role is also played by the coordination number and the symmetry of the environment in the case of multivalent atoms.[36,37]

In the case of a series of molecules of the type $R_{n-1}M^nX$, in which only the substituent R (which is bonded to the central atom M) is changed, the range of NQR frequency shifts for a given M is substantially narrower. Thus, for example, for compounds in the series RCOCl, values for ν^{77} vary in the range 28.36 [R = $(CH_3)_3C$][38] to 36.28 MHz [R = Cl].[29] In Fig. 2 the NQR frequency ranges are depicted for the best studied series of compounds in which X is a substituent with a Taft constant $\sigma_X^* > \sigma_H^*$ having a $-I$ effect, R is a substituent with $\sigma_R^* \leq \sigma_H^*$ having a $+I$ effect, and ϕ is an aromatic or heterocyclic substituent directly bonded to the chlorine atom.

The great precision in the measurement of NQR frequencies ($\pm 10^{-2}$–10^{-5}%) allows us to note the effect of any change in the character of the chemical bond on the indicator atom with great sensitivity. However, it is necessary to allow for a certain ambiguity in the NQR data resulting from crystal field effects caused by van der Waals interactions between the species making up the crystal lattice, and by the structure of the lattice. A shift in the NQR frequency due to crystal effects is usually ≈ 0.3%, but may reach 1.5–2%.[35–37,39]

Besides weak van der Waals forces which account for the crystal field effect, strong intra- or intermolecular interactions may exist in some crystals, which can lead to a shift of the NQR frequency up to 40%.[35–37]

III. THE NATURE OF THE CHEMICAL BOND

The electric field gradient (q_{zz}) or the quadrupole interaction constant (eQq_{zz}) are related to the character of the chemical bond by semiempirical relationships obtained by Townes and Dailey,[3,6,40,41] Gordy,[5,42] Bersohn,[43] and others.[9,12,18,31] On this basis, it is possible, with varying degrees of accuracy, to evaluate, for example, hybridization, ionic character, and π character of the M—X bond (for example, where M = C, Si, Ge, and Sn, and X is a halogen atom). This evaluation, however, has a very approximate character and is useful only for qualitative conclusions.

If a halogen atom is bonded to another atom only by a σ bond, the electric field gradient near the atomic nucleus and the quadrupole interaction constant are determined by the values for sp- (s) and dp- (d) hybridization and the ionic character of the bond (i):

$$eQq_{zz} = [1 - s + d - i(1 - s - d)]eQq_0 \qquad (8)$$

However, d hybridization is usually negligibly small relative to sp hybridization and, therefore, is not taken into consideration.

If p_x or p_y electrons of the halogen atom take part in the formation of the M—X bond, it is necessary to calculate the π character of the bond:

$$eQq_{zz} = (1 - s + d - i - \pi)eQq_0. \qquad (9)$$

NQR data alone do not allow the separate evaluation of these contributions to the value q_{zz} so that either a series of simplifying assumptions is made or data are used which were obtained in other ways. Thus, for example, according to Townes and Dailey[3,40] it is possible to calculate the s hybridization of atom X = Cl at 15% for the condition $\chi_X - \chi_M > 0.25$ (where χ_X and χ_M are electronegativities of the halogen atom and the atom M bonded to it). In reality, however, the degree of hybridization changes depending on the nature of the M—X bond. The bond between atoms which do not significantly differ from one another in electronegativity is preferably formed from pure p electrons.[3,40] The ionic character of the M—X bond may be evaluated from electronegativity differences.[3,9,40,44-46] The absence of a linear dependence of ionic character on electronegativity difference may arise from changes in the bond order, hybridization, etc.[45] NQR frequencies for the series RCH$_2$Cl are linearly related to the electronegativity of the atom located in substituent R.[47]

The electronegativity of the carbon atom varies according to its hybridization, increasing in the transition from the sp^3 to the sp state as a result of the increase in the s character of the bonding orbitals of the carbon atom. This leads to a decrease in the C—X bond length and a reduction of its ionic character and thus, to an increase in the NQR frequency of X.[11,15,31,43]

Thus, for example, in going from $(CH_3)_3CCH_2Cl$ to $(CH_3)_3C\equiv CCl$, the NQR frequency for ^{35}Cl rises from 33.01 to 39.15 MHz with the increase in the s character of the carbon atom bonded to the chlorine atom.

In saturated organic halides (not having halogen atoms in a *gem* position), the C—X bond has very little, if any, π character.[2,47-49] Therefore, in calculating the M—X bond parameters in such molecules, π character can be disregarded. Thus, for example, considering only the ionic character of the C—X bond, and the s hybridization, it was possible to evaluate the s character of the hybrid bonding orbital of the halogen atom in alkyl halides from NQR data. The average value of the s character was found to be 13.6% for C—Cl, 8.6% for C—Br, and 1.8% for C—I.[50]

The value for the π character (π) of the bond of a halogen atom with an sp^2 hybridized carbon atom is not large. From

$$\pi = 2\eta[e^2Qq/e^2Qq_0]/3 \tag{10}$$

it is seen that the π character for the C—Cl bond in compounds containing the

$$\diagdown_{\diagup}C=\overset{|}{C}-Cl$$

group comprises only a few percent[31] (for example, for $CH_2=CHCl$, $\approx 6\%$).[51]

The bond of a halogen atom with an sp-hybridized carbon atom, on the other hand, has very significant π character. Thus, for example, in derivatives of chloracetylene of the type $(CH_3)_3MC\equiv C-Cl$ (M = C, Si, Ge, Sn) and $(CH_3)_2Si(C\equiv C-Cl)_2$, the degree of π bonding of the carbon and chlorine atoms is calculated to be $\approx 70\%$.[52] This conclusion is supported by the very high value of the quadrupole constant of the C—Cl bond in chloroacetylene (79.65 MHz).[53] NQR spectra of chloroacetylenes testify to the significant delocalization of the π electrons of the triple bond which is even further increased if the neighboring atom possesses a vacant d orbital (Si, Ge, Sn).[52]

The presence of vacant nd orbitals in atoms of Si, Ge, and Sn lead to significant π character of the M—X bond which results from the so-called p_π–d_π interaction[2,49,54-56] (in comparison to the C—X bond in saturated organic halides). The π-character of the M—X bond decreases in the group from silicon to lead.[2,55,56]

IV. THE CORRELATION OF NQR FREQUENCIES OF THE TYPE RX, RZX, AND RR'R"MX WITH SUBSTITUENT INDUCTIVE CONSTANTS AND WITH CHEMICAL AND PHYSICAL PROPERTIES

NQR frequencies of ^{35}Cl, ^{79}Br, ^{81}Br, and ^{127}I in compounds in the series RX, RZX, and RR'R"MX, where R, R', and R" are varying substituents,

Z is a divalent atom or group (for example, —CH_2—, —SO_2—, $>P(O)R$, etc.), M = C, Si, Ge, and Sn and X = Cl, Br, and I correlate well with Taft inductive constants for substituents R (σ^*). Similar correlations are observed in the RX series with X = Cl[57-62] and Br,[59,63] and in the series RCH_2Cl,[57-59,61,62] RCH_2CH_2Cl,[57-59] $RCCl_3$,[58,59] $RCOCl$,[37,64] RCH_2Br,[59,63] RCH_2CH_2Br,[59,63] $RSiCl_3$,[57,59,65-68] $R(CH_3)SiCl_2$,[57,65,67] $R(CH_3)_2SiCl$,[57,65,66] $RGeCl_3$,[59,69] $RPOCl_2$, and RSO_2Cl.[70,71] A linear correlation for the value ν^{77} with inductive constants of the RCH_2 group is noted in the series RCH_2Cl,[4,5] and with the sum of the inductive constants for the substituents R, R', and R'' ($\Sigma\sigma^*$) for the series RR'R''MX (X = Cl, M = C,[37,58,59,63,72] Si,[37,57,59,65,66,72-81] Ge[37,57,59,69,72,76,79]; X = Br, M = C,[59,63] Si,[82] Ge,[83] Sn[54]; X = I, M = Si[83]), $RR'SiCl_2$,[57,59,66,73] RR'R''MCH_2Cl (M = C, Si, Ge, Sn)[84] and RR'POCl.[85] Some examples of similar linear correlations are illustrated in Fig. 3. The parameters ν_0^{77} and ρ for the linear equations

$$\nu^{77} = \nu_0^{77} + \rho\sigma^* \quad (11)$$

$$\nu^{77} = \nu_0^{77} + \rho\Sigma\sigma^* \quad (12)$$

where ν_0^{77} is the NQR frequency when σ^* or $\Sigma\sigma^*$ is equal to zero (usually this quantity is somewhat different from the observed value for ν^{77} of the corre-

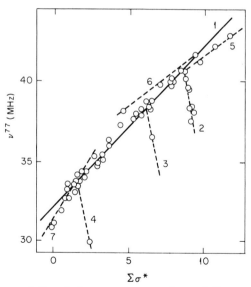

FIG. 3. The correlation between average values of ^{35}Cl NQR frequencies (ν^{77}) for compounds in the series RR'R''CCl and the sum of the Taft constants ($\Sigma\sigma^*$) for substituent R.[58]

sponding compound) and ρ is the coefficient for transmittance of the electronic effect of the substituent to the atom with the resonating nucleus, are presented in Table I.

As an example of such a correlation, it is possible to observe a linear (or close to linear) dependence of the NQR frequency on the number of halogen atoms (n) in molecules of the types

$Y_{4-n}MX_n$ (M = C, Y = H, X = Cl,[11,12,15,31,45,46,48,58,86-91] Br,[89-93] I[89,93];
M = C, Y = F, X = Cl,[12,45,46,58,63,88] Br[63];
Y = CH_3, X = Cl, M = C,[46,58,72,77,89] Si,[65-67,72,75,77,81,89] Ge[72,89];
M = C, Y = NO_2, X = Cl,[58,89-91] Br[89-91];
M = Si, X = Cl, Y = OR,[81] and miscellaneous[45,46,48,58,89])
$R_{3-n}POCl_n$ (R is an electron donating substituent),[86,94]
$ClCH_2(CH_3)_{3-n}SiCl_n$,[77]
$Cl_nSn(CH_2Cl)_{4-n}$,[95]
$Cl_nH_{3-n}HC=CH_2$.[96]

In molecules of the type $Cl_nSn(CH_2Cl)_{4-n}$ the function $\nu^{77}(n)$ has a linear character only for chlorine atoms located on the CH_2Cl group (for chlorine atoms bound to tin, the dependence is nonlinear).[95] An approximately linear dependence of ν^{77} on the number of methyl groups is observed for molecules of the type

$$H_{3-n}(CH_3)_nCCl.^{15,31}$$

The dependence of the NQR frequency on the number of halogen atoms in molecules of the type $Y_{4-n}MX_n$ (and also in the dependence of ν^{77} for molecules of the type $RR'R''MX$ on constants for the R substituents) is found to be linear (or nearly linear) if the following conditions are met[72,86,89-91,94]: (1) The tetrahedral angles between the valence bonds of the central atom M remain practically constant for all substituents; (2) the intramolecular interaction of the substituent through space is minimal or absent (for M = C this is frequently not the case as a result of the small atomic radius of the central carbon atom); and (3) the effect of the substituent is transmitted to the other three bonds of the central atom M equally and independently of the nature of the other substituents bonded to it.

Linear correlations between the NQR frequencies and the sum of the inductive constants of the substituents, and also the number of halogen atoms attached to the central atom M, support the additivity of the inductive contributions of these substituents to the electron density on atom X. This allows the possibility of developing an additive scheme for calculating the NQR frequencies for molecules in the series $RR'R''SiCl$,[75] $Y_{4-n}CX_n$ (Y is a varying

TABLE I

Parameters for Correlation Equations of the Type $\nu^{77} = \nu_0^{77} + \rho\Sigma\sigma^*$

No.	Compound series	ν_0^{77}	ρ	r	Mean deviation	Refs.
1	RCl^{35}	33.120	+2.849	0.93	±0.27	59
2	RCH_2Cl^{35}	33.017	+1.214	0.89	±0.23	58
3	$RCH_2CH_2Cl^{35}$	33.00	+0.425	0.82	±0.21	58,59
4	$RCH_2CH_2CH_2Cl^{35}$	33.09	−0.086	—	±0.130	
5	$RCH_2CH_2CH_2CH_2Cl^{35}$	33.108	+0.005	—	±0.153	
6	$RCCl_3^{35}$	38.045	+0.933	0.96	±0.06	59
7	$RR'CCl_2^{35}$	35.215	+0.959	0.93	±0.27	58,59
8	$RR'R''CCl^{35}$	32.05	+1.019	0.96		58
9	$RSiCl_3^{35}$	18.909	+0.484	0.99		73
10	$RR'SiCl_2^{35}$	17.775	+0.415	0.96	±0.11	66,73
11	$RR'R''SiCl^{35}$	16.632	+0.404	0.96	±0.11	66,73
12	$RSiCl_3^{35}$	18.909	+0.484	0.99	±0.10	66
13	$R(CH_3)SiCl_2^{35}$	17.719	+0.481		±0.060	65
14	$R(CH_3)_2SiCl^{35}$	16.512	+0.447		±0.050	65
15	$RR'R''GeCl^{35}$	17,382	+0.941	0.92		69
16	$RGeCl_3^{35}$	22.747	+1.017	0.93		69
17	RBr^{81}	211.82	+24.353	0.79		63
18	RCH_2Br^{81}	209.37	+9.546	0.87		63
19	$R(CH_2)_2Br^{81}$	209.70	+3.31	0.72		63
20	$RR'R''CBr^{81}$	401.19	−16.833	0.72		63
21	$RR'R''SiBr^{81}$	113.08	+3.953		±0.81	82
22	$RR'R''GeBr^{81}$	120.13	+6.68	0.999	±0.50	83
23	$RR'R''SnBr^{81}$	102.55	+7.48			54
24	$RR'R''SiI^{127}$ ($\frac{1}{2} \leftrightarrow \frac{3}{2}$)	143.09	+7.92	0.998	±0.97	83

substituent and X = Cl, Br, and I),[35,37,89] which would also be applicable to compounds with a number of other elements.

The linear correlations discussed bear witness to the predominantly inductive effect of a substituent on the atom with the resonating nucleus, and they allow the determination of σ^* inductive constants from NQR frequencies.[58,59,61–63,65,66,68,72–74,80] However, correlations between ν^{77} and σ^* are valid only to the first approximation because some deviations from the linear dependence $\nu^{77} = f(\sigma^*)$ are observed. Such deviations result from a change in the nature of the substituent effect on the electron density at atom X (besides such factors as the crystal lattice effect and the nonuniversality of σ^*).

NQR spectra of compounds of the series RR'MXY in which two electronegative atoms X and Y having unshared electron pairs (Y = F, Cl, OR, etc.

are bonded to the central atom M show that the electronegative atoms may interact by the *gem*-coupling mechanism (σ,p coupling) $\overset{\frown}{Y}$—M—$\overset{\frown}{X}$,[12,15,31,37,47,58,59,61-64,78,81,91,97] is depicted in the extreme structure $Y^+ = M - X^-$. The degree of participation in such conjugation by different atoms is found to be directly dependent on their electron affinity. In aliphatic compounds, the ability of a substituent to take part in such conjugation increases in the order $R_2N \geq RO > RS > F > Cl \approx Br$.[64] NQR frequencies for compounds of the type RR'R"MX in which the substituents R are suitable for *gem* coupling with the atom X are linearly related to $\Sigma\sigma^{*}$ [58] (*cf.* Fig. 3, curves 2, 3, and 4) with the sign of p opposite to that found for the main curve (1).[58]

In carboxylic acid chlorides YCOCl, *gem* coupling is found between oxygen and chlorine, $\overset{\frown}{Cl}$—C=$\overset{\frown}{O}$, which substantially lowers the ^{35}Cl NQR frequencies relative to alkyl chlorides.[98,99] It can be assumed that this type of conjugation plays a role for other groups of the type X—M=O in which the halogen atom is separated from the oxygen atom by some atom M (M = S, P, etc.). For example, where Y = RO, Cl, etc., Y may enter conjugation with the chlorine atom as well as the oxygen atom. This effect is important primarily in altering the polarity of the C=O bond, while it is significantly less important in modifying the polarity of the C—Cl bond.[64,70,100]

In molecules of the type CH_3—M—X interaction between the methyl group and the atom with the resonating nucleus (X) is possible as a result of σ-σ coupling (hyperconjugation). This, for example, is the case for ethyl chloride H_3≡C—CH_2—$\overset{\frown}{Cl}$.[38,101,102] Both of these effects leads to an anomalous lowering of the NQR frequency.

The deviation of compounds with unshared electron pairs (of the type Y—M—X) from the correlation of v^{77} and σ^* is predictable and linearly related to the resonance parameters σ_r^0 and σ_r.[61,62]

$$\Delta\sigma^* = (0.57 + 4.164\sigma_r^0) \pm 0.06 \quad (13)$$

$$\Delta\sigma^* = (0.38 + 2.960\sigma_r) \pm 0.10 \quad (14)$$

By calculating the *gem* coupling discussed above, it is possible to compute the σ^* constants for substituents containing atoms with unshared electron pairs according to Eqs. (15) and (16)[61,62]:

$$\sigma^* = (-7.84 + 0.2237v^{77} - 4.164\sigma_r^0) \pm 0.06 \quad (15)$$

$$\sigma^* = (-7.65 + 0.2237v^{77} - 2.960\sigma_r) \pm 0.10 \quad (16)$$

In the series RX, the correlation of v^{77} and σ^* is more exact if taken separately for compounds in which (*i*) R is an alkyl group, (*ii*) R is an electronegative substituent not bonded to X by an atom with an unshared electron pair (CN, CCl_3, COOH, etc.), and (*iii*) R is an electronegative substituent bonded to X by an atom with an unshared electron pair (F, Cl, OR, etc.).[58,60-62]

Deviation from linear dependence of ν^{77} on σ^* in the series RR'R"MX where M = Si, Ge, and Sn may be caused by intra- or intermolecular donor–acceptor interactions with the participation of vacant d orbitals of the atom M, as, for example, is noted in the series $RSiCl_3$ and $R(CH_3)SiCl_2$ when R = $NCCH_2CH_2$.[67]

The possibility of intramolecular interaction between the atom M and a chlorine atom located in the position α to M has been established by the study of NQR spectra for compounds of the series $RR'R"MCH_2Cl$ (M = Si, Ge, and Sn)[84] and in other cases (see below).

In studying the relative importance of inductive and conjugative effects on the electron density of the atom with the resonating nucleus, a two-parameter [Eq. (17)] fit is employed:

$$\nu = \nu_0 + \rho_I \Sigma \sigma_I + \rho_c \Sigma \sigma_c \qquad (17)$$

($\Sigma \sigma_I$ and $\Sigma \sigma_c$ are the summed inductive and conjugative constants found from the NMR shift of ^{19}F-substituted benzenes in inert solvents,[33] ρ_I and ρ_c are the coefficients for the transmittance of the corresponding effects, and ν_0 is the NQR frequency when $\Sigma \sigma_I = \Sigma \sigma_c = 0$).

To a first approximation, $\rho_I \Sigma \sigma_I$ characterizes the change in the number of unbalanced electrons N_z on the p_z orbital of the halogen atom participating in the M—X bond and the term $\rho_c \Sigma \sigma_c$ represents the change for the p_x and p_y orbitals which take part in the formation of the π bond.[37]

The two-parameter correlation $\nu(\sigma_I, \sigma_c)$ has been shown to be unsatisfactory for several series of compounds of the type RR'R"CCl. Therefore, such compounds were divided up into subgroups RCH_2Cl, RR'CHCl, $RHCCl_2$, $RR'CCl_2$, and $RCCl_3$ such that within each subgroup, additivity of the contributions of the substituents R is observed as a result of the close symmetry of the molecules.[37] Parameters for the derived formulas are presented in Table II.

Two-parameter equations are also obtained for molecules of the series RR'R"MX (where M = Si, Ge, Sn, X = Cl, Br, I).[37] In Table III, the two-parameter correlation equations are systematized for halides of group IVB elements for compounds of the type RR'R"MX and the equation parameters are converted to units of electron charge by dividing all terms by $eQq_0/2$, thus allowing the comparison of M—X bonds. For M = Si, Ge, and Sn, ν_0 has a significantly lower value than for M = C as a result of the increased ionic character of the M—X bond in going from M = C to M = Si, Ge, and Sn as well as of the transfer of the unshared electron pair of the halogen atom X to the vacant d orbital of the central atom M = Si, Ge, and Sn (p_π–d coupling), that is, by the reduction in the occupancy of the p_x and p_y orbitals

TABLE II

Parameters for Correlation Equations of the Type
$v^{77} = v_0 + \rho_I \Sigma \sigma_I + \rho_c \Sigma \sigma_c$ for Organic Compounds

Compound series	Equation parameters			
	v_0	ρ_I	ρ_c	r
RCH$_2$Cl	33.37	5.91	4.78	0.954
RR'CHCl	34.35	4.73	5.07	0.981
RHCCl$_2$	36.36	3.88	5.36	0.975
RR'CCl$_2$	34.44	4.37	3.28	0.968
RCCl$_3$	38.63	4.25	2.48	0.889

For a given atom M, the value for p_π–d_π coupling uniformly decreases in going from X = Cl to X = I which agrees with the change observed for v_0 in the correlation equations for molecules of the series RR'R"MX (M = Si and Ge).[37]

Two-parameter correlations $v^{77}(\sigma_I, \sigma_c)$ are found for the series RSCl[105] and RSO$_2$Cl,[71] and they also exist, for example, in the series RR'$_2$Sn^{59}Co(CO)$_4$,[36,37] and (CO)$_3$^{55}MnC$_5$H$_4$X.[36,37] Such correlations are only possible in the case when within a given series the geometry of the environment surrounding the resonating nucleus remains unchanged.[36,37]

NQR frequencies for molecules in the series RX and RR'R"MX are correlated not only with σ constants but also with certain physical and chemical constants. A linear correlation of NQR frequencies with inductive spectroscopic constants calculated from values for the infrared v_{Si-H} has been noted,[65,66,68,74,79,80] as well as with v_{Si-H}[79] and $v_{C=O}$[106] directly, with the half-wave potential for the polarographic reduction of chloronitroalkanes,[90] with chemical shifts in the NMR spectra of halonitroalkanes,[91] with the infrared v_{C-Cl} for molecules in the series CH$_{4-n}$Cl$_n$,[107] with values for the pK_a of RCOOH,[60,61] XCH$_2$COOH, XCH$_2$OH, RNH$_2$, and XCH$_2$NH$_2$,[61] with values for the electronegativity of R groups in compounds of the series RPOCl$_2$,[108] and with dipole moments for C—X bonds (X = Cl, Br) in the series CH$_{4-n}$Cl$_n$.[110] A comparison of NQR frequencies of the tetrahalides of carbon, silicon, germanium, and lead (MX$_4$) with M—X bond parameters (difference in electronegativity, ionic character, etc.)[111] have been related to atomic number,[76] with the electronegativity of atom M (χ_M)[56,76,112] and have been used for evaluating χ_M in the series MX$_4$, H$_3$MX, and (CH$_3$)$_2$MCl$_2$ (where M is an element of group IVB).[112]

The reason for the great variance in electronegativity scales for group IVB elements ("mesoids") lies in the nonuniform transmittance of the electronic effect of substituents by atoms M, and various intramolecular electronic effects which affect the hybridization, intermolecular interactions, etc.[84] In

TABLE III

Parameters for Correlation Equations of the Type $\nu^{77} = \nu_0 + \rho_I \Sigma \sigma_I + \rho_c \Sigma \sigma_c$ for Compounds in the Series $RR'R''MX$ (M = C, Si, Ge, Sn; X = Cl, Br, I)

M	X	$2\nu_0/eQq_0$	$2\rho_I/eQq_0$	$2\rho_c/eQq_0$	ρ_c/ρ_I
C	^{35}Cl	0.712	+0.078	+0.046	+0.68
	^{79}Br	0.887	+0.126	+0.085	+0.67
	^{127}I	0.792	+0.190	+0.172	+0.91
Si	^{35}Cl	0.314	+0.045	−0.004	−0.10
	^{81}Br	0.348	+0.071	−0.041	−0.58
	^{127}I	0.391	+0.112	−0.17	−0.51
Ge	^{35}Cl	0.344	+0.095	−0.017	−0.18
	^{81}Br	0.372	+0.111	−0.065	−0.59
Sn	^{35}Cl	0.183	+0.154	−0.145	−0.94
	^{81}Br	0.323	+0.127	−0.052	−0.41

particular, the apparent low electronegativity (incorrect) of silicon relative to the other mesoids is based on the values for eQq or NQR frequencies for molecules of the type MX_4. It is impossible to conclude from data that cannot be meaningfully compared that the Si—X bond is the most ionic because for different M the M—X bond has varying π character and the halogen atom X has a varying degree of s hybridization. The character of intermolecular interactions in crystals of MX_4 is apparently also nonuniform.

NQR frequencies lead to the determination of inductive and resonance constants for substituents with a sufficient degree of accuracy, as well as to certain chemical and physical constants of the compounds studied, including reactivity and the presence of intra- and intermolecular interactions.

In a series of isostructural compounds, intermolecular association has practically no effect on the linear dependence of ν^{77} on σ^* (chlorides of group IVB elements are associated even as liquids).[113,114] The linear correlation for NQR frequencies of alkylmercuric bromides, RHgBr, in crystals in which intermolecular Hg ← Br coordination exists is also explained by their like structures, and on the basis that the effect of the substituent on the bromine atom occurs primarily by an induction mechanism.[115]

V. TRANSMISSION OF ELECTRONIC EFFECTS ACROSS SATURATED CARBON AND HETEROATOMIC SYSTEMS

NQR data support the proposition that in compounds of the type $X(CH_2)_nY$ (Y is a substituent and X is the atom with the resonating nucleus or a moiety containing such an atom) the mutual interaction of the terminal

5. NQR in Organic and Metalloorganic Chemistry   185

atoms of groups X and Y occurs by an inductive mechanism [with the exception of those special cases when the effect of $\sigma,\sigma,^{38,99,101,102}$ σ,p, or σ,π coupling (see above) is superimposed on the interaction of substituents through space]. With an increase in the number of methylene or other similar groups (n) which separate X and Y, the inductive effect rapidly diminishes. In the NQR spectra of compounds of the type $X(CH_2)_nY$ this fall-off is seen as a gradual decrease or increase in the NQR frequency with an increase in n until a certain, approximately constant value, $\nu_\infty{}^T$, is reached which is characteristic for all members of the series in which the inductive interaction between X and Y is not significant. If the inductive constant for substituent Y, $\sigma_y^* \geq \sigma^*_{-(CH_2)_n-} = -0.13$,[101] the NQR frequency falls with an increase in n. In the case of electron donating substituents for which $\sigma_y^* < \sigma_{-(CH_2)_n-} = -0.13$, the value for ν^T grows for an increase in n. The NQR frequency falls for an increase in n in the series

$Cl(CH_2)_nCl$[11,31,89,98,101,116–118] (Fig. 4, curve 1)

$Cl(CH_2)_nCCl_3$[72,101,119]

$Cl(CH_2)_nC\!\!=\!\!\!=\!\!CH$ [101,120] (Fig. 4, curve 2)
  $\diagdown\diagup$
  $B_{10}H_{10}$

$ClCO(CH_2)_nCOCl$[98,101]
$I(CH_2)_nI$[101,121]
$Br(CH_2)_nBr$[11,101] $I(CF_2)_nI$[101,121]
$H(CH_2)_nCl$[11,31,38,45,87,101] (Fig. 4, curve 3)
$H(CH_2)_nBr$[11,31,92,101,122,123]
$H(CH_2)_nI$[92,101,122,123]
$H(CH_2)_nCOCl$[99,101]
$Cl(CH_2)_nGeCl_3$[72,101]
$Cl(CH_2)_nSiCl_3$[72,101]
$Cl_3Si(CH_2)_nSiCl_3$[68]
$H(CH_2)_nSiCl_3$
$H(CH_2)_nOSiCl_3$.[81]

On the other hand, the NQR frequency increases with an increase in n for the series $(CH_3)_2CH(CH_2)_nCl$[101] (Fig. 4, curve 4) and $(CH_3)_3C(CH_2)_nCl$.[38]

Figure 4 shows that the number of methylene groups, through which the inductive effect may be transmitted, depends on the character of the substituents X and Y in accordance with the value of their inductive constants and usually does not exceed 4.[101] Thus, for example, the effect of a hydrogen atom is only transmitted through one methylene group[38,101]; the effect of a

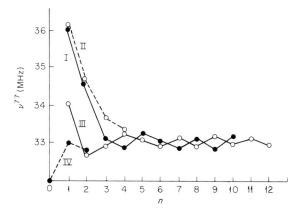

FIG. 4. The dependence of ^{35}Cl NQR frequencies (ν^{77}) for compounds in the series Y(CH$_2$)$_n$Cl on the number of methylene groups (n) for Y = Cl (curve I),

HC≡≡C— (II), H (III), and (CH$_3$)$_2$CH (IV).
  \\ /
  B$_{10}$H$_{10}$

chlorine atom, COCl group,[98,101] and SiCl$_3$[68] group is transmitted through two methylene groups, while the effect of the

HC≡≡C—
  \\ /
  B$_{10}$H$_{10}$

group is apparently transmitted through three methylene groups.

When the inductive effect or other manifestations of proximate interactions of substituents are completely eliminated in compounds of the type X(CH$_2$)$_n$Y, an oscillation of the NQR frequency is observed about an average ν_∞^T (Fig. 4) whose value does not depend on the character of Y and is determined only by the nature of the atoms or groups neighboring X, that is, by the electron density of the fragment XZ(CH$_2$)$_n$— (where XZ = ClCH$_2$—, ClCF$_2$—, ClS—, ClCH$_2$O—, ClCO—, etc.).[99,101,102] Thus, for example, for compounds in the series Y(CH$_2$)$_n$Cl, Y(CH$_2$)$_n$COCl, and Y(CH$_2$)$_n$Br, values for ν_∞^{77} are equal to 33,[11,89,101,116] 29,[98,99,101] and 250 MHz,[89] respectively.

The deviation of NQR frequencies of the first members of series of the type X(CH$_2$)$_n$Y from the average ν_∞^T may serve as a measure of the electronic effect of the substituent Y.[101] Comparatively insignificant regular periodic deviations in the NQR frequencies of group members with $n = 4\text{--}6$ from the average ν_∞^T are caused by the "even–odd" oscillation. This effect arises, apparently, as the result of hyperconjugation of the substituents X and Y through the σ-bond system.[101,124]

The irregular character of the oscillation in ν^T for certain homologous compounds [for example, $H(CH_2)_nCl$ and $Cl(CH_2)_nCl$] for $n = 4$–6 is possibly caused by the interaction of substituents through space and not through the σ-bond system.

The change in NQR frequencies for homologous compounds in a series of the type $X(CH_2)_nY$ with n indicates that the inductive constants of all substituents $Y(CH_2)_n$— at sufficiently high values of n must have a constant value which does not depend on the nature of Y and is determined by the electronic properties of the polymethylene chain, $[\sigma_{-(CH_2)_n-} \approx -0.13]$.[101] This observation invalidates the formula for calculating the inductive constants for substituents $Y(CH_2)_n$—

$$\sigma^*_{Y(CH_2)_n-} = \xi^{-n}\sigma^*_Y \tag{18}$$

according to which the quantity $\sigma_{Y(CH_2)_n-}$ in the case of an electron accepting substituent ($\sigma^*_Y > 0$) cannot fall to less than zero, no matter how large the value for n becomes. The correctness of this relationship had earlier become an object of some doubt.[34]

For evaluating the constant $\sigma^*_{Y(CH_2)_n-}$, Eq. (19), which reproduces the data well, is recommended[101]:

$$\sigma^*_{Y(CH_2)_n-} = -0.13 + 2.22^{-n}(\sigma^*_Y + 0.13) \tag{19}$$

According to NQR data, the transmittance of certain systems —M—, in compounds of the type X—M—Y, is characterized by the angle made by the slope of the correlation line (φ) or by the parameter $\rho = tg\varphi$ (transmittance coefficient) in linear equations which relate the values for NQR frequencies (ν^T) to σ constants for substituents $Y(\sigma)$, as for example in Eq. (6).

The linear correlation between ν^{77} and σ^* in molecules of the series $Y(CH_2)_nCl$ for varying values of n [Fig. 5, Eqs. (1–5) in Table I] leads to the conclusion that the transmittance of the system (ρ) has a value which falls with increasing n: $\rho = 2.8$ ($n = 0$), 1.2 ($n = 1$), 0.4 ($n = 2$), 0.0 ($n = 3$), and 0.0 ($n = 4$).[101,125,126] From these values for ρ, the inductive effect of substituents is completely eliminated for $n \geq 3$.

The subtle peculiarities of the interaction of substituents in series of the type $X(CH_2)_nY$ when $n \geq 3$ may be found by an analysis of the deviations from the linear correlation between ν^T and σ.* Comparison of the correlation equations for compounds in the series $R(CH_2)_3Cl$ [Eq. (4) in Table I] and RAdCl (Ad = adamanthyl)

$$\nu^{77} = 31.337 + 0.192\sigma^* \tag{20}$$

shows that the transmission of the inductive effect through the spatially fixed

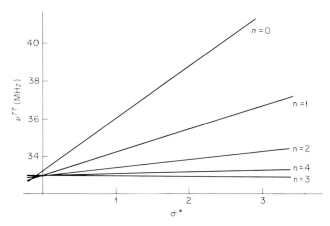

FIG. 5. The dependence of ^{35}Cl NQR frequencies (v^{77}) of compounds in the series $Y(CH_2)_nCl$ on Taft constants (σ^*) for the substituent Y when $n = 0$–4.

three-carbon system in 3-substituted 1-chloroadamantanes ($\rho = 0.192$) is significantly higher than through the corresponding acyclic chain ($\rho \approx 0$).[125]

A comparative study of NQR spectra of compounds with group IVB elements[84] shows that the transmittance of the inductive effect of substituents across the mesoid atom M (M = C, Si, Ge, and Sn) does not remain constant in different systems, and instead, even the order of transmittances of atoms M changes. From the value for ρ in correlation equations of the type in Eq. (11) for compounds in the series of type RR'R"MCl, where M = C, Si, and Ge (Table I), the sensitivity of the chlorine atom to a change in the nature of the substituents R, R', and R" falls according to the following order of a change in M: C > Ge > Si.[37] Such an order for the change in transmittance apparently is caused by the possibility of d_π–p_π coupling of the halogen atom with the central atom Si and, to a certain extent, with Ge.[31,37] The NQR frequencies of compounds of the type $(CH_3)_{4-n}MCl_n$ (M = C, Si, Ge, and Sn), on the other hand, show that the transmittance of the inductive effect of substituents falls—Sn > Ge ≥ C > Si[77,84]—while in compounds RR'R"MCH$_2$Cl the order is Sn ≥ Ge ≥ C > Si.[84] Likewise, the transmittance in compounds $^{35}Cl(CH_2)_nMCl_3$ ($n = 0$–2, M = C, Si, and Ge) depends on the nature of atom M as well as on the value of n. The ^{35}Cl NQR frequency for these compounds falls for a change in M in the following order: Ge > Si > C ($n = 0$), C > Ge > Si ($n = 1$), and C > Si > Ge ($n = 2$).[72] The change in the transmittance of the effect of chlorine atoms in this series is explained by the difference in the relative contributions of the effects of σ,σ coupling and of the field depending on the nature of atom M.[72]

5. NQR in Organic and Metalloorganic Chemistry

It is possible to use Eq. (21) for comparative evaluation of the transmittance of the electronic effect of substituents through any system [125,127]:

$$\gamma = tg\phi/tg\phi' = \rho/\rho' \tag{21}$$

where ϕ and ϕ' are the slopes of the respective correlation lines and ρ and ρ' are the transmittance coefficients in the correlation equations of the type represented by Eq. (6). Thus, for example, by considering the values for ρ, the transmittance of the adamantane system from the 1 to the 3 position is found to be less than half that of the system XCH_2CH_2Cl ($\gamma = 0.425/0.192 = 2.2$) and less than one sixth than the transmittance for the system XCH_2Cl ($\gamma = 1.214/0.192 = 6.3$).[125]

With a rigorously identical mechanism for the interaction of substituents, it is possible to determine values of γ from the linear equations which directly relate values of NQR frequencies for the two series of compounds under consideration. From the equations for the lines

$$\nu = \nu_0 + \rho\sigma^*$$

$$\nu' = \nu'_0 + \rho'\sigma^*$$

which relate values for ν and σ^* in the two series, it follows that

$$\nu = (\nu_0 + \gamma\nu'_0) + \gamma\nu' = \alpha + \gamma\nu' \tag{22}$$

Here γ is the tangent of the slope of the line which represents the linear dependence between NQR frequencies of the two series being compared and may be determined without recourse to any of the substituent σ constants.[125] Significant deviations from Eq. (22) are taken to mean that different types of interactions are operating.

Use of the correlation equations for 3-substituted 1-chloroadamantanes and the corresponding compounds in the series XCH_2Cl and XCH_2CH_2Cl leads to $\gamma = 5.84$ and 1.93, respectively, i.e., to values close to those shown above (6.3 and 2.2) which were found by using σ^*.[125]

Equations of the type represented by Eq. (7) allow comparison of the transmittance of inductive and conjugation effects. Thus, for example, the linear equation which relates NQR frequencies of sulfenyl chlorides, RSCl, with σ constants of substituents R

$$\nu^{77} = 38.40 + 3.65\sigma_I + 5.31\sigma_c \tag{23}$$

indicates that divalent sulfur is better in the transmittance of the conjugation effect than of the inductive effect of substituents (the ratio of the transmittance coefficients $\rho_I/\rho_c = 0.69$).[105]

Equations of the type represented by Eq. (12) show that in compounds of the series $RR'R''CX$ ($X = Cl$, Br, and I), $\rho_I > 0$ and $\rho_c > 0$, that is, that both σ- and π-donor substituents affect the NQR frequency in the same direction.[36,37] This is possible if the substituents cause a change only in the occupancy of the p_z orbitals of the halogen atom. In such a case the π system of the substituent interacts with the halogen atom by $p(\pi)$–σ coupling.[36,37,47]

In series of the type RR'R"MX (where M = Si, Ge, and Sn; X = Cl, Br, and, in the case of Si, I), $\rho_I > 0$, $\rho_c < 0$ because of the transfer of electron density from the p_x and p_y orbitals of the halogen atom to the vacant d orbitals of atom M.[36,37] With a change in X the value for ρ_c increases in the order Cl > Br > I, which results, apparently, from a change in the conditions for overlap of the corresponding d and p orbitals of the atoms M and X.[37] One of the reasons leading to the consistent growth in ρ_I in the series RR'R"MX in going from X = Cl to X = I is the change in the polarizability of the M—X bonds (I > Br > Cl).[37]

From the transmittance coefficients ρ_I and ρ_c in Eq. (24) for cyclopentadienylmanganesetricarbonyls, $(CO)_3{}^{55}MnC_5H_4Y$,

$$\nu_{(\pm 3/2 \leftrightarrow \pm 5/2)} = 19.12 + 0.65\rho_I - 4.32\rho_c \tag{24}$$

it follows that the transfer of the effect of substituents Y to the metal atom takes place primarily by way of conjugation.[36,37] The negative sign for the coefficient ρ_c arises from participation of the vacant orbitals of the metal atom in interaction with the substituent.[37]

In Eq. (25) which relates eQq_{zz}, ρ_c, and ρ_I in the series $RR'_2Sn^{59}Co(CO)_4$,

$$eQq_{zz} = 74.99 + 39.65\Sigma\rho_I - 71.35\Sigma\rho_c \tag{25}$$

the value of the coefficient of ρ_c is almost twice as large as that for ρ_I, owing to the very short Sn—Co bond.[37]

In conclusion, it is worthwhile to mention a number of inaccuracies in the comparative evaluation of the transmittance of the inductive effect from NQR spectroscopic data. For example, for the evaluation of the transmittance of the carborane system [120,128] the differences between the Cl NQR frequencies of compounds of the types $ClCH_2CB_{10}H_{10}CH$ and $ClCH_2CB_{10}H_{10}CCH_2Cl$ ($\Delta\nu^{77} = 1.07$ MHz) were compared with the differences in the series $ClCH_2C{\equiv}CH$ and $ClCH_2C{\equiv}CCH_2Cl$ ($\Delta\nu^{77} = 0.56$ MHz). It was concluded that the transmittance of the carborane system was greater than for —C≡C—. However, if one compares the difference in NQR frequencies of the pairs of compounds $ClCH_2CB_{10}H_{10}CCH_3$ and $ClCH_2CB_{10}H_{10}CCH_2Cl$ ($\Delta\nu^{77} = 0.23$ MHz) and $ClCH_2C{\equiv}CCH_3$ and $ClCH_2C{\equiv}CCH_2Cl$ ($\Delta\nu^{77} = 1.01$ MHz), it is possible to arrive at the opposite conclusion. Such contradictory conclusions are the result of the fact that the mechanisms for the electronic interactions between substituents in pairs are not identical. While the transfer of substituent effects to the chlorine atom in acetylenic compounds takes place primarily along the valence bonds by virtue of the linearity of the structure, the determining factor in the o-carborane system is the substituent interaction with the chlorine atom through space.[126] Thus, it is only possible to evaluate transmission when comparing the NQR frequencies of isostructural compounds and when the transmission mechanisms are identical.[126]

VI. ELECTRONIC EFFECTS OF ALKYL GROUPS

The question of the electronic effects of alkyl groups arose with the development of the electronic theory of the structure of organic compounds. However, this question has still not been answered unambiguously. Furthermore, this question has recently become significantly more complex. Not long ago, alkyl groups were considered to be electron donors as their electron repelling properties grow with an increase in the length and branching of the radical. A different order for the change in the effect of alkyl groups was noted for molecules of the type $H_{3-n}(CH_3)_nCC_6H_4CH_2Cl$,[127] in the series of 1-chloromethyl-2-substituted carboranes,[126] 2-p-alkylbenzylideneindane-1,3-diones[124,129] etc. Furthermore, experimental data were explained by the ability of alkyl groups to exhibit properties of electron acceptors (see, for example, Kwart and Miller,[130] Schubert et al.,[131] and Fort and Schleyer[132]). Along with these approaches, there exists a view that all alkyl groups possess a fixed and equal inductive effect.[129,133] Opinions regarding the ability of alkyl substituents to participate in hyperconjugation are quite contradictory.[34,134–138] It is often postulated that C—H bonds in alkyl groups are capable of entering into hyperconjugation with π-electron systems (see, for example, Palm[34] and Baker and Nathan[134]). However, evidence has appeared which invalidates such interactions in the molecular ground state,[135–137] leading to the conclusion that hyperconjugation is a dynamic effect and exists only in the excited state.[34,137,138]

The inductive effect of alkyl groups may be easily studied by the NQR method because if, as a result of the $+I$ effect of these substituents, the electron density on the atom with the resonating nucleus would increase, the NQR frequency would, in turn, decrease.

The classic example of the $+I$ effect of alkyl groups is the lowering of Cl NQR frequency in the series $H_{3-n}(CH_3)_nCCl$ with an increase in the number of methyl groups (n) on the carbon atom attached to the chlorine atom.[45,87] In this way, ν is correlated with the Taft inductive constants of the corresponding alkyl groups (σ^*).[32] A satisfactory correlation of NQR frequencies for halides RX (where X = Cl, Br, and R is an alkyl group) and σ^* (Fig. 6)[57–62] supports the assumption, to a first approximation, that an inductive effect for substituents R is operative.

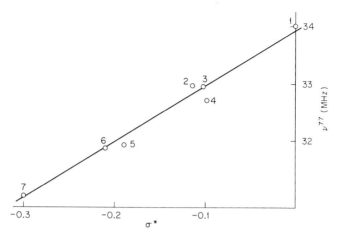

FIG. 6. The dependence of ^{35}Cl NQR frequencies (ν^{77}) of compounds in the series RCl on Taft constants (σ^*) for substituents R when R = CH_3 (1), $(CH_3)_3CCH_2$ (2), $(CH_3)_2CHCH_2$ (3), CH_3CH_2 (4), $(CH_3)_2CH$ (5), $CH(CH_3)C_2H_5$ (6), and $(CH_3)_3C$ (7).

Nevertheless, NQR data indicate[38] that in the series $H_{3-n}(CH_3)_nCCl$, in addition to the positive inductive effect of alkyl groups CH_3CH_2, $(CH_3)_2CH$, and $(CH_3)_3C$, σ,σ coupling of the type $H_3\equiv C\!-\!C\!-\!Cl$ is also an important factor. Such a conclusion is derived from a consideration of the dependence of NQR frequency of compounds of the series $H_{3-n}(CH_3)_nCCOCl$ and $p\text{-}[H_{3-n}(CH_3)_nC]C_6H_4CH_2Cl$ on σ^* of the substituents $H_{3-n}(CH_3)_nC$ (Fig. 7). In these series, NQR frequencies observed for $n = 0$ have anomalous values (deviating from a linear dependency) which may be explained by assuming that σ,σ coupling between the C—H and C—Cl bonds, which acts to lower the NQR frequency, is superimposed on the positive inductive effect of the alkyl groups. In the series $H_{3-n}(CH_3)_nCCl$, conjugation of this type is evident for $n = 1\text{-}3$, but not for $n = 0$, and thus in the latter case the NQR frequency has a value higher than expected by considering only the positive inductive effect of the methyl group. In the series $H_{3-n}(CH_3)_nCOCl$ and $p\text{-}[H_{3-n}(CH_3)_nC]C_6H_4CH_2Cl$, on the other hand, σ,σ coupling is seen only for $n = 0$, and the NQR frequency of the methyl derivatives is lower than for $n = 1$ (if in each of the series only the positive inductive effect of the alkyl substituents is operative, the relationship of the frequencies for $n = 0$ and $n = 1$ would be the reverse).

The presence of σ,σ coupling in molecules of the type CH_3CH_2X is supported by the relationship of NQR frequencies in the series $CH_3(CH_2)_nX$ (X = Cl, Br, and I; $n = 0\text{-}2$), which have the greatest value for $n = 1$ and not

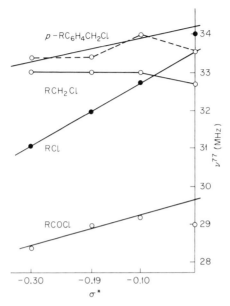

FIG. 7. The dependence of values of ^{35}Cl NQR frequencies (ν^{77}) on Taft constants (σ^*).

for $n = 2$. If in these series the $CH_3(CH_2)_n$ groups were acting only by the positive inductive effect, the NQR frequency would steadily decrease in going from $n = 0$ to $n = 2$.[38]

Evidence for the presence of hyperconjugation in the molecule CH_3CHCl_2 is found in the lower NQR frequency of this compound (35.41 MHz)[29] than that for $(CH_3)_3CCHCl_2$ (36.80 MHz).[29] If only the positive inductive effect of the substituents were operative, the NQR frequency of the latter compound should be lower than that of the former. The hyperconjugative effect is composed of additive contributions of the separate methyl groups so as to permit a good correlation of NQR frequencies with σ^* constants.[38] The hyperconjugation from NQR indicates that this effect is present in the molecular ground state[38] and is not restricted to the excited state.

The contribution of σ,σ coupling to the overall electron-donating effect of alkyl groups in compounds of the series $H_{3-n}(CH_3)_nCCl$ does not change the usual order of the increase in the positive inductive effect of alkyl substituents—$CH_3 < CH_3CH_2 < (CH_3)_2CH < (CH_3)_3C$—and the alkyl iodides and bromides of the type $H_{3-n}(CH_3)_nCX$ (X = I, Br; $n = 0$–2) are similar.[92]

The same order in the inductive effects of groups $H_{3-n}(CH_3)_nC$ is also confirmed in the series RCOCl,[38] if the contribution of σ,σ coupling which acts to lower the NQR frequency is taken into account only for R = CH_3 in this series (Fig. 7).

However, in the series p-[$H_{3-n}(CH_3)_nC$]$C_6H_4CH_2Cl$,[38] NQR frequencies, as shown in Fig. 7, do not change in conformity with values of the $+I$ effect of alkyl groups [$CH_3CH_2 > CH_3 > (CH_3)_3C \approx (CH_3)_2CH$]. The reduction in NQR frequency for R = CH_3 ($n = 0$), in comparison with p-$CH_3CH_2C_6H_4CH_2Cl$ ($n = 1$), may be caused by the hyperconjugative effect according to the scheme

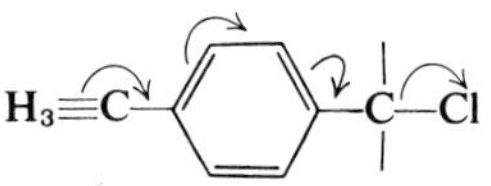

$$H_3\equiv C\!-\!\!\bigcirc\!\!-\!\!C\!-\!Cl$$

The similar NQR frequencies for compounds with $n = 2$ and 3 are as yet difficult to explain.

NQR spectra in the series RCl and RCOCl[38] allow the evaluation of the contribution of σ,σ coupling of the methyl C—H bonds to the NQR frequency of the corresponding compound of this type (≈ 0.57 MHz) and show that in this case only methyl group C—H bonds (and not those of CH_2 or CH groups) are capable, apparently, of participating in σ,σ coupling.

The recent proposal of equal inductive effect and electron accepting behavior for alkyl groups (for example, in 3-alkyl-substituted 1-bromoadamantanes)[132] is convincingly refuted by NQR spectra of 1-chloroadamantane, 3-methyl-, and 3-isopropyladamantanes, where the electron density on the chlorine atom in compounds of this series increases (Fig. 8)[125]:

$$H < CH_3 < (CH_3)_2CH$$

In compounds of the type $R(CH_2)_nX$, the inductive effect of substituents R as seen above rapidly diminishes as n is increased and approaches a constant value which is determined by the electron-donating character of the polymethylene chain ($\sigma^*_{-(CH_2)_n-} = -0.13$). However the monotonic diminution of the inductive effect of alkyl substituents $R(CH_2)_n$ with an increase in n may be changed by σ,σ coupling, thus explaining the anomalous order for the change in their electron donating effect.[101] For example, in the series $CH_3(CH_2)_nX$ (X = Cl, Br, and I; $n = 0$–2) the order of the electron donating ability is $C_2H_5 > n$-$C_3H_7 > CH_3$.[101] A similar order in many classes of organic and heteroorganic compounds is also supported by other physical and chemical methods.[101] NQR frequencies have been used to establish that the electron-donating effect of the methyl group is less than that of the ethyl group.[11] In this connection the assumption has been made that the methyl group, when separated from the chlorine atom by three carbon atoms, does not affect the chlorine atom.[11]

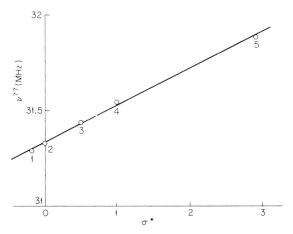

FIG. 8. The dependence of ^{35}Cl NQR frequencies (ν^{77}) of compounds in the adamantane series YAdCl on Taft constants (σ^*) for substituents Y for Y = $(CH_3)_2CH$ (1), CH_3 (2), H (3), CH_2Br (4), and CO_2H (5).

A comparison of NQR frequencies of compounds in the series $R(CH_2)_nCl$ [R = CH_3, $(CH_3)_2CH$,[101] and $(CH_3)_3C$[38]] allows the conclusion that the effect of substituents R is already diminished for $n = 1$. Similarly, the inductive effects of the alkyl groups $R(CH_2)_n$ for $n \geq 1$ are practically identical.

VII. ALICYCLIC COMPOUNDS

Chlorine-35 NQR frequencies for compounds of the cyclopropane series are significantly higher than for the corresponding acyclic analogs. Thus, for example, the NQR frequencies of 2-chloropropane and 2,2-dichloropropane (31.939 and 34.883 MHz, respectively) are lower than their cyclic analogs, chlorocyclopropane, and 1,1-dichlorocyclopropane (ν^{77} = 34.063 and 36.610 MHz, respectively).[139] Such a difference in NQR frequencies for cyclic and acyclic compounds is the result of the greater s character of the carbon atoms in the cyclopropane ring. Thus, NQR frequencies of chlorocyclohexane, 1,1-dichloropentane, and 1,1-dichlorocyclohexane (ν^{77} = 32.698, 35.022, and 34.947 MHz, respectively)[29] are lower than in their cyclopropane analogs.

Values for the ionic character of the C—Cl bond and the s hybridization of the chlorine atom in a chlorine derivative of the cyclopropane series, as well as for the effective orbital electronegativity of the ring carbon atoms bonded to chlorine, have been calculated.[31,139] Electronegativity values correlate with the NQR frequencies of the corresponding cyclopropane derivatives.[139] NQR

TABLE IV
Chlorine-35 NQR Frequencies (v^{77}) for Substituted Chlorobenzenes with Hammett Substituent Constants

No.	Compound	v^{77}(MHz)	Position of chlorine atom	Substituent	$\Sigma\sigma$
53	1-Cl-2-CH$_3$C$_6$H$_4$	34.19		o-CH$_3$	+0.292
54	1-Cl-4-CH$_3$C$_6$H$_4$	34.53		p-CH$_3$	−0.17
55	1-Cl-4-IC$_6$H$_4$	34.77		p-I	+0.276
56	1-Cl-4-BrC$_6$H$_3$	34.80		p-Br	+0.232
57	1-Cl-4-NO$_2$C$_6$H$_4$	34.88		p-NO$_2$	+0.778
58	1-Cl-4-FC$_6$H$_4$	35.103[a]		p-F	+0.062
59	1-Cl-2,6-(NO$_2$)$_2$C$_6$H$_3$	39.383		2(o-NO$_2$)	+4.060
60	1,3-Cl$_2$-2-CH$_3$C$_6$H$_3$	34.771[a]		o-CH$_3$ + m-Cl	+0.665
61	1,3-Cl$_2$-5-NH$_2$C$_6$H$_3$	35.021[a]		m-Cl + m-NH$_2$	+1.083
62	1,3-Cl$_2$-4-OHC$_6$H$_3$	35.668[a]	1	m-Cl + p-OH	+0.016
		35.075[a]	3		
63	1,3-Cl$_2$-4-COHC$_6$H$_3$	35.986	1	m-Cl + p-COH	+0.589
		35.461	3		
64	1,2-Cl$_2$-4-COHC$_6$H$_3$	36.762	2	o-Cl + m-COH	+1.615
		35.685	1	o-Cl + p-COH	+1.476
65	1,2,3-Cl$_3$-5-NH$_2$C$_6$H$_2$	37.380	2	2(o-Cl) + p-NH$_2$	+1.860
		36.493[a]	1,3	o-Cl + m-Cl + m-NH$_2$	+1.472

5. NQR in Organic and Metalloorganic Chemistry

66	1,2,3-Cl$_3$-5-NO$_2$C$_6$H$_2$	37.720	2	2(o-Cl) + p-NO$_2$	+3.298
		37.359	1,3	o-Cl + m-Cl + m-NO$_2$	+2.343
67	1,2,4-Cl$_3$-5-NO$_2$C$_6$H$_2$	38.364[a]	4	o-NO$_2$ + m-Cl + p-NO$_2$	+2.630
		37.386[a]	2	o-Cl + m-Cl + p-NO$_2$	+2.411
		36.651[a]	1	o-Cl + m-NO$_2$ + p-Cl	+2.197
68	1,3,5-Cl$_3$-2,6(NO$_2$)$_2$C$_6$H	39.165	1	2(o-NO$_2$) + 2(m-Cl)	+4.806
		37.922	3,5	o-NO$_2$ + 2(m-Cl) + p-NO$_2$	+3.554
69	1,2,4-Cl$_3$-3,5(NO$_2$)$_2$C$_6$H	39.917	4	2(o-NO$_2$) + m-Cl + p-Cl	+4.660
		38.741	2	o-Cl + o-NO$_2$ + m-Cl + p-NO$_2$	+3.551
		38.245	1	o-Cl + 2(m-NO$_2$) + o-Cl	+2.907
70	1,2,4,5-Cl$_4$-3,6-(CH$_3$)$_2$C$_6$	36.60	1,2,4,5	o-Cl + o-CH$_3$ + m-Cl + m-CH$_3$ + p-Cl	+2.083
71	1,2,3,5-Cl$_4$-4,6-(CH$_3$)$_2$C$_6$	37.57	2	2(o-Cl) + 2(m-CH$_3$) + p-Cl	+2.607
		36.83	1,3	o-Cl-o-CH$_3$ + 2(m-Cl) + p-CH$_3$	+2.128
		35.69	5	2(o-CH$_3$) + 2(m-Cl) + p-Cl	+1.557
72	1,2,3,5-Cl$_4$-6-NH$_2$C$_6$H	37.708	2	2(o-Cl) + m-NH$_2$ + p-Cl	+2.586
		36.961	3	o-Cl + 2(m-Cl) + p-NH$_2$	+1.346
		36.461	1		
		35.618	5		
73	1,2,4,5-Cl$_4$-3,6-(NO$_2$)$_2$C$_6$	39.110	1,2,4,5	o-Cl + o-NO$_2$ + m-Cl + m-NO$_2$ + p-Cl	+4.600
74	1,2,3,4,5-Cl$_5$-6-NHCH$_3$C$_6$	37.975	2,4	2(o-Cl) + m-Cl + m-NHCH$_3$ + p-Cl	+2.818
		37.695	3	o-Cl + 2(m-Cl) + p-NHCH$_3$	+2.674
		37.148	1,5		

[a] Mean value.

data support the proposal by Coulson concerning the absence of π orbitals in cyclopropane.[31,139]

NQR methods have been employed for the study of α, β, γ, δ isomers of hexachlorocyclohexane.[24,140-145] This study of chlorosubstituted cyclohexanes has shown the possibility of using NQR for conformational analysis.

The ^{35}Cl NQR spectrum of the β isomer of 1,2,3,4,5,6-hexachlorocyclohexane (a single line at 36.9 MHz) corresponds to the molecular conformation in which all the chlorine atoms are located in equivalent positions.[24,140,141,144] Evidence has been presented that this position is equatorial.[140,141,144] In the spectrum of the γ isomer, four resonance lines are observed (ν^{77} = 35.9, 36.1, 36.5, and 36.9 MHz) with the relative intensities of 1:2:2:1, respectively. The frequency observed at 36.9 MHz may be assigned to an equatorial chlorine atom adjacent to two other chlorine atoms in the equatorial position since it has the same value as the frequency observed for the β isomer. The lowest frequency belongs to an axial chlorine atom adjacent to two other axial chlorine atoms and so on. The assignment of NQR frequencies of the α and δ isomers of hexachlorocyclohexane has also been made.[140,141,145] NQR Zeeman spectra of single crystals of the α[145] and δ isomers[143] of hexachlorocyclohexane allow the evaluation of the asymmetry parameter of the electric field gradient and the value of the angles between the C—Cl bonds which deviate substantially from tetrahedral.

The quadrupole constant of the δ-hexachlorocyclohexane isomer bond calculated with the average NQR frequency is linearly dependent on values for eQq and C—Cl bond dipole moments in chlorosubstituted hydrocarbons.[141,146]

NQR data allow the establishment of the structure of dichlorocyclohexanones obtained upon chlorination of α-chlorocyclohexanone. A single line (ν^{196} = 35.501 MHz) in the ^{35}Cl NQR spectrum of one of the isomers establishes its structure as 2,6-dichlorocyclohexanone, which possesses chlorine atoms in the equatorial position.[147] Another isomer has the structure 2,2-dichlorocyclohexanone in which the chlorine atoms are situated in both axial and equatorial positions (two lines are observed in its spectrum: ν^{196} = 36.475 and 36.703 MHz).[147]

NQR spectra have also been studied for a number of chloro-substituted unsaturated alicyclic compounds (tetrachlorocyclopropene[148,149] and its derivatives,[149] and chlorinated polycyclic hydrocarbons[140]). In the spectrum of tetrachlorocyclopropene the high-frequency group of lines is assigned to the chlorine atoms in the CCl$_2$ group and the low-frequency group is assigned to the chlorine atoms adjacent to the double bond. This conclusion is in agreement with the NQR frequency observed for cis-1,2-dichloroethylene.[148] However, in studying the NQR spectra of tetrachlorocyclopropene derivatives, it turned out that such an assignment of frequencies was incorrect.[149] In halosubstituted cyclopropenes, the halogen atoms adjacent to the double bond resonate at much higher frequencies than the geminal atoms (in contradistinction to chlorinated four-, five-, six-, and seven-membered rings).[149] Thus, for example, in hexachlorocyclopentadiene the chlorine atoms bonded to an sp^2-hybridized carbon atom resonate in the range 36.9–37.5 MHz, while those bonded to a saturated carbon atom resonate in the range 38.1–39.1 MHz.[11,149] On the other hand, in the spectrum of tetrachlorocyclopropene, chlorine atoms bonded to a saturated carbon atom are assigned the frequencies 36.41 and 36.73 MHz, while those bonded to an unsaturated carbon atom are found in the range 38.24–38.74 MHz.[149] In the spectrum of 1,2-dichloro-3,3-difluorocyclopropene, four resonance lines are observed in the range 38.14–38.75 MHz, while in the case of 1,2-diphenyl-3,3-dichlorocyclopropene, two resonance lines are found at 34.78 and 35.02 MHz.[149]

Further information on the NQR spectra of a number of alicyclic compounds and the assignment of NQR frequencies is found in Hooper and Bray,[11] Roll and Biros,[140] and Lucken and Mazeline.[148]

NQR permits the establishment of the structure of octachlorocycloheptatriene obtained by treating octachlorobicyclo[3.2.0]hepta-2,6-diene with two equivalents of aluminum chloride.[150]

VIII. AROMATIC SYSTEMS

NQR spectroscopy was first used for the correlation of organic compounds in 1952 when a linear dependence of ^{35}Cl NQR frequencies of *meta*- and *para*-substituted chlorobenzenes on Hammett σ constants for substituents on the aromatic nucleus was discovered.[151,152] It was later shown that the linear correlation between v^{77} and σ is not only observed for chlorobenzene derivatives[127,151,152] but also for bromo- and iodobenzene derivatives.[12,93,153]

In substituted halobenzenes as in the aliphatic series the contribution of individual substituents on the aromatic ring to the electron density of the halogen atom, and thus, on the NQR frequency, has an additive character.[11,93,153] Further evidence on this point testifies to the linear correlation of NQR frequencies with the sum of substituent σ constants[10–12,15,31,93,120,154–161] (Fig. 9).

Data obtained for 52 chlorobenzene derivatives are described by the linear equation (26).[154] Equation (27) for bromobenzenes[11] is derived from a smaller body of data.

$$v^{77}{}_{35_{Cl}} = 1.024\Sigma\sigma + 34.826 \qquad (26)$$

$$v^{77}{}_{81_{Br}} = 7.693\Sigma\sigma + 226.932 \qquad (27)$$

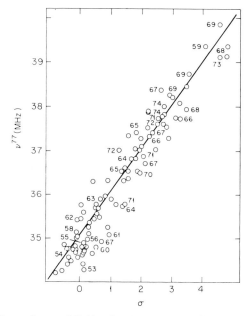

FIG. 9. The dependence of ^{35}Cl NQR frequencies (v^{77}) of substituted benzenes on the sum of Hammett constants of the ring substituents. The numbering of points 53–74 is the same as that in Table IV; for the remaining points, see Bray and Barnes.[154]

The assumption of additivity for the change in electron density on the atom with the resonating nucleus (and thus, in the NQR frequency) leads to the calculation of incremental NQR frequencies of *ortho*, *meta*, and *para* substituents on the benzene ring.[31,162–169] However, in many cases a significant discrepancy from linear dependency of v and σ is observed. Such discrepancies may result from crystal effects,[11,12,152,153,155,156,161,170] ambiguities in the determination of constants,[106,152,153] steric effects[153,163] (according to other data,[156] steric effects have little influence on NQR frequencies), formation of hydrogen bonds,[152,163] or other types of intramolecular interactions.[151]

One of the reasons considered for deviations from the linear dependence of v^{77} on σ has been the comparatively low sensitivity of NQR frequency to change in the C—X bond order.[60,63,171] On the other hand, it has been proposed[31] that change in the C—Cl bond order cannot account for the significant scattering of points in the correlation of NQR frequencies with Hammett constants. It was proposed[156] that the mesomeric effect results only in an insignificant shift in NQR frequency (on the same order as crystal effects). Thus, substituent effects on NQR frequency in the spectra of aromatic compounds is determined primarily by their inductive effect[60,63,156,164,171] (the inductive effect of *ortho* substituents is twice as large as for *meta* substituents and three times greater than for *para* substituents).[156] Nevertheless, σ constants serve as a measure of not only the inductive effect, but of conjugative interactions as well.

Substituent effects on a chlorine atom in the *meta* position on the benzene ring are transmitted primarily by an inductive mechanism. Thus, for *meta*-substituted chlorobenzenes, a linear correlation of NQR frequencies with σ_m constants is observed:

$$\sigma_m = -26.71 + 0.7735\nu_m \quad (28)$$

$$\nu_m = 34.55 + 1.223\sigma_m \quad (29)$$

Chlorine-35 NQR frequencies for *para*-substituted compounds of the type $p\text{-RC}_6\text{H}_4\text{Cl}$ in which the substituent R does not conjugate with the chlorine atom obey the same correlation equation as that for the frequencies of the *meta* derivatives. Deviations of NQR frequencies from this correlation observed for compounds which contain a strongly conjugating element of the first row or any substituent of the second row have a well-defined character. These deviations are linearly related to Taft resonance parameters σ_r and thus allow calculation of σ_r from NQR frequencies.[171]

Deviations from a linear dependence of the NQR frequencies of *ortho*-substituted chlorobenzenes are most clearly observed, and a satisfactory linear correlation between ν^{77} and σ_o is not possible. Anomalous values for ^{35}Cl NQR frequencies in compounds of this type are explained, primarily, by the "*ortho* effect",[157,160,172] though the nature of this effect has not been specified.

An increase in NQR frequencies for aromatic molecules containing chlorine atoms in the *ortho* position (i.e., $o\text{-C}_6\text{H}_4\text{Cl}_2$ and C_6Cl_6) has been explained by the fact that the C—Cl bonds are forced out of the ring plane, thus partial double-bond character is reduced.[173] However, it was later shown[120] that by removing C—Cl bonds from the ring plane, bond order actually increases because the decrease in π-orbital overlap is more than compensated by an increase in σ-orbital overlap. Thus, the increase in ^{35}Cl NQR frequency in the case of two chlorine atoms in the *ortho* position must be the result mainly of the reduction in the ionic character of the C—Cl bonds.[10,157] X-ray structure analysis of hexachlorobenzene indicates that the molecule is planar in the crystal.[10,174]

Step-wise substitution of hydrogen atoms in chlorobenzenes by chlorine atoms[12,157,173,175] and in bromobenzenes by bromine[158] results in a linear increase in NQR frequency. This linear dependence is of three types: (1) for compounds in which the halogen atom does not have neighbors in the *ortho* position, (2) for compounds in which the halogen atom has one such neighbor, and (3) for compounds in which the halogen atom has two such neighbors. This trichotomy is explained primarily by the inductive effect of the halogen atoms, although the resonance frequency is also influenced by a change in the C—X bond order, especially in the case of *ortho* substitution.[12,157,158] An increase in the number of chlorine atoms on the ring leads to an increase in the π character of the C—X bond.[43,176] The correlation between NQR frequencies and σ constants allows the calculation of ν^T from known constants or, vice versa, to determine values of σ from NQR frequencies.[11,12,120,154,160,161] The constant K_i, which is determined from the NQR frequency by $\nu = \nu_0 + \sum_i k_i$, where k for hydrogen is zero and additivity is obeyed, has been proposed as a measure of the substituent effect on the distribution of electron density on the atom with the resonating nucleus in the molecular ground state.[156] On the basis of a correlation equation derived either from σ constants or from substituent inductive constants, it has been possible to assign NQR frequencies in aromatic compounds containing a number of halogen atoms.[11,152,156,157,158,172,177–179] It has also been found that σ constants correlate with C—Cl bond order (or with the asymmetry parameter)[176] determined from NQR data. As the electron-accepting character of the substituents becomes greater, the degree of double-bond character increases.[180] A similar correlation has been found for iodobenzene derivatives.[181] NQR frequencies of *p*-dihalobenzenes change according to the electronegativity of the C—X bond.[111]

Comparison of the NQR frequencies of compounds RC_6H_4—OC(O)—C_6H_4Cl and RC_6H_4—C(O)O—C_6H_4Cl with σ constants of the substituents R led to the conclusion[182] that the effect of substituents occurs mainly by an inductive mechanism since the NQR frequencies are best correlated with the inductive constants σ_I [Eqs. (30) and (31), respectively].

$$\nu^{77} = 34.69 + 0.94\sigma_I, \quad r = 0.97 \quad (30)$$

$$\nu^{77} = 34.85 - 0.24\sigma_I, \quad r = 0.94 \quad (31)$$

On the basis of the coefficients in Eqs. (30) and (31), the transmittance of the inductive effect of substituents R in opposite directions through the same system is fundamentally different.[182] Nevertheless, the fact that correlations between ν^{77} and σ_I are observed is not convincing proof of a purely inductive effect between the substituents R and Cl. In the case under discussion, it is unlikely that the inductive effect of the substituents can be transmitted across such a great distance.

NQR frequencies of chloro derivatives of naphthalene, anthracene, phenanthrene, and other polynuclear aromatic systems[29] fall in approximately the same range as those observed for chloro-substituted benzenes. This indicates the similarity in the electronic structure of the C—Cl bond in these aromatic systems.[183] The NQR frequencies for chlorobenzene and 1-chloronaphthalene are almost identical (the frequency for 2-chloronaphthalene is significantly higher). However, upon further chlorine substitution, the quadruple interactions in the two systems begin to differ substantially. Apparently the ionic and π character of the C—Cl bond changes in such a manner that the resultant effect appears identical in chlorobenzene and 1-chloronaphthalene. With increasing number of chlorine atoms, the change in the ionic character in both systems is approximately the same. The effect of the change in π character of the C—Cl bond on the NQR frequency in chloro derivatives of naphthalene increases more rapidly than in polychlorobenzenes.[184]

Upon increasing chlorosubstitution in naphthalenes, the NQR frequency, as in the case of polychlorobenzenes, increases linearly[11,184-187] (Fig. 10). However, the increase is slower.[184-186] (The ratio of the slopes of their correlation lines is ≈ 1.36.)

The mutual interaction of substituents in naphthalene derivatives occurs mainly by an induction mechanism. The π character of the C—Cl bond has a

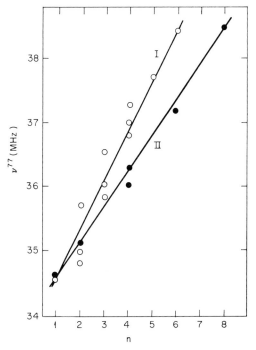

FIG. 10. The dependence of ^{35}Cl NQR frequencies (ν^{77}) of chlorosubstituted benzenes (I) and naphthalene (II) on the number of chlorine atoms on the aromatic ring.

secondary effect on the NQR frequency. In chloroderivatives of naphthalene, it is greater than in chloroderivatives of benzene.[184]

Among other aromatic systems, chloro-substituted quinones[159] and anthraquinones[11] have been studied. The increase in the NQR frequency for 2-chloranthraquinone relative to chlorobenzene is explained by the inductive effect of the oxygen atom. In dichloro-substituted anthraquinones the NQR frequency is higher than in 2-chloroanthraquinone but lower than in 1-chloroanthraquinone. This anomaly is ascribed to differing values for the asymmetry parameters of the C—Cl bonds. NQR data for chloro-substituted quinones[159] are in good agreement with the distribution of charge in quinones calculated by the LCAO MO method.[188]

IX. ARALIPHATIC COMPOUNDS

Among the arylalkanes containing halogen atoms in their side chain, eso-substituted benzyl chlorides are the best studied.[11,127,135,189,190] The electron density on the chlorine atom in the molecule $C_6H_5CH_2Cl$ is significantly higher than for the chlorine atom in C_6H_5Cl,[11,127,135] presumably because of the differing hybridizations of the carbon atom bonded to the chlorine atom.[11] It was later shown, however, that the increase in electron density on the chlorine atom in benzyl chloride (relative to chlorobenzene) is caused mainly by a lessening of the negative inductive effect of the phenyl group by the methylene bridge.[127] A certain contribution to the increase in electron density on the chlorine atom is also made by geminal conjugation of the type

$$\langle\!\!\!\!\bigcirc\!\!\!\!\rangle\!-\!CH_2\!-\!Cl.\,^{127,135}$$

In the diphenylchloromethane molecule $(C_6H_5)_2CHCl$, such σ,π coupling is apparently hindered by the homoconjugation of the phenyl groups.[127]

Changes in the NQR frequency for molecules in the series $YC_6H_4CH_2Cl$ are found, in general, to follow the Hammett constants for the substituents Y. Thus, as is the case for σ constants, the NQR frequencies of $m\text{-}YC_6H_4CH_2Cl$ are always higher than for $p\text{-}YC_6H_4CH_2Cl$.[127]

A comparison of the difference in NQR frequencies (Δv) for a number of substituted and unsubstituted chlorobenzenes and benzyl chlorides[11] led to the conclusion that the substituent effect on the electron density in $YC_6H_4CH_2Cl$ is approximately $\frac{3}{4}$ as great as in YC_6H_4Cl. However, on the basis of a greater amount of data, it was shown[127] that the greater transmittance of the substituent effect in YC_6H_4Cl relative to $YC_6H_4CH_2Cl$ is observed only in the case of *ortho-* and *meta*-substituted derivatives. In the *para* case the transmittance of p-$YC_6H_4CH_2Cl$ is greater than for p-YC_6H_4Cl.

In ^{35}Cl NQR spectra of *o*-, *m*-, and *p*-chlorobenzyl chlorides, assignments were made for chlorine atoms on the ring and on the side chain,[11,152,189] and on the basis of these values, an attempt was made to evaluate σ constants for the CH_2Cl group.[11,154,189]

In molecules of the type $(Cl_3C)_nC_6H_{6-n}$ the sensitivity of the chlorine atoms in the CCl_3 group to the substituent effect on the aromatic ring is significantly lower than for substituted chlorobenzenes.[191] NQR frequencies of these compounds are linearly related to the value for n.

X. HETEROCYCLIC COMPOUNDS

NQR has been employed primarily for the study of heterocyclic compounds containing one (pyridine derivatives[159,192–194]) or two (pyrimidines[161,193,195]) nitrogen atoms in the ring. Among oxygen heterocycles, NQR spectra have been studied only for chloro-substituted cyclic anhydrides of dicarboxylic acids[11,196]; among sulfur heterocycles NQR spectra have been studied for chloroderivatives of thiophene,[11,16,29,196] including silicon organic compounds.[197] Furthermore, NQR spectra have been investigated for heterocyclic compounds containing the heteroatoms silicon[78] and boron (in carboranes).[120,126,198,199]

In going from tetrachlorophthalic acid to its anhydride there is almost no change of the electron density on the chlorine atoms because the change in molecular structure occurs far from these atoms.[11]

The oxygen atom in the ring of chloromaleic anhydride has a stronger effect on the electron density of the chlorine atom than does another chlorine atom in 2,3-dichloronaphthoquinone.[11] NQR spectra have led to the conclusion that in mono- and dichloromaleic anhydrides, the π character of the C—Cl bond is higher than in vinyl chloride and *cis*-dichloroethylene, respectively.[196] Assignment of NQR frequencies in the spectrum of the spirolactone

was made on the basis of NQR data for three- and five-membered rings.[149]

In the NQR spectrum of 2-chloroethyl chlorosulfinate

only one frequency was observed close to the frequency for 2-chloroethyl p-toluenesulfonate. This frequency may be assigned to the chlorine atom of the 2-chloroethyl group, although it cannot be excluded that it belongs to the chlorine atom in the S—Cl bond.[11]

NQR frequencies for chlorothiophenes[11,16,29,196] support the notion that the sulfur atom, which transfers one $3p$ electron to a $3d$ orbital, has greater electronegativity than a carbon atom[200] (in their ground states, the electronegativities of the sulfur and carbon atoms are about equal).[11,201]

The sulfur atom in 2-chlorobenzothiazole behaves similarly. However, the NQR frequency of this compound is approximately 0.9 MHz lower than for dichlorothiophene, and this fact is explained by a change in the π character of the C—Cl bond under the influence of the nitrogen atom.[11] NQR frequencies of a number of organosilicon derivatives of thiophene[197] indicate that the values for σ^* for the 2-thienyl and phenyl groups are approximately equal.

The substitution for a carbon atom in the benzene ring by the more electronegative nitrogen atom should raise the NQR frequency of the corresponding haloderivatives. Thus, for example, NQR frequencies for 3- and 4-chloropyridines (35.24 and 34.89 MHz, respectively) are higher than for chlorobenzene (34.62 MHz), while the frequencies for 6- and 7-chloroquinolines (34.63 and 34.69 MHz, respectively) are higher than for 2-chloronaphthalene (34.69 MHz). However, this inductive effect is not always observed. In particular, the NQR frequencies for 2-chloropyridine (34.19 MHz) and 2-chloroquinoline (33.27 MHz) are significantly lower than for chlorobenzene and 2-chloronaphthalene, respectively. This anomaly results from the increase in π character of the C—Cl bond in these heterocyclic compounds which leads to a lowering of the NQR frequency.[15,31,193,194] A comparison of the NQR frequencies shows that the partial double-bond character of the C—Cl bond in 2-chloropyridine is less than in the case of 2-chloroquinoline.[193] A rough calculation shows that the partial double-bond characters of C—Cl bond in chloroderivatives of the heterocycles discussed are 8.9 and 12%, respectively.[16]

However, since the contribution of the π character of the C—Cl bond to the NQR frequency is usually insignificant,[60,63] the sharp reduction in NQR frequencies as the result of an increase in the bond order of the C—Cl bond is unlikely. The anomalous lowering of NQR frequency in such cases is the result, apparently, of the geminal coupling between the nitrogen and chlorine atoms according to the scheme $\overset{\curvearrowright}{N}-\overset{|}{\underset{|}{C}}-\overset{\curvearrowleft}{Cl}$.

The NQR spectrum of cyanuric chloride $(CNCl)_3$[202-204] shows that only two of the three chemically equivalent chlorine atoms of this molecule occupy equivalent positions in the crystal. The partial ionic and π characters of the C—Cl bonds in $(CNCl)_3$ have been calculated from the quadrupole constants and asymmetry parameters.[202] The increase in the π character of the C—Cl bond in this molecule is compensated by the lowering of its ionic character by the three nitrogen atoms (the NQR frequency of cyanuric chloride is somewhat higher than that observed for 1,3,5-trichlorobenzene).[203]

For analysis of the NQR spectra of heterocyclic compounds, correlation equations, obtained from derivatives of chlorobenzene [Eq. (26)][154] and bromobenzene [Eq. (32)][159] are employed

$$\nu^{77} = 227.02 + 7.866\Sigma\sigma \tag{32}$$

in which the constant σ represents, in the present case, a measure not only of the substituent effect but also of the effect of the heteroatomic ring. Correlation equations (26) and (32) take the form of Eqs. (33) and (34) when applied to heterocyclic compounds

$$\text{for } {}^{35}\text{Cl:} \quad \nu^{77} = 34.826 + 1.024(\Sigma\sigma + \Sigma\sigma_r) \tag{33}$$

$$\text{for } {}^{81}\text{Br:} \quad \nu^{77} = 227.02 + 7.866(\Sigma\sigma + \Sigma\sigma_r) \tag{34}$$

where σ represents the corresponding Hammett constant and σ_r characterizes the change in electron density caused by the substitution of a carbon atom by a heteroatom in the benzene ring.[159,195]

Values for the constant σ_r calculated from these equations agree satisfactorily with the values determined by other methods only in the case of pyridines[161] and pyrimidines[159,195] which do not have a halogen atom in the α position to the heteroatom. For quinolines, no agreement is found. Anomalous values for NQR frequencies of compounds having chlorine atoms adjacent to a nitrogen atom are explained on the basis of an increase in the π character of the C—Cl bond.[31,159,195] In support of this explanation, the asymmetry parameter for the C—Cl bond in these compounds was found to have an increased value.[159,192,195] However, the explanation is not convincing if we consider that the partial double-bond character of the C—Cl bond, which is characterized by the asymmetry parameter, has an insignificant effect on the NQR frequency.

The partial double-bond character of the C—Cl bond in 2-, 4-, and 6-chloro-substituted pyrimidines was calculated[195] from NQR frequencies upon the assumption that nitrogen atoms have a weak inductive effect on adjacent atoms in the ring.

The NQR frequency for N-chlorosuccinimide is only 0.3% lower than that observed in the chlorine molecule.[194] Such a similarity is admittedly coincidental because the nitrogen molecule in this molecule must be positively charged. It is assumed that a transfer of electrons from chlorine to oxygen occurs, thus increasing the p-electron deficiency on the chlorine atom and, hence, the NQR frequency.[194]

The chlorine atom in 1-chloropiperidine has a partial negative charge because the quadrupole interaction constant, calculated according to the NQR frequency, under the assumption that the N—X bond has insignificant π character, is less that for the quantity eQq for the chlorine atom. On the other hand, bromine and iodine atoms bonded to nitrogen have a partially positive character because the value for eQq of these compounds is greater than that for the corresponding halogen atom.[205] Bromine atoms in N-bromosuccinimide and N-bromophthalimide are considerably more positive than in N-bromoacetamide, 1-bromo-2-pyrrolidinone, and N-bromo-ε-caprolactam. A similar relationship in the polarity of N—X bonds (X = Cl and Br) is the result of the inductive effect of the carbonyl group (in N-chlorosuccinimide and N-bromophthalimide there are two carbonyl groups adjacent to the nitrogen atom, in N-bromoacetamide, 1-bromo-2-pyrrolidinone, and N-bromo-ε-caprolactam, only one, while in 1-chloropiperidine, none at all).[205]

The ionic character of the N—X bond of nitrogen containing heterocyclic compounds may be calculated, assuming that it is possible to ignore π, s, and d character. On the basis of values for the ionic character of the N—X bond (X = Cl, Br, and I), dipole moments have been calculated for N-chloro-, N-bromo, and N-iodosuccinimides.[205] The direction of the N—X moment in N-chlorosuccinimide coincides with the direction of the molecular moment, while in N-bromo and N-iodosuccinimides the direction of the N—X moment is opposite.[205]

An investigation by NQR of silacycloalkanes of the type

$$H_2C\text{——}SiR$$
$$(CH_2)_n\quad Cl$$

has shown[78] that the polarity of the Si—Cl bond in such compounds varies according to the overall inductive effect of the substituent R and the polymethylene fragment. NQR frequencies of these compounds are linearly related to the overall inductive constants of the substituents R and R' [R' = —(CH$_2$)$_n$CH$_2$—] at the $\diagdown$Si—Cl group. Upon increasing the size of the ring the polarity of the Si—Cl bond falls and, finally, does not differ significantly from the polarity of the Si—Cl bond in the analogous acyclic trialkylsilyl chlorides.

In a study of the NQR spectra of compounds in the carborane series, the Hammett (σ)[120,198] and inductive (σ^*) constants of the carborane system were evaluated, as were the inductive constants of a number of groups containing this system.[120,199] The transmittance of the substituent effect through the

$$-C\!=\!\!=\!\!CCH_2\!-$$
$$B_{10}H_{10}$$

system was determined.[120,126,128,199] In 2-substituted 1-chloromethyl-*o*-carboranes, NQR frequencies do not completely correlate with σ^*, σ_I, or, for that matter, with any σ constants. The anomalous substituent effect on the NQR frequency in this series of compounds is possibly related to intraspatial interaction of the *ortho* substituents.[126]

The introduction of two halogen atoms in positions 9 and 12 in the carborane system, as might be expected, lowers the electron density on the chlorine atom in the chloromethyl group, located in the 1 position relative to that for 1-chloromethyl-*o*-carborane. However, the effect in this direction decreases: Br > Cl > I. This type of transmittance is not in accord with the electronegativity or the constants σ^* and σ for these atoms. The unusual behavior of halogens was also discovered using other means,[206-208] including examples in the carborane series[209] and ascribed to a possible $p \to d$ coupling effect with participation of the vacant d orbitals of chlorine, bromine, or iodine adding to the negative inductive and positive mesomeric effects of the halogen.[126] NQR values in *ortho*- and *meta*-carboranes were compared with the reactivity of C—Cl and B—Cl bonds contained by such systems.[199]

Analysis of NQR spectra allowed further verification of the structure of derivatives of 2-hydroxy-3-chloroisoxazoline-4,[210] and 2-methoxycarbonyl-2-(2-oxo-1-cyclohexyl)-4-chloro-2H1-benzopyran.[211]

XI. PHOSPHORUS COMPOUNDS

NQR data show that the effect of substituents on the chlorine atom in compounds in the series $RPOCl_2$, $RCOCl$, RCH_2Cl, and RSO_2Cl have, at first glance, similar character.[108] However, a more detailed comparison of the NQR frequencies in these series has led to the discovery of large differences in the substituent effects in alkyl chlorides and carboxylic, phosphonic, and sulfonic acid chlorides related to conjugative effects.

The correlation of NQR frequencies for the series $RPOCl_2$ with Taft (σ^*)[70] and Kabachnik (σ_p)[212] constants or with the electronegativities of the R groups[108] indicates a predominantly inductive effect of substituents R on the electron density on the chlorine atom.

In phosphinous chlorides RR'PCl and phosphinic chlorides RR'POCl, in addition to the inductive substituent effect, the conjugative effect involving the vacant $3d$ orbitals of phosphorus is significant. The participation of these orbitals leads to a smaller overall conjugative effect relative to alkyl halides on the change in NQR frequencies.[64,100]

The correlation between NQR frequencies for compounds in the series RR'POCl and the overall σ^* constant for the substituents R and R'[85] and the linear dependence of the NQR frequencies on the number of chlorine atoms in the molecules $R_{3-n}P(O)Cl_n$,[85,94] $R_{3-n}P(S)Cl_n$[94] (where R is an electron donating substituent), and $Y_{3-n}P(O)Cl_n$ (Y = OCH_3 and OC_2H_5)[94,213] establishes the additivity of the inductive effects of the substituents of the tetrahedral phosphorus atom on the electron density on the chlorine atom.[85,94,213] (In Lucken and Whitehead,[108] the absence of additivity at tetrahedral phosphorus was postulated.) Thus, the condition of additivity for the NQR frequencies for tetrahedral molecules is upheld.[85,94,213] If substituents of the type R and OR are both found on the tetrahedral phosphorus atom, additivity for the NQR frequencies is not maintained, however.[94,213] NQR frequencies for compounds of trivalent phosphorus are likewise not additive—the result of a change in the degree of sp hybridization upon deformation of the pyramidal angle $P{\diagup \atop \diagdown}$.[94,213,214]

A comparatively large line splitting in spectra of compounds in the series $ROPCl_2$ and $RSPCl_2$ which have large symmetrical [i-C_3H_7, $N(CH_3)_2$, and $N(C_2H_5)_2$] and a number of unsymmetrical substituents (OC_6H_5, $OC_6H_4CH_3$, and CH_2Cl) is explained by the asymmetric conformation of these molecules.[94,213]

In the series RR'PYCl, NQR data indicate a reduction in the electron density on the chlorine atom for Y = S relative to analogous compounds with Y = O.[215]

The relationship between NQR frequencies and the reactivity of organophosphorus compounds has been considered.[64,70,100,108,212] Satisfactory linear correlations, generally were not found.[64,212]

XII. SULFUR COMPOUNDS

The correlation of ^{35}Cl NQR frequencies for p-YC_6H_4SCl with σ constants for the substituents Y found in the following equation

$$\nu^{77} = 37.25 + 2.04\sigma_I + 1.23\sigma_c, \quad r = 0.97 \tag{35}$$

testifies[216] that the C_6H_4S system transmits the inductive substituent effect better than the conjugative despite the considerable contribution of the latter. This conclusion is supported by the lowering of the value for r in correlating the NQR frequencies with inductive constants σ^* as in Eq. (36)[216]:

$$\nu^{77} = 36.17 + 1.84\sigma_y^*, \quad r = 0.93 \tag{36}$$

In compounds of the series YSCl, on the other hand, divalent sulfur is a better transmitter of the conjugative than the inductive effect of substituents. This is seen from the comparison of the two transmittance coefficients in Eq. (23) which represents this series.[105] Thus, the comparison of the correlation equations for the series p-YC$_6$H$_4$SCl and YSCl indicates that the mechanisms for the transfer of the substituent effect through the —C$_6$H$_4$S— and —S— systems are different.[105] NQR frequencies of some *para*-substituted arylsulfenyl chlorides are linearly related with their hydrolysis rates.[216]

The ^{35}Cl NQR spectrum of o-O$_2$NC$_6$H$_4$SCl points to the existence of intramolecular coordination between the sulfur and oxygen atoms. Intramolecular coordination is apparently found in o-chloro- and o-bromophenylsulfenyl chlorides as well. The NQR spectrum of 2,4-(NO$_2$)$_2$C$_6$H$_3$SCl indicates that, in addition to intramolecular coordination, an additional intermolecular interaction exists between sulfur and the oxygen atom of the nitro group of the neighboring molecule.[216]

NQR frequencies for sulfochlorides of the series RSO$_2$Cl correlate with σ^* constants of substituents R and with ν_{SO_2} in their infrared spectra. Upon increasing the negative inductive effect of the substituent, the polarity of the S—Cl and S=O bonds is reduced.[64,70] The p–π coupling of the substituents has a slight effect on the polarity of these bonds explained by the acceptance of substituent electrons by sulfur.[70]

NQR data for substituted benzosulfochlorides and -fluorides indicate that the SO$_2$Cl group is a better acceptor than SO$_2$F.[217] For these two series of compounds an additive scheme has been proposed for calculating NQR frequencies. The relationship of the NQR incremental frequencies for the SO$_2$Cl and SO$_2$F groups ($\alpha_o \gg \alpha_p \sim \alpha_m$) is confirmed by the greater lability of chlorine atoms in the position *ortho* to SO$_2$Cl and SO$_2$F groups to nucleophilic exchange. This same order for incremental change for Cl and F ($\alpha_o > \alpha_p > \alpha_m$) in substituted benzosulfochlorides indicates the significant role of conjugation in the transmittance of substituent effects on the electron density of chlorine in the SO$_2$Cl group.[217]

In going from RR'POCl to RR'PSCl the NQR frequency falls, although this result would not be expected on the basis of the electronegativities for sulfur and oxygen. This anomaly is explained by the lesser ability of sulfur to enter conjugation.[64,100,215]

XIII. INTRA- AND INTERMOLECULAR COORDINATION IN METALLOORGANIC COMPOUNDS

NQR spectroscopy is becoming one of the most valuable methods in the investigation of intra- and intermolecular coordination interactions in inorganic and metalloorganic compounds. The formation of coordination (donor–acceptor) bonds results from the transfer of valence electrons (such as the p) of the donor atom to vacant orbitals in the acceptor molecule. Coordination interactions in NQR spectra of metalloorganic compounds may be seen not only in a frequency shift, but in an increase in the asymmetry parameter and splitting of the resonance line, which frequently is greater than the crystallographic line splitting (1.5–2%).[36,37,218]

Coordination interactions are sometimes detected by anomalous effects of substituents found in comparing NQR spectra for series of organic and metalloorganic compounds of like structure. Thus, for example, the decrease in the NQR frequency of 2-$O_2NC_6H_4SCl$ (36.15 MHz) relative to that of C_6H_5SCl (37.016 MHz) observed, instead of the expected increase due to the electron accepting properties of the nitro group, is explained by intramolecular coordination of the sulfur atom with the oxygen atom of the nitro group.[37,216]

The low value for the NQR frequency ($v^{77} = 33.864$ MHz) of the ring chlorine in o-ClC_6H_4SCl is also explained by the intramolecular interaction of the chlorine with the *ortho* sulfur.[37,216]

In pentachlorophenylmercury derivatives C_6Cl_5HgY (Y = Cl, C_6H_5, C_6Cl_5, and CH_3), the NQR frequency shifts for the chlorine in the *ortho* position, relative to the *meta*- and *para*-chlorine atoms are significantly greater than in cases of other substituents attached to the C_6Cl_5 group. This observation is explained on the basis of intramolecular coordination.[36,37,219]

Intramolecular coordination between atoms M and Cl in compounds in the series RR'R"MCH$_2$Cl when M = Si, Ge, and Sn (the α effect) is further verified by comparing their NQR frequencies with those of the corresponding organic molecules (M = C). NQR frequencies in metalloorganic molecules are higher than for the analogous organic molecules, although the electronegativities of the atoms M would require the opposite. This anomaly was explained by the intramolecular interaction of the p_z electrons of the chlorine atom which participate in the C—Cl bond with the vacant d orbitals of atom M (M = Si, Ge, and Sn) by the following scheme[84]:

$$\begin{array}{c} Cl \\ \diagdown\diagup\diagdown \\ M\!-\!CH_2 \\ \diagup \end{array}$$

The existence of intramolecular coordination in Cl_3SnCH_2Cl was also proposed on the basis of the large NQR line splitting for the $SnCl_3$ group.[37,95]

The linear correlation of NQR frequencies of compounds RR'R"MCH$_2$Cl (M = Si, Ge, and Sn) with σ^* constants indicates that the contribution of the α effect on the electron density of the chlorine atom is a constant value or a value proportional to the inductive substituent constant for a given M.[84]

In the spectra of Cl_3CHgY (Y = Cl, Br, and CCl_3)[36,37,218,220] a significant splitting of the NQR frequencies for the chemically equivalent chlorine atoms in the CCl_3 group is observed, indicating intramolecular coordination of one of them with mercury. The intramolecular character of the coordination in these compounds is confirmed by X-ray structure analysis.[220]

In the NQR spectra of molecules in the series 2,6-$Cl_2C_6H_3$OMR (M = Hg, Sn, and Pb), two resonance frequencies are observed with a shift greater than that usually encountered in cases characteristic of crystallographic splitting. This apparently results from the intramolecular coordination of one of the chlorine atoms with the atom M. The value for the observed splitting depends on the nature of atom M. If we assume that the values for the splitting and the coordination interaction are proportional, it is possible to conclude that in the compounds investigated, mercury has a greater propensity towards coordination than tin or lead.[22,36,37] NQR spectra for these mercury compounds also indicate that the intramolecular coordination between the mercury and chlorine atoms is practically independent of the electronic properties and size of substituents R.[36,37,221]

Interesting results were obtained in the study of the effect of complex formation on intramolecular coordination. NQR spectra of solutions of Cl_3CHgBr and Cl_3CHgCl in 1,2-dimethoxyethane show that there is no intramolecular coordination in these compounds (the NQR line splitting is significantly less than for neat Cl_3CHgBr or Cl_3CHgCl).[37,222] Such splitting in solutions of $(Cl_3C)_2$Hg in dimethoxyethane and tetrahydrofuran is also small (does not exceed that characteristic of crystals). This indicates that in a medium composed of an electron-donating solvent, intramolecular coordination of the type C—Cl···Hg, if not entirely eliminated, is significantly reduced.[37,222] The solvent effect on the magnitude of the intramolecular coordination falls in the order $(CH_3OCH_2)_2 > (CH_2)_4O > (C_2H_5)_2O$.[37,222]

The NQR spectrum of cis-β-chlorovinylmercuric chloride confirms the existence of intramolecular coordination in a molecule of the type

```
      H         H
       \       /
        C=C
       /     \
      Cl·········Hg—Cl
```

This has an effect on the electron density distribution in the molecule (as does the intermolecular coordination in the *trans* isomer) no less than that resulting from the *cis* or *trans* arrangement of the substituents.[37,223]

Intermolecular coordination is seen in NQR spectra in the same fashion as intramolecular coordination. Therefore, it is often difficult to determine the nature of the coordination without incorporating data from X-ray structure analysis.

The unusually high multiplicity of the NQR spectrum of CCl_4 (15 signals at 77°K)[29] was explained on the basis of the coordination of the molecule in the crystal lattice resulting from "interhalogen bonds" (the transfer of the unshared electron pair of a chlorine atom in one molecule to the vacant $3d$ orbital of a chlorine atom in another). The existence of similar coordination in liquid carbon tetrachloride at 5°C has been proved by cryoscopic methods.[224]

The introduction of a second nitro group in the *para* position in 2-$O_2NC_6H_4SCl$ leads to a sharp decrease in the NQR frequency to 34.752 MHz. This decrease is apparently caused by the intermolecular coordination of sulfur with oxygen in the *para*-nitro group in the neighboring molecule.[37,216]

The large deviations from the correlation relating ν^{77} and σ^* for the series $RSiCl_3$ and $R(CH_3)SiCl_2$ found for $R = NCCH_2CH_2$ are explained by the intermolecular interaction of silicon with the nitrole group.[67]

Intermolecular coordination of the type C—Cl···H between two molecules (of the three which make up the unit cell of the crystal) of *trans*-ClCH=CHHgCl is confirmed by X-ray structural data. The NQR spectrum consists of three lines, one of which is found significantly higher than the other two and is assigned to the chlorine atom which does not participate in coordination.[36,37,220] In this molecule coordination is also found of the type Hg—Cl···Hg which becomes predominant for *trans*-ClCH=CHHgBr (H—Br···H).[36,37]

The comparatively large ^{81}Br NQR resonance line splitting in the case of Cl_3CHgBr indicates the existence of intermolecular coordination of the type Br···Hg which is confirmed by X-ray structural studies.[37,220] The large resonance line splitting (to 15%) in the NQR spectra of ethoxytitanium bromides $(C_2H_5O)_nTiBr_{4-n}$ ($n = 1$–3) is explained by intermolecular coordination and the difference in electron density on the bromine atoms located in the equatorial and axial positions as has been confirmed by X-ray structure analysis in the case of $(C_2H_5O)_2TiBr_2$.[36,37]

Anomalously large splitting of the NQR frequencies of the chemically equivalent chlorine atoms in the CCl_3 group is observed in salts of Cl_3CCO_2H such as Cl_3CCO_2Ag and $(Cl_3CCO_2)_2Hg_2$. The splitting is ascribed to the intermolecular coordination of one of the chlorine atoms with mercury, Cl···Hg.[36,37,218,225]

A convenient object for the study of coordination interactions has been found in cyclopentadienyltricarbonyl compounds of manganese and rhenium, because the ^{55}Mn, ^{185}Re, and ^{187}Re NQR spectra (nuclear spin $I = \frac{5}{2}$) of these compounds allow the determination of both eQq_{zz} and η. The asymmetry parameter for compounds of this type is not zero [not even for $(CO)_3MnC_5H_5$ or $(CO)_3ReC_5H_5$, which is explained by the difference in the symmetry of substitution] but has a small value for both substituted and unsubstituted compounds. However, the introduction of substituents of the type $COCH_3$, COC_6H_5, COCl, etc., on the cyclopentadiene ring significantly raises the value of η. This effect is ascribed to the coordination interaction between the substituent and the metal atom.[36,37] The value for η in rhenium compounds is greater than for the corresponding derivatives of manganese. It may thus be postulated that rhenium has a greater propensity towards coordination (resulting from the greater size of this atom) than manganese.[36,37] The large values for η for $Co_2(CO)_8$ and $Re_2(CO)_{10}$ are confirmed by X-ray data indicating a coordination interaction in the crystal lattices of these compounds.[37]

Intermolecular coordination may also be found by comparing values for eQq_{zz} in the gas phase (eQq_{gas}) where there is no coordination and in the solid state (eQq_{sol}). In a crystal with only van der Waals forces the ratio of the quadrupole constants $eQq_{sol}/eQq_{gas} \approx 0.9$. A significant deviation from this value indicates coordination interaction.[36,37,115]

The large differences in the quadrupole coupling constants for the gas and solid states confirms the existence of intermolecular coordination in compounds RHgBr (R is an alkyl group). For example, the values of these constants differ by approximately 40% for CH_3HgBr[81] (the ratio $eQq_{sol}/eQq_{gas} = 0.638$).[36,37,115]

Complex formation hinders intermolecular coordination (association). This can be seen, for example, in NQR spectra of $(Cl_3CCO_2)_2Hg_2$ in the solid and in dimethoxyethane solution. Tetrahydrofuran apparently only weakens the intermolecular coordination.[220] In complexes of trans-ClCH=CHHgCl with pyridine, tetrahydrofuran, and dimethylformamide, intermolecular coordination also disappears.[37]

The existence of a linear correlation between $\nu^{77\ 81}Br$ and σ_R^* in the series RHgBr (where R = CH_3, C_2H_5, and n-C_4H_9), in which there is strong intermolecular interaction, indicates that the relative changes in this interaction in the series are small. This is only possible for isostructural compounds where substituents capable of coordinating with a metal atom are absent.[36,37,115]

In the series 1,4-RC_6H_4HgX (X = Cl, Br, and I), NQR frequencies do not correlate with any of the σ constants as a result of the large variation in solid structural types. The range of frequencies in this series for X = Cl is significantly broader than in the series 1,4-ClC_6H_4R since the change in the latter is primarily due to the intermolecular coordination of mercury with the substituent and not to the nature of the substituent as for the series 1,4-ClC_6H_4R.[37,115]

The electronic properties of the HgX group vary markedly in different series of organomercury compounds. Thus, for example, in Cl_3CHgX, the HgCl and HgBr groups are weak electron donors relative to H. In C_6Cl_5HgX these groups are, on the other hand, weak electron acceptors, while in p-ClC_6H_4HgX, they are strong electron acceptors. The variance in the electronic properties of the HgX group arises from the varying character of the intra- and intermolecular interactions involving mercury and halogen atoms in different compounds.[37,115]

The hydrogen bond is a frequent example of a coordination interaction;

therefore, it is natural that it affects NQR spectra by the same criteria. Thus, for example, the NQR spectrum of 2,6-dichloro-4-nitrophenol indicates that the chlorine atoms are found in nonequivalent positions, and the splitting is greater than that characteristic for a crystallographic effect. This may be explained by an intramolecular hydrogen bond.[36,37,218] The same may be said for 2,6-diiodo-4-nitrophenol. The iodine atom participating in the hydrogen bond in this molecule should have the lower frequency and asymmetry parameter.[218] Upon the substitution of the hydrogen atom in the OH group by potassium, the large NQR line splitting disappears (in the NQR spectrum of potassium 2,6-diiodo-4-nitrophenylate, two lines are observed at 35.57 and 35.67 MHz).[37,218] The NQR spectrum of $Cl_3CCH(OH)_2$ is supported by X-ray data showing intermolecular hydrogen bonds in the crystal.[37,226]

The NQR spectrum of the 1:1 complex of chloroform with ether retains the planes of symmetry found for pure $CHCl_3$ (the NQR spectrum consists of two lines with the intensity ratio 1:2). This is possible only upon formation of the hydrogen bond of type $Cl_3CH\cdots O(C_2H_5)_2$.[227]

Intermolecular hydrogen bonds are also postulated for C_6Cl_5OH,[228] $ClCH_2CO_2H$, and $ClCH_2CO_2D$. The NQR spectra of the latter two compounds are similar (two resonance lines). However, the NQR frequencies for $ClCH_2CO_2D$ are found at 0.2 MHz higher. This shift is explained by the difference between the bonds O$\cdots$H—O and O$\cdots$D—O in the dimers of the acids.[106]

XIV. MOLECULAR COMPLEXES

The investigation of intermolecular complexes by NQR began in 1953.[229] NQR data indicate not only the existence and mechanism of complex formation and the composition and structure of the complex formed, but also permit the evaluation of the activity of various donors and acceptors and the clarification of the role of electronic and steric aspects in complex formation. Complex formation leads to a shift in the NQR frequency ($\Delta\nu$) from that in the component molecules. The magnitude and direction of the shift are determined by the amount and distribution of transferred charge (the change in the electric field gradient upon complex formation is directly proportional to the amount of transferred charge).[36,230] Significant effects may also result from steric hindrance which prevents close contact between the donor and acceptor, changes the character of the bonds due to the distortion of molecular geometry, and/or deforms the electron cloud of the atoms with a resonating nucleus owing to polarization. Electronic and steric factors may be seen in the NQR spectra of complexes in varying degrees depending on the strength of the donor–acceptor interaction. In stable complexes, electronic factors are apparently predominant, while in weak complexes, steric effects are more significant. A substantial contribution arises from the difference in the crystal structure of the component molecules and that of the complex.[36]

Complex formation may also be indicated by NQR line splitting which frequently exceeds the crystallographic effect, a change in the value for η, and the temperature dependency of the NQR frequency.

NQR frequency shifts in complexes of π acceptors are relatively small since the degree of charge transfer is minor (1–6%).[231] An increase in the π-electron density of the acceptor upon complex formation leads to an increase in the electric field gradient (and, hence, the NQR frequency) as a result of an increase in the occupancy of the p_x orbitals in the indicator atom.[36] However, the predominant effect on $\Delta\nu$ in complexes of this type results from steric interactions.[36,232,233] Therefore, the upfrequency shift is far from always observed. Thus, for example, the NQR frequency for 1,3,5-$(CH_3)_3C_6H_3 \cdot C_6H_5Cl$ ($\nu^{77} = 34.265$ MHz) is slightly lower than for C_6H_5Cl ($\nu^{77} = 34.622$ MHz).[234] In spectra of π complexes of halosubstituted benzenes, a slight downfrequency shift is also observed, the magnitude of which is on the order of crystallographic splitting. The small $\Delta\nu$ confirms the weak electron accepting properties for halobenzenes,[235] and the X-ray structural data indicate that the distance between the components of the complex is not less than the sum of their van der Waals radii.[232,233,236] This explains the lack of dependence of $\Delta\nu$ on the electronic nature of the donors, and the significant dependence on their geometry. For complexes of C_6F_5Cl and 1,3,5-$C_6F_3Cl_3$ with the planar or centrosymmetric donor molecules $C_6H_5NH_2$, 1,3,5-$(CH_3)_3C_6H_3$, and $C_6(CH_3)_6$, values for $\Delta\nu^{77}$ are about the same (≈ 730 and ≈ 460 KHz for C_6F_5Cl and 1,3,5-$C_6F_3Cl_3$, respectively). Values of $\Delta\nu^{77}$ for C_6F_5Cl complexes with nonplanar or noncentrosymmetric donor molecules ($C_6H_5CH_3$ and $N(C_2H_5)_3$) differ strongly.[232,233]

In the chloroaniline complex with methylene a slight upfrequency shift is observed. The formation of this complex is also confirmed by a decrease in the transverse relaxation time (T_2). In going from pure chloroaniline ($T_2 = 100$ mseconds) to the complex, T_2 is decreased by 2.5 times.[234] A small upfrequency shift is also observed in the spectrum of the chloroaniline complex with hexamethylbenzene.[237] In the spectrum of the picryl chloride complex with naphthalene, the upfrequency shift is 560 KHz at 77°K.[238] The temperature dependences of the NQR frequencies of picryl chloride and its complex with naphthalene are different.[238]

The effect of steric and electronic factors may be separated by comparing a series of complexes in which one of the components is invariant and the symmetry type of the other component remains constant (it is assumed that the geometries of the mutual orientation of the components of the complex are identical).[36,239,240] Thus, for example, in complexes of picryl chloride with monosubstituted benzenes, the upfrequencies shifts (Δv^{77}) correlate with the ionization potential of the donors which, in turn, are linearly related to the degree of charge transfer. The linear correlation between these quantities may be used for evaluating the ionization potential of the donors.[36,239,240]

The doublet spectrum of the picryl chloride complex with toluene, which is different from the spectra of other complexes of similar structure, is explained by the possible existence of two structures with the *syn* and *anti* orientation of the methyl group relative to the chlorine atom. This assumption is also supported by the spin-lattice relaxation time T_1, which in the case of the *syn* isomer is the same as for other picryl chloride complexes with monosubstituted benzenes, and in the case of the *anti* isomer, the same as in the picryl chloride complex with *p*-xylene. The doublet spectrum of the picryl chloride complex with *p*-chlorotoluene is also apparently the result of the existence of *anti* and *syn* configurations.[239,240]

If *para*-substituted toluenes are taken as the donors in complexes of picryl chloride, the NQR frequency shift is mainly determined by the distance between the components of the complex which depends on the size of substituents. Thus, a linear correlation is observed between the NQR frequency shift and the van der Waals radii of the substituent Y in the donor molecule p-$YC_6H_4CH_3$ (an increase in substituent size leads to a decrease in the NQR frequency shift).[239,240] It has not been possible to discover any systematic relationship in the change of NQR frequency shifts for picryl chloride complexes with di- and higher substituted benzenes and condensed aromatic systems.[239,240]

In complexes of σ acceptors, charge transfer to the antibonding σ orbital of the acceptor leads to a downfrequency shift in the NQR spectrum. In most complexes of this type, the degree of charge transfer is slight and the steric effect has a significant influence on the frequency shifts.[241] Among the complexes of this type studied by NQR are complexes of donor molecules with Cl_2,[242] Br_2,[243–247] I_2, ICl, IBr,[248] CCl_4, CBr_4,[36,227,234,243,249–251] $CHCl_3$,[36,227,234,250,252] CH_2Cl_2,[36,227] and Cl_3CCO_2H.[253]

The lowering of the NQR frequency (in most cases greater than that characteristic for crystallographic splitting) upon formation of complexes of Cl_2 with various donors indicates the transfer of charge from the donor molecule to the chlorine atoms. Complex formation in this case is accompanied by an increase in the multiplicity of the spectrum and a decrease in the transverse relaxation time T_2.[242]

In the three-component spectrum of the complex $Cl_2 \cdot C_6H_5NO_2$, one of the lines has a higher value than the frequency for the Cl_2 molecule. This is explained by the effect of the strongly electron-accepting nitro group on the charge distribution in the complex.[242]

In the spectra of most complexes of bromine with various n and π donors, a slight downfrequency shift (0.01–4%) is observed, while bromine complexes with benzene and ether have an upfrequency shift (0.03 and 0.3%, respectively).[241,245]

The complex $Br_2 \cdot C_6H_5Br$ has bromine atoms in both the donor and acceptor molecules.[247] The degree of charge transfer calculated from the ^{81}Br NQR $\Delta \nu$ differs strongly in the two cases. This anomaly may be explained by the participation of vacant d orbitals of the bromine atom in the acceptor molecule, Br_2, upon complex formation.[241] NQR spectra of all complexes with bromine indicate the complete equivalence of the bromine atoms in the molecule.[244]

Upon complex formation involving I_2, ICl, and IBr with various donors, the value of eQq_{zz} for ^{127}I, contrary to expectation, decreases. It is assumed that charge transfer in these complexes is accompanied by sp^3d hybridization of the iodine atom.[241]

The small NQR frequency shifts of the molecular complexes of trichloroacetic acid with various oxygen-containing molecules (aldehydes, ketones, acids, ethers, alcohols, etc.) were explained on the basis of the crystal effect,[253] but the possibility of charge transfer in these complexes was ignored.[241]

The frequency shift in the spectra of CBr_4 and CCl_4 complexes with donor molecules is minor in most cases (as in the case of π complexes); $\Delta \nu$ is less than the downfrequency shift in the spectra of analogous $CHCl_3$ complexes, which indicates the stronger acceptor properties of $ChCl_3$.[234,250] The average $\Delta \nu$ of chloroform complexes are linearly correlated with the ionization potential of the donor molecules.[250] NQR spectra allow conclusions about the mechanism of charge transfer in chloroform complexes[234,250] and about the structure of chloroform complexes with ethers.[227]

In the spectrum of $I(CF_2)_2I \cdot NH(C_2H_5)_2$ a comparatively large downfrequency shift is observed,[121] indicating that the degree of charge transfer is $\approx 3\%$.[36] The single component nature of the spectrum of this complex and symmetry considerations allow the assumption that the complex has a layer structure with equivalent iodine atoms.[36,121]

The complexes studied by NQR include the acceptors $HgCl_2$,[36,254] $HgBr_2$,[36] Cl_3CHgY (Y = Cl, Br, and CCl_3),[36,222] $AlCl_3$,[229] $AlBr_3$,[36,255] $GaCl_3$,[256,257] $SnCl_4$,[36,229,254,258,259] $SbCl_5$,[259] $SbCl_3$ and $SbBr_3$,[250,254,261–270] $BiCl_3$,[36] and $AsCl_3$.[254,271]

The effect of complex formation on intramolecular coordination of the type C—Cl⋯H was studied for complexes of CCl_3HgX (see above).[36,222]

In complexes of $HgCl_2$, the downfrequencies shifts are 7–15%. By lowering the ionization potential of the donor, the average $\Delta \nu$ for $HgCl_2$ increases. The absence of a rigorous linear dependence between these quantities is explained by the significant steric effect on the frequency shift.[36]

In the spectra of $HgBr_2$ complexes, both upfrequency and downfrequency shifts are observed. The comparatively large upfrequency shift in the spectrum of $HgBr_2 \cdot (CH_3OCH_2)_2$ (8%) indicates a substantial change in the Hg—Br bond order upon complex formation.[241]

In the spectra of $AlBr_3$ complexes with ethers, sulfides and pyridines, a significant downfrequency shift is observed (from 3.3% for $AlBr_3 \cdot O(C_6H_5)_2$ to 10% for $AlBr_3 \cdot NC_5H_5$). The decrease in the frequency shift upon stepwise substitution of ethyl groups by more bulky aryl groups is explained on the basis of an increase in steric interactions between these groups and the bromine atoms. In this series of complexes, strong splitting is observed, indicating a significant role for the steric effect. Assignment of NQR frequencies in multiplet spectra of these complexes has been accomplished.[25,36]

Changes in the NQR frequencies for gallium chloride complexes with a number of donors are quite large (for ^{35}Cl, 16–18.5 MHz and for ^{69}Ga, 0–25 MHz). The correlation of ^{35}Cl and ^{69}Ga NQR frequencies with values for the enthalpy of formation for the donor–acceptor bond ($-\Delta H$) in $GaCl_3$ complexes[257] indicates that for an increase in the donor–acceptor bond strength, ^{35}Cl and ^{69}Ga NQR frequencies move to lower frequencies. The existence of these correlations testifies to the predominant effect of charge transfer, and not steric interactions, on Δv.[241] The linear correlation between ^{35}Cl and ^{69}Ga NQR frequencies for $GaCl_3$ complexes[257]

$$\nu_{Cl} = 16.326 + 0.826\nu_{Ga} \tag{37}$$

allows evaluation of the distribution of transferred charge between chlorine and gallium. If we assume that the sp^3 hybridization of the gallium atom does not change, it is possible to conclude that $\frac{2}{7}$ of the transferred charge is localized on the central gallium atom and $\frac{5}{7}$ shared among the three chlorine atoms.[241]

The formation of hexacoordinate complexes of stannic chloride with oxygen-containing donors leads to an increase in the ionic character of the Sn—Cl bond (and a decrease in the NQR frequency).[254] The relationship between nature of the NQR spectrum and the structure of the complex is considered in Semin[36] and Kravchenko et al.[259] Molecular orbital calculations indicate that shifts in the average NQR frequencies for *cis* and *trans* isomers are not necessarily identical; NQR frequencies are also not identical for *axial*- and *equatorial*-halogen atoms in complexes with *cis* configuration. It has been assumed that the NQR spectra for *trans* forms of the hexacoordinate complexes of $SnCl_4$ will be composed of a single line, narrow doublet, or quartet. For the *cis* isomers, we expect a doublet with large splitting, two greatly separated narrow doublets, or a doublet and a greatly separated singlet of double intensity.[259] Comparison of the splitting in NQR spectra of $SnCl_4$ complexes with the splittings proposed for *cis* and *trans* isomers allows conclusions about the structure of a series of octahedral complexes (SnCl$_4$ complexes with 2C$_2$H$_5$OH, 2POCl$_3$, C$_6$H$_5$NO$_2$, 2C$_6$H$_5$CH$_2$Cl, and 2CH$_3$OCH$_2$Cl have *cis* configurations and those with 2(C$_2$H$_5$)$_2$O, CH$_3$O(CH$_2$)$_2$CH$_3$, and CH$_3$O(CH$_2$)$_4$OCH$_3$ have *trans* configurations).[36,259] The increase of 4% in the ^{35}Cl NQR frequency for POCl$_3$ in the complex SnCl$_4\cdot$2POCl$_3$ indicates a decrease in the σ-electron density on the chlorine atom of the donor.[259]

The best-studied NQR spectra of complexes are those of SbCl$_3$ and SbBr$_3$, for which ^{121}Sb, ^{123}Sb, ^{37}Cl, ^{79}Br, and ^{81}Br as well as ^{35}Cl NQR frequencies have been measured for a number of donors, values of η have been calculated, and the temperature dependence of the frequency determined. In most complexes of this type, $\Delta\nu$ of Sb is significantly greater than those for Cl or Br, indicating participation of antimony and not the halogen atoms in the donor–acceptor interaction.[250,264,266] In a number of cases, however, the halogen atoms may serve as the acceptor (e.g., in 2SbCl$_3\cdot$C$_{10}$H$_8$).[265] Comparison of the ^{121}Sb and ^{35}Cl (in C$_6$H$_5$Cl) NQR frequencies for the complex SbCl$_3\cdot$C$_6$H$_5$Cl with those of the separate donor and acceptor molecules indicates that π electrons of the benzene ring take part in complex formation.[264]

The type of donor–acceptor interaction in complexes of SbX$_3$ (X = Cl and Br) with aromatic and oxygen-containing donors differs because different acceptor orbitals are used for accepting the π electrons of the aromatic ring and for the unshared pair of oxygen.[262,266]

Distortion of the pyramidal structure of SbCl$_3$ and SbBr$_3$ has a significant effect on the value for the change in η and NQR frequencies.[250,262,264–266]

Single-component ^{121}Sb and ^{123}Sb NQR spectra are characteristic for SbX$_3\cdot$D complexes, while for 2SbX$_3\cdot$D, a number of resonance lines are frequently observed. This indicates the nonequivalence of the SbX$_3$ molecules in the 2:1 complexes. NQR spectra for the Cl or Br nuclei of these complexes have three times the number of lines in the ^{121}Sb and ^{123}Sb spectra.[16] The average value for eQq_{zz} of the 2:1 complexes is always higher than for the 1:1 complexes.[250] The value of eQq_{zz} for ^{121}Sb in 1:1 SbCl$_3$ complexes correlates with Hammett constants for *meta* and *para* substituents in the aromatic ring of the donor.[250] The change in the constant (ΔeQq_{zz}) for 2:1 and 1:1 SbCl$_3$ complexes with C$_6$H$_5$Y (Y is a varying substituent) relative to Y = H is also linearly tied to the Hammett constants for substituents Y.[265]

The similarity of the splitting in the NQR spectra of 2AsCl$_3\cdot$D and 2SbCl$_3\cdot$D allows the conclusion that the structures of these complexes are identical.[254]

The ^{121}Sb, ^{123}Sb, ^{35}Cl, and ^{37}Cl NQR spectra for SbCl$_5\cdot$POCl$_3$ are in agreement with the established structure of this complex. Information derived from NQR indicates that in the SbCl$_5\cdot$CH$_3$CN complex, four of the five chlorine atoms are equivalent and are located in a position *cis* to CH_3CN; the fifth is unique and *trans* to CH_3CN.[260] The degree of charge transfer in these complexes is ≈ 40-45%, and $\approx 35\%$ of the transferred charge is localized on the chlorine atoms.[241]

The NQR spectrum of the trichlorogermanium etherate $2(C_2H_5)_2O \cdot HGeCl_3$ shows a strongly split triplet (the splitting magnitude is 3-4%) significantly shifted to the downfrequency region (13.4-14.4 MHz) relative to that of $HGeCl_3$ (23.0-23.3 MHz). This shift indicates a sharp increase in the ionic character of the Ge—Cl bond in the etherate, and the large splitting in the NQR spectrum is due to the significant nonequivalence of the chlorine atoms in the complex. The short relaxation times T_1 and T_2 indicate an intensive energy exchange in the system. However, the narrow NQR lines do not agree with the small values for T_1 and T_2 and may be related to rapid collective motion in the system. This information, together with results obtained by other methods of investigation, allow the conclusion that the crystal structure of trichlorogermanium etherate consists of polymeric chains with tetrahedral unit cells in which rapid proton exchange occurs.[227]

As complexes of $BiCl_3$ with various donors are relatively unstable, the wide range of values for eQq_{zz} and η observed is explained by a marked steric effect which significantly influences the degree of sp^3d hybridization of the Bi atom.[36]

The use of NQR for the study of the complexes discussed here is described in greater detail in Lucken[31] and Maksyutin.[241]

XV. POLYMERIC COMPOUNDS

For the calculation of NQR frequencies for halogen atoms (^{79}Br or ^{35}Cl) located on an aliphatic chain at a primary (ν_I), secondary (ν_{II}) or tertiary (ν_{III}) carbon atom, the following empirical equations were proposed[116]:

$$\nu_I = \nu_{0I} + \Delta_1 \nu \Sigma \xi^{n-1} \qquad (38)$$

$$\nu_{II} = \nu_{0II} + \Delta_1 \nu \Sigma \xi^{n-1} \qquad (39)$$

where ξ is the extinction coefficient for effect transmittance through a carbon atom chain, and $\Delta_1 \nu$ is the difference between the average characteristic frequency (ν_{0I} = constant) and the halogen NQR frequency for compounds of the type $X(CH_2)_nX$ for $n = 1$.

Comparison of the observed NQR frequencies of a number of halogen-containing polymers with those calculated according to Eqs. (38) and (39) allows determination of the structure of the polymer fragment. The eight-component NQR spectrum of polyvinylidene chloride[272,273] is found in the range 36.6–38.04 MHz. Calculation by the above equations leads to about the same NQR frequencies (36.54 and 38.30 MHz for the extreme lines), if the assumption is made that in an eight-membered fragment of polyvinylidene chloride, for seven monomeric units joined head to tail, one unit is joined head to head:

$[-CH_2Cl_2-CH_2Cl_2-CH_2Cl_2-CH_2Cl_2-CH_2Cl_2-CH_2Cl_2-CH_2Cl_2-Cl_2CH_2-]_m$

This structure is confirmed by NMR.[273]

Similarly, the structure of the polymeric fragment $[-CH_2CHCl-CCl_2-]$ of the polymerization product of 3,3,3-trichloro-1-propene ($Cl_3C-CH=CH_2$) was determined.[272–274] The NQR spectra of the amorphous polymerization product of 3,3,3-trichloro-1-propene polytrifluorochloroethylene and polychloral were also studied.[273]

XVI. OTHER AREAS OF APPLICATION OF THE NQR METHOD

By analyzing NQR spectra of organic or metalloorganic compounds, it is possible to draw conclusions concerning their reactivities, possible reaction mechanisms, and the effect of static and dynamic factors on reactivity. Thus, for example, the reactivity of the chlorine atom in molecules in the series $RR'R''MCH_2Cl$ where $M = Si$ and Ge) towards electrophilic reagents relative to the corresponding organic chlorides ($M = C$) is found to agree completely with NQR data for these compounds.[84]

The reactivity of compounds of the type $(CH_3)_3MCH_2Cl$ relative to nucleophilic reagents increases in going from $M = C$ to $M = Si$, Ge, and Sn, while, on the other hand, the electron density on the chlorine atom (from NQR) is substantially greater when $M = C$ than when $M = Si$, Ge, and Sn. This suggests that in this case, the site of nucleophilic attack is not only on the carbon atom in the CH_2 group but also on the group IV atom M so that the reaction takes place through the transition state

$$\diagdown_{\diagup}M\cdots\cdots CH_2 \atop \phantom{\diagdown_{\diagup}M}X^{Cl}$$

Since the electron densities at the chlorine atom in $(CH_3)_3MCH_2Cl$ ($M = Si$, Ge, and Sn) are very similar, although the reactivity increases significantly in going from $M = Si$ to $M = Sn$, it follows that the reactivity must be determined by the stability of the transition complex gained by participation of the vacant d orbitals of the atom M (the dynamic factor).[84]

The NQR frequency of 2-chloro-*para*-carborane (23.94 MHz) obtained in various chlorination reactions ($Cl_2/AlCl_3$ or $CCl_4/AlCl_3$) indicates that in both instances chlorination occurs at the B—H bond of the *para*-carborane system. NQR spectra of chlorosubstituted carboranes containing a B—Cl bond allow conclusions to be drawn concerning possible reaction mechanisms. If the NQR frequency for a compound in this series is greater than 24 MHz, nucleophilic reactions are expected, if less, electrophilic and ion–radical reactions. Compounds with NQR frequencies close to 24 MHz may be expected to have dual reactivity.[275]

NQR is also used in the study of the products of the radical telomerization of propylene with CCl_4[276] and of vinyl chloride with chloroform,[277] for the effect of various anions on the properties of diazonium cations and the influence of the diazonium cation on the complexed metallohalide anion in *p*-, *m*-, and *o*-chlorophenyldiazonium double salts,[278] and for the study of phase transitions in organic crystals.[187,279–285]

XVII. CONCLUSION

This review shows that NQR is an extremely effective method for the study of the nature of the chemical bond, electronic effects in molecules, intra- and intermolecular coordination, polymeric structure, features of the crystal lattice, and phase transitions. NQR spectroscopy may be successfully employed in the structure determination of unknown compounds. Information derived from NQR allows conclusions to be drawn about the reactivity of organic and metalloorganic compounds and reaction mechanisms and the evaluation of the role of static and dynamic factors in various chemical processes.

The area of problems which may be solved with the aid of NQR is quite large. Therefore, on the basis of the information obtained, this method can become no less important than such widely used physical methods as NMR and IR. Judging from the increasing number of publications in the area of NQR spectroscopy, it may be predicted that in the near future, this method will find as wide an application as these other methods, provided that suitable equipment becomes commercially available.

References

[1] H. Dehmelt and H. Krüger, *Naturwissenschaften* **37**, 111 (1950).
[2] B. P. Dailey, *J. Phys. Chem.* **57**, 490 (1953).
[3] C. H. Townes and A. L. Schawlow, "Microwave Spectroscopy." McGraw-Hill, New York, 1955.
[4] W. J. Orvill-Thomas, *Quart. Rev., Chem. Soc.* **11**, 162 (1957); *Usp. Khim.* **27**, 731 (1958).

5. NQR in Organic and Metalloorganic Chemistry 225

[5] W. Gordy, *in* "Chemical Applications of Spectroscopy" (W. West, ed.). New York, 1956; *in* "Primenenie spektroskopii v khimii" (W. West, ed.), p. 90. IL, Moskva, 1959.

[6] T. P. Das and E. L. Hahn, "Nuclear Quadrupole Resonance Spectroscopy." Academic Press, New York, 1958.

[7] J. W. Smith, *Sci. Progr. (London)* **46**, 293 (1958).

[8] V. S. Grechishkin, *Usp. Fiz. Nauk* **19**, 189 (1959).

[9] G. K. Semin and E. I. Fedin, *Zh. Strukt. Khim.* **1**, 252 (1960).

[10] E. I. Fedin and G. K. Semin, *Zh. Strukt. Khim.* **1**, 464 (1960).

[11] H. O. Hooper and P. J. Bray, *J. Chem. Phys.* **33**, 334 (1960).

[12] C. T. O'Konski, *in* "Determination of Organic Structures by Physical Methods" (F. C. Nachod and W. D. Phillips, eds.), Vol. 2, p. 661. Academic Press, New York, 1961.

[13] G. Bondoris, *J. Phys. Radium* **23**, 43 (1962).

[14] S. L. Segel and R. G. Barnes, "Catalog of Nuclear Quadrupole Interactions and Resonance Frequences in Solids," Parts 1 and 2. 1962 and 1965.

[15] E. A. C. Lucken, *Tetrahedron* **19**, Suppl. 2, 123 (1963).

[16] E. A. C. Lucken, *Phys. Methods Heterocycl. Chem.* (1963); *in* "Fizicheskie metody v khimii geterotsyklicheskikh soedinenii" (A. R. Katrizky, ed.), p. 398. Khimiya, Moskva, 1966.

[17] V. S. Grechishkin and G. B. Soifer, *Tr. Estestvenno-nauch. Inst. Perm. Gos. Univ.* **11**, 3 (1964).

[18] G. K. Semin and E. I. Fedin, *in* "Mössbauer Effect and Its Applications in Chemistry" (V. I. Goldanskii, ed.), p. 68. Consultants Bureau, New York, 1964.

[19] R. S. Drago, "Physical Methods in Inorganic Chemistry." Van Nostrand: Reinhold, Princeton, New Jersey, 1965; R. Drago, "Fizicheskie metody v neorganicheskoii khimii." Mir, Moskva, 1967.

[20] F. Mairinger, *Oesterr. Chem. Ztg.* **67**, 279 (1966).

[21] D. Nakamura, *Kagaku To Kogyo (Tokyo)* **19**, 816 (1966).

[22] M. Kubo and D. Nakamura, *Advan. Inorg. Chem. Radiochem.* **8**, 257 (1966).

[23] J. M. Lehn, *Angew. Chem.* **79**, 1001 (1967).

[24] E. G. Brame, *Anal. Chem.* **39**, 918 (1967).

[25] F. Mairinger, *Allg. Prakt. Chem.* **18**, 71 (1967).

[26] E. Wendling, *Bull. Soc. Chim. Fr.* [5] p. 181 (1968).

[27] J. M. Lehn, *Z. Anal. Chem.* **235**, 10 (1968).

[28] E. Brown, *Ann. Chim. (Paris)* [14] **3**, 323 (1968).

[29] I. P. Biryukov, M. G. Voronkov, and I. A. Safin, "Tablitsy chastot yadernogo kvadrupol'nogo rezonansa." Khimiya, Leningrad, 1968.

[30] V. A. Afanas'ev, "Fizicheskie metody issledovaniya stroeniya molekul organicheskikh soedinenii." Ilim, Frunze, 1968.

[31] E. A. C. Lucken, "Nuclear Quadrupole Coupling Constants." Academic Press, New York, 1969.

[32] R. W. Taft, *in* "Steric Effects in Organic Chemistry" (M. S. Newman, ed.), p. 556. Wiley, New York, 1956; *in* "Prostranstvennye effekty v organicheskoi khimii" (A. N. Nesmeyanov, ed.), p. 562. IL, Moskva, 1960.

[33] Yu. A. Zhdanov and V. I. Minkin, "Korrelyatsionnyi analiz v organicheskoi khimii." Izd. Rostovskogo Universiteta, Rostov na Donu, 1966.

[34] V. A. Pal'm, "Osnovy kolichestvennoi teorii organicheskikh reaktsii." Khimiya, Leningrad, 1967.

[35] G. K. Semin, "Spektry yadernogo kvadrupol'nogo rezonansa Cl^{35} khlororganicheskikh soedinenii." Avtoreferat kand. diss., Moskva, 1968.
[36] G. K. Semin, "Nekotorye primeneniya yadernogo kvadrupol'nogo rezonansa v khimii." Avtoreferat doktorskoi diss., Moskva, 1970.
[37] E. V. Bryukhova, "Issledovanie elektronnykh effektov i koordinatsionnogo vzaimodeistviya v elementoorganicheskikh soedineniyakh metodom YaKR." Avtoreferat kand. diss., Moskva, 1969.
[38] M. G. Voronkov, V. P. Feshin, and E. P. Popova, *Teor. Eksp. Khim.* **7**, 40 (1971).
[39] T. A. Babushkina, "Vliyanie kristallicheskogo polya na spektry yadernogo kvadrupol'nogo rezonansa molekulyarnykh kristallov." Avtoreferat kand. diss., Moskva, 1968.
[40] C. Townes and B. Dailey, *J. Chem. Phys.* **17**, 782 (1949).
[41] C. H. Townes and B. P. Dailey, *Phys. Rev.* **78**, 346A (1950).
[42] W. Gordy, *J. Chem. Phys.* **19**, 792 (1951).
[43] R. Bersohn, *J. Chem. Phys.* **22**, 2078 (1954).
[44] B. P. Dailey and C. H. Townes, *J. Chem. Phys.* **23**, 118 (1955).
[45] R. Livingston, *J. Phys. Chem.* **57**, 496 (1953).
[46] R. Livingston, *Rec. Chem. Progr.* **20**, 173 (1959).
[47] E. A. C. Lucken, *J. Chem. Soc.*, London p. 2954 (1959).
[48] J. D. Graybeal and C. D. Cornwell, *J. Phys. Chem.* **62**, 483 (1958).
[49] E. D. Swiger and J. D. Graybeal, *J. Amer. Chem. Soc.* **87**, 1464 (1965).
[50] B. P. Dailey, *J. Chem. Phys.* **33**, 1641 (1960).
[51] J. H. Goldstein, *J. Chem. Phys.* **24**, 106 (1956).
[52] W. Zeil and B. Haas, *Z. Naturforsch. A* **23**, 2011 (1968).
[53] A. A. Westenberg, J. H. Goldstein, and J. B. Wilson, *J. Chem. Phys.* **17**, 1319 (1949).
[54] E. V. Bryukhova, G. K. Semin, V. I. Goldanskii, and V. V. Khrapov, *Chem. Commun.* p. 491 (1968).
[55] J. M. Mays and B. P. Dailey, *J. Chem. Phys.* **20**, 1695 (1952).
[56] E. A. V. Ebsworth, in "Organometallic Compounds of the Group IV Elements" (A. G. MacDiarmid, ed.), Vol. 1, Part I, p. 76. New York, 1969.
[57] I. P. Biryukov and M. G. Voronkov, *Izv. Akad. Nauk Latv. SSR, Ser. Khim.* p. 115 (1965).
[58] I. P. Biryukov and M. G. Voronkov, *Collect. Czech. Chem. Commun.* **32**, 830 (1965).
[59] I. P. Biryukov and M. G. Voronkov, *Izv. Akad. Nauk Latv. SSR* **10**, 39 (1966).
[60] E. N. Tsvetkov, G. K. Semin, D. I. Lobanov, and M. I. Kabachnik, *Dokl. Akad. Nauk SSSR* **161**, 1102 (1965).
[61] E. N. Tsvetkov, G. K. Semin, D. I. Lobanov, and M. I. Kabachnik, *Teor. Eksp. Khim.* **4**, 452 (1968).
[62] E. N. Tsvetkov, G. K. Semin, D. I. Lobanov, and M. I. Kabachnik, *Tetrahedron Lett.* p. 2933 (1967).
[63] I. P. Biryukov, M. G. Voronkov, and V. T. Danilkin, *Teor. Eksp. Khim.* **2**, 533 (1966).
[64] A. A. Neimysheva, G. K. Semin, T. A. Babushkina, and I. L. Knunyants, *Dokl. Akad. Nauk SSSR* **173**, 585 (1967).
[65] I. P. Biryukov, M. G. Voronkov, and I. A. Safin, *Teor. Eksp. Khim.* **1**, 373 (1965).

5. NQR in Organic and Metalloorganic Chemistry 227

[66] I. P. Biryukov, M. G. Voronkov, and I. A. Safin, *Izv. Akad. Nauk Latv. SSR, Ser. Khim.* p. 706 (1965).
[67] I. P. Biryukov, M. G. Voronkov, B. N. Pavlov, and D. Ya. Shtern, *Izv. Akad. Nauk Latv. SSR, Ser. Khim.* p. 501 (1965).
[68] I. P. Biryukov, M. G. Voronkov, and I. A. Safin, *Izv. Akad. Nauk Latv. SSR, Ser. Khim.* p. 153 (1965).
[69] I. P. Biryukov, M. G. Voronkov, V. F. Mironov, and I. A. Safin, *Dokl. Akad. Nauk SSSR* **173**, 381 (1967).
[70] A. A. Neimysheva, V. I. Savchuk, and I. L. Knunyants, *Izv. Akad. Nauk SSSR, Ser. Khim.* p. 2730 (1968).
[71] G. K. Semin, A. A. Neimysheva, and T. A. Babushkina, *Izv. Akad. Nauk SSSR, Ser. Khim.* p. 486 (1970).
[72] G. K. Semin, T. A. Babushkina, V. I. Robas, G. Ya. Zueva, M. A. Kadina, and V. I. Svergun, in "Radiospektroskopicheskie i kvantovokhimicheskie metody v strukturnykh issledovaniyakh," p. 225. Nauka, Moskva, 1967.
[73] I. P. Biryukov, M. G. Voronkov, and I. A. Safin, *Dokl. Akad. Nauk SSSR* **165**, 857 (1965).
[74] I. P. Biryukov, M. G. Voronkov, and I. A. Safin, in "Radiospektroskopiya tverdogo tela," p. 252. Atomizdat, Moskva, 1967.
[75] M. G. Voronkov and I. P. Biryukov, *Teor. Eksp. Khim.* **1**, 122 (1965).
[76] M. G. Voronkov and I. P. Biryukov, *Teor. Eksp. Khim.* **1**, 124 (1965).
[77] I. P. Biryukov, M. G. Voronkov, G. V. Motsarev, V. R. Rozenberg, and I. A. Safin, *Dokl. Akad. Nauk SSSR* **162**, 130 (1965).
[78] I. P. Biryukov, M. G. Voronkov, E. D. Babich, T. M. Arkhipova, V. M. Vdovin, and N. S. Nametkin, *Dokl. Akad. Nauk SSSR* **161**, 1336 (1965).
[79] M. G. Voronkov and I. P. Biryukov, *Izv. Akad. Nauk Latv. SSR, Ser. Khim.* p. 170 (1968).
[80] I. P. Biryukov, M. G. Voronkov, V. F. Mironov, and I. A. Safin, *Izv. Akad. Nauk Latv. SSR, Ser. Khim.* p. 287 (1968).
[81] I. P. Biryukov, M. G. Voronkov, E. Ya. Lukevits, and I. A. Safin, *Teor. Eksp. Khim.* **6**, 566 (1970).
[82] I. P. Biryukov, E. Ya. Lukevits, M. G. Voronkov, and I. A. Safin, *Izv. Akad. Nauk Latv. SSR, Ser. Khim.* p. 754 (1967).
[83] E. V. Bryukhova, T. A. Babushkina, V. I. Svergun, and G. K. Semin, *Uch. Zap. Mosk. Obl. Ped. Inst.* **222**, 64 (1969).
[84] M. G. Voronkov, V. P. Feshin, V. F. Mironov, S. A. Mikhaiilyants, and T. K. Gar, *Zh. Obshch. Khim.* **41**, 2211 (1971).
[85] A. A. Neimysheva, V. A. Pal'm, G. K. Semin, N. A. Loshadkin, and I. L. Knunyants, *Zh. Obshch. Khim.* **37**, 2255 (1967).
[86] R. Livingston, *J. Chem. Phys.* **19**, 1434 (1951).
[87] R. Livingston, *J. Chem. Phys.* **20**, 1170 (1952).
[88] L. H. Meyer and H. S. Gutowsky, *J. Phys. Chem.* **57**, 481 (1953).
[89] G. K. Semin, in "Radiospektroskopiya tverdogo tela," p. 205. Atomizdat, Moskva, 1967.
[90] G. K. Semin and A. A. Fainzil'berg, *Zh. Strukt. Khim.* **6**, 213 (1965).
[91] G. K. Semin, A. V. Kessenikh, L. V. Okhlobystina, A. A. Fainzil'berg, N. N. Shapet'ko, and L. A. Kurkovskaya, *Teor. Eksp. Khim.* **3**, 233 (1967).
[92] H. Zeldes and R. Livingston, *J. Chem. Phys.* **21**, 1418 (1953).
[93] J. Hatton and V. Y. Rollin, *Trans. Faraday Soc.* **50**, 358 (1954).
[94] G. K. Semin and T. A. Babushkina, *Teor. Eksp. Khim.* **4**, 835 (1968).

[95] G. K. Semin, T. A. Babushkina, A. K. Prokofiev, and R. G. Kostyanovskii, *Izv. Akad. Nauk SSSR, Ser. Khim.* p. 1401 (1968).
[96] G. K. Semin and A. A. Boguslavskii, *Uch. Zap. Mosk. Obl. Ped. Inst.* **222**, 74 (1969).
[97] A. V. Kessenikh, L. V. Okhlobystina, V. M. Khutoretskii, A. A. Fainzil'berg, T. A. Babushkina, and G. K. Semin, *Teor. Eksp. Khim.* **5**, 426 (1969).
[98] M. G. Voronkov, S. A. Giller, I. N. Goncharova, L. I. Mironova, and V. P. Feshin, *Izv. Akad. Nauk Latv. SSR, Ser. Khim.* p. 250 (1969).
[99] M. G. Voronkov, V. P. Feshin, I. N. Goncharova, and L. I. Mironova, *Izv. Akad. Nauk Latv. SSR* **2**, 35 (1970).
[100] A. A. Neimysheva and I. L. Knunyants, *Dokl. Akad. Nauk SSSR* **181**, 888 (1968).
[101] M. G. Voronkov and V. P. Feshin, *Teor. Eksp. Khim.* **7**, 444 (1971).
[102] M. G. Voronkov, V. P. Feshin, and E. P. Popova, *Izv. Akad. Nauk Latv. SSR* **2**, 33 (1970).
[103] V. A. Pal'm, *Usp. Khim.* **30**, 1069 (1961).
[104] C. D. Rithie, *J. Phys. Chem.* **65**, 2091 (1961).
[105] T. A. Babushkina, V. S. Levin, M. I. Kalinkin, and G. K. Semin, *Izv. Akad. Nauk SSSR, Ser. Khim.* p. 2340 (1969).
[106] H. C. Allen, *J. Phys. Chem.* **57**, 501 (1955).
[107] P. Machmer, *Nature (London)* **217**, 61 (1968).
[108] E. A. C. Lucken and M. A. Whitehead, *J. Chem. Soc. London* p. 2459 (1961).
[109] S. G. Vul'fson, I. P. Biryukov, and A. N. Vereshchagin, *Izv. Akad. Nauk SSSR, Ser. Khim.* p. 1008 (1970).
[110] V. I. Svergun, A. E. Borisov, N. V. Novikova, T. A. Babushkina, E. V. Bryukhova, and G. K. Semin, *Izv. Akad. Nauk SSSR, Ser. Khim.* p. 484 (1970).
[111] A. L. Schawlow, *J. Chem. Phys.* **22**, 1211 (1954).
[112] A. L. Allred and E. G. Rochow, *J. Inorg. Nucl. Chem.* **5**, 269 (1958); **20**, 167 (1961).
[113] M. G. Voronkov and A. Ya. Deich, *Dokl. Akad. Nauk SSSR* **168**, 337 (1966).
[114] M. G. Voronkov, M. V. Pozdnyakova, and L. A. Zhagata, *Zh. Obshch. Khim.* **40**, 6 (1970).
[115] A. N. Nesmeyanov, O. Yu. Okhlobystin, E. V. Bryukhova, V. I. Bregadze, D. N. Kravtsov, V. A. Faingor, L. S. Golovchenko, and G. K. Semin, *Izv. Akad. Nauk SSSR, Ser. Khim.* p. 1928 (1969).
[116] G. K. Semin, *Dokl. Akad. Nauk SSSR* **158**, 1169 (1964).
[117] N. D. Sokolov and S. Ya. Umanskii, *Teor. Eksp. Khim.* **2**, 171 (1966).
[118] N. D. Sokolov, *Usp. Khim.* **36**, 2195 (1967).
[119] G. K. Semin and V. I. Robas, *Zh. Strukt. Khim.* **7**, 117 (1966).
[120] G. K. Semin, V. I. Robas, V. I. Stanko, and V. A. Brattsev, *Zh. Strukt. Khim.* **6**, 305 (1965).
[121] G. K. Semin, T. A. Babushkina, S. P. Khrlakyan, E. Ya. Pervova, V. V. Shokina, and I. L. Knunyants, *Teor. Eksp. Khim.* **4**, 275 (1968).
[122] S. Kojima, K. Tsukada, S. Ogawa, and A. Shimauchi, *J. Chem. Phys.* **21**, 1415 (1953).
[123] S. Kojima, K. Tsukada, S. Ogawa, and A. Shimauchi, *J. Chem. Phys.* **21**, 2237 (1953).
[124] L. P. Zalukaev, "Gomolizatsiya organicheskikh molekul." Izd. Voronezhskogo Universiteta, Voronezh, 1968.

5. NQR in Organic and Metalloorganic Chemistry 229

[125] M. G. Voronkov, V. P. Feshin, and Ya. Yu. Polis, *Teor. Eksp. Khim.* **7**, 555 (1971).
[126] M. G. Voronkov, V. P. Feshin, A. P. Snyakin, V. N. Kalinin, and L. I. Zakharkin, *Khim. Geterotsikl. Soedin.* 565 (1970).
[127] M. G. Voronkov, V. P. Feshin, and E. P. Popova, *Teor. Eksp. Khim.* **7**, 356 (1971).
[128] L. I. Zakharkin, V. I. Stanko, V. A. Bratcev, and Yu. A. Chapovskii, *Dokl. Akad. Nauk SSSR* **157**, 1149 (1964).
[129] I. K. Anokhina, "Elektronnye effekty alkil'nykh grupp." Kand. diss., Voronezh, 1968.
[130] H. Kwart and L. I. Miller, *J. Amer. Chem. Soc.* **83**, 4552 (1961).
[131] W. M. Schubert, R. B. Murphy, and J. Robins, *Tetrahedron* **17**, 199 (1962).
[132] R. C. Fort and P. R. Schleyer, *J. Amer. Chem. Soc.* **86**, 4194 (1964).
[133] E. Heilbronner, *Tetrahedron* **19**, Suppl. 2, 289 (1963).
[134] J. Baker and W. Nathan, *J. Chem. Soc., London* p. 1844 (1935).
[135] M. J. S. Dewar, "Hyperconjugation." Ronald Press, New York, 1962.
[136] B. M. Mikhailov, *Izv. Akad. Nauk SSSR, Otd. Khim. Nauk* p. 1379 (1960).
[137] Ya. K. Syrkin, *Zh. Vses. Khim. Obshchest.* **1**, 461 (1962).
[138] T. I. Temnikova, "Kurs teoreticheskikh osnov organicheskoi khimii." Khimiya, Leningrad, 1968.
[139] J. E. Todd, M. A. Whitehead, and K. E. Weber, *J. Chem. Phys.* **39**, 404 (1963).
[140] D. B. Roll and F. J. Biros, *Anal. Chem.* **41**, 407 (1969).
[141] Y. Morino, I. Miyagawa, T. Chiba, and T. Shimozava, *J. Chem. Phys.* **25**, 185 (1956).
[142] I. Tatsuzaki, *J. Phys. Soc. Jap.* **14**, 578 (1959).
[143] Y. Morino, M. Toyama, and K. Itoh, *Acta Crystallogr.* **16**, 129 (1963).
[144] J. Duchesne, A. Monfils, and J. Depireux, *C. R. Acad. Sci.* **243**, 144 (1956).
[145] G. Soda, M. Toyama, and Y. Morino, *Bull. Chem. Soc. Jap.* **38**, 1965 (1965).
[146] I. Miyagawa, *J. Chem. Soc. Jap.* **75**, 1061 (1954).
[147] K. Kozima and S. Saito, *J. Chem. Phys.* **31**, 560 (1959).
[148] E. A. C. Lucken and C. Mazeline, *J. Chem. Soc., A* p. 153 (1968).
[149] R. M. Smith and R. West, *Tetrahedron Lett.* p. 2141 (1969).
[150] R. West and K. Kusuda, *J. Amer. Chem. Soc.* **90**, 7354 (1968).
[151] E. B. Wilson, Jr., *Proc. N.Y. Acad. Sci.* **55**, 943 (1952).
[152] H. C. Meal, *J. Amer. Chem. Soc.* **74**, 6121 (1952).
[153] G. K. Semin, *Zh. Strukt. Khim.* **3**, 292 (1962).
[154] P. J. Bray and R. G. Barnes, *J. Chem. Phys.* **27**, 551 (1957).
[155] P. J. Bray and R. G. Barnes, *J. Chem. Phys.* **22**, 2023 (1954).
[156] D. Biedenkapp and A. Weiss, *J. Chem. Phys.* **49**, 3933 (1968).
[157] P. J. Bray and R. G. Barnes, *J. Chem. Phys.* **25**, 813 (1956).
[158] P. A. Casabella, P. J. Bray, S. L. Segel, and R. G. Barnes, *J. Chem. Phys.* **25**, 1230 (1956).
[159] P. J. Bray, S. Moskowitz, H. O. Hooper, R. G. Barnes, and S. L. Segel, *J. Chem. Phys.* **28**, 99 (1958).
[160] G. W. Ludwig, *J. Chem. Phys.* **25**, 159 (1956).
[161] P. J. Bray, *J. Chem. Phys.* **22**, 1787 (1954).
[162] E. Scrocco, P. Bucci, and M. Maestro, *J. Chim. Phys. Physicochim. Biol.* **56**, 623 (1959).

[163] A. I. Kitaigorodskii, G. K. Semin, and G. G. Yakobson, in "Radiospektroskopiya tverdogo tela," p. 202. Atomizdat, Moskva, 1967.
[164] G. K. Semin, L. S. Kobrina, and G. G. Yakobson, *Izv. Sib. Otd. Akad. Nauk SSSR, Ser. Khim. Nauk* p. 84 (1968).
[165] A. I. Kitaigorodskii and G. K. Semin, *Tezisy Dokl. Vses. Soveshch. Primen. Fiz. Metod. Org. Khim.* (1962).
[166] G. K. Semin, T. A. Babushkina, L. S. Kobrina, and G. G. Yakobson, *Izv. Sib. Otd. Akad. Nauk SSSR, Ser. Khim. Nauk* p. 73 (1968).
[167] G. K. Semin, T. A. Babushkina, L. S. Kobrina, and G. G. Yakobson, *Izv. Sib. Otd. Akad. Nauk SSSR, Ser. Khim. Nauk* p. 63 (1968).
[168] G. K. Semin, T. A. Babushkina, L. S. Kobrina, and G. G. Yakobson, *Izv. Sib. Otd. Akad. Nauk SSSR, Ser. Khim. Nauk* p. 69 (1968).
[169] T. A. Babushkina, A. P. Zhukov, L. S. Kobrina, G. K. Semin, and G. G. Yakobson, *Izv. Sib. Otd. Akad. Nauk SSSR, Ser. Khim. Nauk* p. 93 (1969).
[170] P. J. Bray, *J. Chem. Phys.* **22**, 950 (1954).
[171] E. N. Tsvetkov, G. K. Semin, D. I. Lobanov, and M. I. Kabachnik, *Tetrahedron Lett.* p. 2521 (1967).
[172] Y. Morino, M. Toyama, K. Itoh, and S. Kyono, *Bull. Chem. Soc. Jap.* **35**, 1667 (1962).
[173] J. Duchesne and A. Monfils, *J. Chem. Phys.* **22**, 562 (1954).
[174] Yu. T. Struchkov and I. N. Strel'tsova, *Zh. Strukt. Khim.* **2**, 3 (1961).
[175] A. Monfils and J. Duchesne, *J. Chem. Phys.* **22**, 1275 (1954).
[176] M. E. Ainbinder, V. S. Grechishkin, and A. N. Osipenko, *Zh. Strukt. Khim.* **7**, 111 (1966).
[177] G. K. Semin, V. I. Robas, L. S. Kobrina, and G. G. Yakobson, *Zh. Strukt. Khim.* **5**, 915 (1964).
[178] T. L. Weatherly and W. Quitman, *J. Chem. Phys.* **21**, 2073 (1953).
[179] P. J. Bray and D. Esteva, *J. Chem. Phys.* **22**, 570 (1954).
[180] M. Suhara, T. Yonemitsu, and T. Tonomura, *Bull. Chem. Soc. Jap.* **38**, 2205 (1965).
[181] M. Kubo and Y. Kurita, *Kagaku (Kyoto)* **12**, 566 (1957).
[182] V. V. Korshak, S. V. Vinogradova, V. A. Vasnev, E. V. Bryukhova, and G. K. Semin, *Izv. Akad. Nauk SSSR, Ser. Khim.* p. 681 (1970).
[183] G. K. Semin, T. A. Babushkina, and V. I. Robas, *Zh. Fiz. Khim.* **40**, 2564 (1966).
[184] V. S. Grechishkin and G. B. Soifer, *Tr. Estestvennonauch. Inst. Perm. Gos. Univ.* **11**, vyp. 4, p. 3 (1966).
[185] V. S. Grechishkin and G. B. Soifer, *Zh. Strukt. Khim.* **4**, 763 (1963).
[186] V. S. Grechishkin and G. B. Soifer, *Zh. Strukt. Khim.* **5**, 914 (1964).
[187] V. S. Grechishkin and G. B. Soifer, in "Radiospektroskopiya tverdogo tela," p. 242. Atomizdat, Moskva, 1967.
[188] C. A. Coulson, "Valence." Oxford Univ. Press, London and New York, 1961.
[189] S. Saito, *J. Chem. Phys.* **36**, 1397 (1962).
[190] H. C. Meal, *J. Chem. Phys.* **24**, 1011 (1956).
[191] V. S. Grechishkin, I. V. Izmest'ev, and G. B. Soifer, *Zh. Fiz. Khim.* **43**, 757 (1969).
[192] P. Bucci, P. Cecchi, and A. Colligiani, *J. Amer. Chem. Soc.* **87**, 3027 (1965).
[193] M. J. S. Dewar and E. A. C. Lucken, *J. Chem. Soc., London* p. 2653 (1958).
[194] S. L. Segel, R. G. Barnes, and P. J. Bray, *J. Chem. Phys.* **25**, 1286 (1956).
[195] H. O. Hooper and P. J. Bray, *J. Chem. Phys.* **30**, 957 (1959).

5. NQR in Organic and Metalloorganic Chemistry   231

[196] M. J. S. Dewar and E. A. C. Lucken, *J. Chem. Soc., London* p. 426 (1959).
[197] I. P. Biryukov, M. G. Voronkov, and I. A. Safin, *Izv. Akad. Nauk Latv. SSR, Ser. Khim.* p. 638 (1966).
[198] L. I. Zakharkin, V. I. Stanko, V. A. Brattsev, and Yu. A. Chapovskii, *Dokl. Akad. Nauk SSSR* **155**, 1119 (1964).
[199] E. V. Bryukhova, V. I. Stanko, A. I. Klimova, N. S. Titova, and G. K. Semin, *Zh. Strukt. Khim.* **9**, 39 (1968).
[200] H. C. Longuet-Higgins, *Trans. Faraday Soc.* **45**, 173 (1949).
[201] L. Pauling, "Nature of the Chemical Bond." Cornell Univ. Press, Ithaca, New York, 1948.
[202] Y. Morino, T. Chiba, T. Shimozawa, and M. Toyama, *J. Phys. Soc. Jap.* **13**, 869 (1958).
[203] H. Negita and S. Satoy, *J. Chem. Phys.* **27**, 602 (1957).
[204] H. Negita, S. Satoy, T. Yonezawa, and K. Fukui, *Bull. Chem. Soc. Jap.* **30**, 721 (1957).
[205] H. Kashiwagi, D. Nakamura, and M. Kubo, *Tetrahedron* **21**, 1095 (1965).
[206] Ya. S. Bobovich and V. V. Perekalin, *Dokl. Akad. Nauk SSSR* **127**, 1239 (1959).
[207] Yu. P. Egorov, L. A. Leites, I. D. Kravtsova, and V. F. Mironov, *Izv. Akad. Nauk SSSR, Otd. Khim. Nauk* p. 1114 (1963).
[208] A. E. Lutskii, *Zh. Strukt. Khim.* **2**, 640 (1968).
[209] L. A. Leites, L. E. Vinogradova, V. N. Kalinin, and L. I. Zakharkin, *Izv. Akad. Nauk SSSR, Ser. Khim.* p. 1016 (1968).
[210] T. A. Babushkina, S. D. Sokolov, and G. K. Semin, *Izv. Akad. Nauk SSSR, Ser. Khim.* p. 2065 (1969).
[211] E. K. Orlova, I. D. Tsvetkova, V. S. Troitskaya, V. G. Vinokurov, and V. A. Zagorevskii, *Khim. Geterotsikl. Soedin.* p. 429 (1969).
[212] E. N. Tsvetkov, G. K. Semin, T. A. Babushkina, D. I. Lobanov, and M. I. Kabachnik, *Izv. Akad. Nauk SSSR, Ser. Khim.* p. 2375 (1967).
[213] G. K. Semin, T. A. Babushkina, and V. I. Svergun, *Uch. Zap. Mosk. Obl. Ped. Inst.* **222**, 78 (1969).
[214] D. Purdela, *Rev. Roum. Chem.* **10**, 949 (1965).
[215] A. A. Neimysheva, V. I. Savchuk, M. V. Ermolaeva, and I. L. Knunyants, *Izv. Akad. Nauk SSSR, Ser. Khim.* p. 2222 (1968).
[216] T. A. Babushkina and M. I. Kalinkin, *Izv. Akad. Nauk SSSR, Ser. Khim.* p. 157 (1969).
[217] G. K. Semin, T. A. Babushkina, V. M. Vlasov, and G. G. Yakobson, *Izv. Sib. Otd. Akad. Nauk SSSR, Ser. Khim. Nauk* p. 99 (1969).
[218] G. K. Semin and V. I. Robas, *in* "Radiospektroskopiya tverdogo tela," p. 229. Atomizdat, Moskva, 1967.
[219] V. I. Bregadze, T. A. Babushkina, O. Yu. Okhlobystin, and G. K. Semin, *Teor. Eksp. Khim.* **3**, 547 (1967).
[220] T. A. Babushkina, E. V. Bryukhova, F. K. Velichko, V. I. Pakhomov, and G. K. Semin, *Zh. Strukt. Khim.* **9**, 207 (1968).
[221] D. N. Kravtsov, A. P. Zhukov, V. A. Faingor, E. M. Rokhlina, G. K. Semin, and A. N. Nesmeyanov, *Izv. Akad. Nauk SSSR, Ser. Khim.* p. 1703 (1968),
[222] E. V. Bryukhova, F. K. Velichko, and G. K. Semin, *Izv. Akad. Nauk SSSR, Ser. Khim.* p. 960 (1969).
[223] E. V. Bryukhova, T. A. Babushkina, O. Yu. Okhlobystin, and G. K. Semin, *Dokl. Akad. Nauk SSSR* **183**, 827 (1968).
[224] M. G. Voronkov and L. A. Zhagata, *Dokl. Akad. Nauk SSSR* **194**, 847 (1970).

[225] F. K. Velichko, L. A. Nikonova, and G. K. Semin, *Izv. Akad. Nauk SSSR, Ser. Khim.* p. 84 (1967).
[226] H. C. Allen, *J. Amer. Chem. Soc.* **74**, 6074 (1952).
[227] T. A. Babushkina, S. P. Kolesnikov, O. M. Nefedov, V. I. Svergun, and G. K. Semin, *Izv. Akad. Nauk SSSR, Ser. Khim.* p. 1055 (1969).
[228] I. Sakurai, *Acta Crystallogr.* **15**, 1164 (1962).
[229] H. G. Dehmelt, *J. Chem. Phys.* **21**, 380 (1953).
[230] Yu. K. Maksyutin, E. N. Gur'yanova, and G. K. Semin, *Zh. Strukt. Khim.* **9**, 701 (1968).
[231] G. Briegleb, "Electron-Donator-Acceptor-Complexe." Springer-Verlag, Berlin and New York, 1961.
[232] G. K. Semin, V. I. Robas, G. G. Yakobson, and V. D. Shteingarts, *in* "Radiospektroskopiya tverdogo tela," p. 239. Atomizdat, Moskva, 1967.
[233] G. K. Semin, V. I. Robas, V. D. Shteingarts, and G. G. Yakobson, *Zh. Strukt. Khim.* **6**, 160 (1965).
[234] V. S. Grechishkin and I. A. Kyuntsel', *Zh. Strukt. Khim.* **7**, 119 (1966).
[235] N. E. Ainbinder and A. N. Osipenko, *Tr. Estestvennonauch. Inst. Perm. Gos. Univ.* **11**, vyp. 4, p. 101 (1966).
[236] A. I. Kitaigorodskii and A. A. Frolova, *Izv. Sekt. Fiz.-Khim. Anal. Inst. Obshch. Neorg. Khim., Akad. Nauk SSSR* **19**, 307 (1949).
[237] D. C. Douglass, *J. Chem. Phys.* **32**, 1882 (1960).
[238] A. I. Kitaigorodskii and E. I. Fedin, *Zh. Strukt. Khim.* **2**, 216 (1961).
[239] Yu. K. Maksyutin, T. A. Babushkina, E. N. Gur'yanova, and G. K. Semin, *Zh. Strukt. Khim.* **10**, 1025 (1969).
[240] Yu. K. Maksyutin, T. A. Babushkina, E. N. Gur'yanova, and G. K. Semin, *Theor. Chim. Acta* **14**, 48 (1969).
[241] Yu. K. Maksyutin, E. N. Gur'yanova, and G. K. Semin, *Usp. Khim.* **39**, 727 (1970).
[242] V. S. Grechishkin, A. D. Gordeev, and Yu. A. Galishevskii, *Zh. Strukt. Khim.* **10**, 743 (1969).
[243] H. O. Hooper, *J. Chem. Phys.* **41**, 599 (1964).
[244] P. Cornil, M. Read, J. Duchesne, and R. Cahay, *Bull. Cl. Sci., Acad. Roy. Belg.* **50**, 235 (1964).
[245] P. Cornil, J. Duchesne, M. Read, and R. Cahay, *Bull. Soc. Belge Phys.* **2**, 89 (1964).
[246] M. Read, R. Cahay, P. Cornil, and J. Duchesne, *C. R. Acad. Sci.* **257**, 1778 (1963).
[247] R. M. Keefer, and L. J. Andrews, *J. Amer. Chem. Soc.* **72**, 4677 (1950).
[248] L. T. Jones, *Diss. Abstr. B* **27**, 130 (1968).
[249] D. F. R. Gilson and C. T. O'Konski, *J. Chem. Phys.* **48**, 2767 (1968).
[250] V. S. Grechishkin and I. A. Kyuntsel', *Tr. Estestvennonauch. Inst. Perm. Gos. Univ.* **12**, vyp. 1, p. 15 (1966).
[251] F. G. Strieter and D. H. Templeton, *J. Chem. Phys.* **39**, 1225 (1963).
[252] R. A. Bennet and H. O. Hooper, *J. Chem. Phys.* **47**, 4855 (1967).
[253] D. Biedenkapp and A. Weiss, *Ber. Bunsenges. Phys. Chem.* **70**, 788 (1966).
[254] D. Biedenkapp and A. Weiss, *Z. Naturforsch. A* **19**, 1518 (1964).
[255] Yu. K. Maksyutin, E. V. Bryukhova, G. K. Semin, and E. N. Gury'anova, *Izv. Akad. Nauk SSSR, Ser. Khim.* p. 2658 (1968).
[256] T. S. Srivastava, *Curr. Sci.* **37**, 253 (1968).
[257] D. A. Tong, *J. Chem. Soc., London* p. 790 (1969).

5. NQR in Organic and Metalloorganic Chemistry 233

258 M. Rogers and J. A. Ryan, *J. Phys. Chem.* **72**, 1340 (1968).
259 E. A. Kravchenko, Yu. K. Maksyutin, E. N. Gur'yanova, and G. K. Semin, *Izv. Akad. Nauk SSSR, Ser. Khim.* p. 1271 (1968).
260 R. F. Schneider and J. V. DiLorenzo, *J. Chem. Phys.* **47**, 2343 (1967).
261 V. S. Grechishkin and I. A. Kyuntsel', *Zh. Strukt. Khim.* **4**, 269 (1963).
262 V. S. Grechishkin and I. A. Kyuntsel', *Tr. Estestvennonauch. Inst. Perm. Gos. Univ.* **11**, 119 (1964).
263 V. S. Grechishkin and I. A. Kyuntsel', *Opt. Spektrosk.* **15**, 832 (1963).
264 V. S. Grechishkin and I. A. Kyuntsel', *Opt. Spektrosk.* **16**, 161 (1964).
265 V. S. Grechishkin and I. A. Kyuntsel', *Zh. Strukt. Khim.* **5**, 53 (1964).
266 V. S. Grechishkin and I. A. Kyuntsel', *Tr. Estestvennonauch. Inst. Perm. Gos. Univ.* **11**, vyp. 2, p. 9 (1966).
267 V. S. Grechishkin and A. D. Gordeev, *Tr. Estestvennonauch. Inst. Perm. Gos. Univ.* **12**, 29 (1966).
268 M. Negita, T. Okuda, and M. Kashima, *J. Chem. Phys.* **45**, 1076 (1966).
269 M. Negita, T. Okuda, and M. Kashima, *J. Chem. Phys.* **46**, 2450 (1967).
270 V. S. Grechishkin, S. I. Gushchin, and V. A. Shishkin, *Zh. Strukt. Khim.* **11**, 145 (1970).
271 D. Biedenkapp and A. Weiss, *Z. Naturforsch.* B **23**, 174 (1968).
272 T. A. Babushkina, V. I. Robas, and G. K. Semin, *Dokl. Akad. Nauk SSSR* **159**, 164 (1964).
273 G. K. Semin, V. I. Robas, and T. A. Babushkina, in "Radiospektroskopiya tverdogo tela," p. 218. Atomizdat, Moskva, 1967.
274 A. P. Suprun, A. S. Shashkov, T. A. Soboleva, G. K. Semin, T. T. Vasil'yeva, G. P. Lopatina, T. A. Babushkina, and R. Kh. Freidlina, *Dokl. Akad. Nauk SSSR* **173**, 1356 (1967).
275 V. I. Stanko, E. V Bryukhova, Yu. V. Gol'tyapin, and G. K. Semin, *Zh. Strukt. Khim.* **10**, 745 (1969).
276 B. A. Englin, B. N. Osipov, V. A. Valovoi, T. A. Babushkina, G. K. Semin, V. B. Bondarev, and R. Kh. Freidlina, *Izv. Akad. Nauk SSSR, Ser. Khim.* p. 1251 (1968).
277 B. A. Englin, T. A. Onishchenko, V. A. Valovoi, T. A. Babushkina, G. K. Semin, L. G. Zelenskaya, and R. Kh. Freidlina, *Izv. Akad. Nauk SSSR, Ser. Khim.* p. 332 (1969).
278 A. V. Upadysheva, T. A. Babushkina, E. V. Bryukhova, V. I. Robas, L. A. Kazitsyna, and G. K. Semin, *Izv. Akad. Nauk SSSR, Ser. Khim.* p. 2068 (1969).
279 C. Dean and R. V. Pound, *J. Chem. Phys.* **20**, 195 (1952).
280 T. Kushida, G. B. Benedekand, and N. Boembergen, *Phys. Rev.* **104**, 1364 (1956).
281 V. S. Grechishkin, *Zh. Struct. Khim.* **6**, 162 (1965).
282 T. A. Babushkina, V. I. Robas, and G. K. Semin, in "Radiospectroscopija Tverdogo Tela," p. 221. Atomizdat, Moskva, 1967.
283 T. A. Babushkina and D. F. Baisa, *Fiz. Tverd. Tela* **6**, 2663 (1964).
284 T. A. Babushkina, V. I. Robas, I. A. Safin, and G. K. Semin, *Fiz. Tverd. Tela* **7**, 924 (1965).
285 V. S. Grechishkin and G. B. Soifer, *Tr. Estestvennonauch. Inst. Perm. Gos. Univ.* **12**, vyp. 1, p. 3 (1966).

6 Mössbauer Spectra of Organometallics

N. W. G. DEBYE AND J. J. ZUCKERMAN

I. Theory	235
A. Nuclear Processes and Chemistry	235
B. The Resonance Fluorescence of γ Rays	236
C. Mösssbauer's Discovery	238
D. Mössbauer Spectroscopy	239
E. Parameters of Mössbauer Spectra	240
II. Applications	250
A. Composition	250
B. Solution Interactions	251
C. Kinetics	253
D. Surface Studies	255
E. Polymerization	255
F. Radiation Damage	257
G. Spin States of Iron	259
H. The Sign of the Quadrupole Coupling Constant	260
I. Oxidation State	261
J. Applications to Biological Systems	263
K. Structural Determinations	264
References	278

I. THEORY*

A. Nuclear Processes and Chemistry

The law which states that nuclear processes are independent of the chemical environment of the nucleus has been quietly repealed. Until recently, radioactive decay processes were thought to be entirely independent of chemical

---

* This section is based in part on an earlier review.[1]

state. This view was based upon the fundamentally sound precept that the energies involved in nuclear processes are so much larger than the energies of chemical binding that the nucleus can almost always be thought of as being part of a free atom. On the other hand, nuclear properties (aside from mass and electric charge) were considered to be of little interest to the chemist. Today, however, a broad range of phenomena is known which arise from interactions of nuclei with their electron shells. Among the latest additions to this list is the observation of chemical effects upon γ-ray energies. These energies are studied by the technique of resonance fluorescence.

B. The Resonance Fluorescence of γ Rays

A pair of systems, an emitter and an absorber of energy with nearly the same characteristic frequencies, can give rise to the resonance phenomenon. The two frequencies need not be exactly equal, however, since every resonance is characterized by a plot of response against frequency peaked at the resonant energy E but with a characteristic width Γ at half-maximum. A sharply tuned system gives a narrow resonance line relative to the resonance frequency. Radio engineers speak of this sharpness of "tuned" electrical circuits as the quality, or Q factor, and when the energy added is exactly E, the probability of excitation of the level will be maximum; the excitation probability at any other energy E' will be something less.

Resonance absorption for optical processes in atomic systems was predicted by Lord Rayleigh and demonstrated by R. W. Wood in 1904 for sodium emission. Wood allowed yellow atomic sodium radiation to pass into a sample of sodium vapor at low pressure in a glass bulb. A yellow glow emanating from the bulb was the result of the sodium atoms absorbing and reradiating the incident yellow light; other atoms, not "tuned" to the incident radiation, were effectively transparent. Rayleigh's classical predictions may be understood in quantum terms: The emission of radiation arises from transitions between discrete electronic energy levels with both the emitting and the absorbing sodium atoms having the same level separation. For a typical atom the resonance energy is of the order of 1 eV, and the half-height width Γ about 10^{-8} eV so that the Q factor, E/Γ, is 10^8. The atom is thus a finely tuned oscillator, and the lines in atomic spectra are very sharp. In an optical laser, where an ensemble of atoms emits in unison, the ratio of E to Γ can be much higher (up to $\approx 10^{13}$).

The nuclear analog to the optical process is the resonant absorption of γ rays. The search begun in the late twenties by W. Kuhn for an example of this nuclear resonance, however, remained unsuccessful for two decades and the reasons for this can be readily understood. The energies involved in the nuclear transitions are much larger than those in atoms, but the natural widths of the spectral lines are about the same. Thus the E/Γ ratio can exceed even 10^{15}. The resonance between γ rays and nuclei is thus one of the most sharply tuned systems in nature and is sensitive to the smallest departures from resonant conditions. The part played by recoil effects now becomes very important. In accord with Newton's third law, some of the energy of the emitted γ ray will be taken up in the recoil of the emitting nucleus; this energy dissipation destroys the sharply tuned resonance. Recoil processes are present in all such resonant systems, but the higher energies involved in nuclear transitions make recoil a much more important consideration here.

The recoil energy E_R is proportional to the square of the transition energy E, and for a free atom this is computed (from momentum conservation) to be $E^2/2Mc^2$, where M is the mass of the emitting nucleus and c is the speed of light. For the tin-119 nucleus, for example, the recoil energy is only about 10^{-3} eV, tiny compared with the transition energy (23,800 eV) but large compared with Γ (10^{-8} eV). The energy of a γ ray emitted by a freely recoiling tin nucleus is thereby reduced by an amount E_R, and the energy carried by the γ ray is less than that necessary for resonance. Most of the early experiments, therefore, failed to detect nuclear resonance fluorescence. In optical transitions, on the other hand, the recoil energy is much less than the natural linewidth, and resonance is readily observed despite the fact that the emission line is displaced $2E_R$ from the energy required by the absorber. For example, in the case of yellow sodium radiation, E_R is only 10^{-10} eV, less than $\frac{1}{100}$ of Γ.

The linewidths for the γ rays are also broadened considerably by Doppler effects. An emitted γ ray of energy E produced by a nucleus receding with velocity V appears to have energy $E - EV/c$ while the energy associated with

TABLE I

Transition Characteristics

	Atomic: sodium	Nuclear: tin-119m
Transition energy, E (eV)	2.1	23,800
Natural linewidth of excited level, Γ (eV)	4.4×10^{-8}	2.4×10^{-8}
Resonance wavelength (cm)	5.89×10^{-5}	5.3×10^{-9}
Doppler width at room temperature (eV)	3.13×10^{-6}	1.6×10^{-2}
Recoil energy of a single nucleus, E_R (eV)	10^{-10}	2.5×10^{-3}
Ratio E/Γ	5×10^7	10^{12}

a γ ray produced by an approaching nucleus is $E + EV/c$. The definition or frequency spread of the γ rays is thus made larger than the natural linewidth. The velocities of the nuclei, and hence the line broadening, are proportional to the square root of the temperature in degrees absolute. Since the Doppler width is also proportional to the transition energy E, the broadening of the nuclear transition energies will be much greater than the broadening of the electronic transition energies. Some of the transition characteristics for the atomic sodium line and an analogous nuclear transition in tin-119m are shown in Table I.

C. Mössbauer's Discovery

Rudolf L. Mössbauer simplified the whole problem considerably by demonstrating in 1958 that the loss of energy by nuclear recoil could be avoided altogether in solids. To understand Mössbauer's discovery it is necessary to distinguish three cases of the scattering of γ rays by atoms bound in solids (see Fig. 1).

Case 1. If E_R is large compared with the strength of the chemical forces binding the atom in the solid, then the atom will be ripped out of its lattice site [Fig. 1(b)].

Case 2. If E_R is larger than the characteristic vibrational energy of the solid (the phonon energy), but not great enough to dislodge the atom completely from its site, then the atom will oscillate at its site, and the recoil energy will be dissipated as heat [Fig. 1(c)].

Case 3. If the recoil energy is less than the phonon energy, a new effect arises because the lattice is a quantized system which cannot be excited in an arbitrary manner. In this case it becomes possible for some fraction of the emission and absorption events to occur with no loss in recoil energy at all. For these events it is possible to consider that the effective recoiling mass is now the mass of the whole lattice rather than just the mass of the emitting nucleus. The recoil energy would then drop to a negligible value.

This production of monochromatic γ rays is the Mössbauer effect. Conventional techniques of spectroscopy are hopelessly inadequate for resolving the tiny differences in energy in these Mössbauer effect γ rays due to changes in chemical state. The nuclear transition energies themselves are quite apart from those encountered in other physical techniques useful to organometallic chemists, as shown in Table II. The γ-ray energies are huge, but the precise values are a measure of small changes in the atom. Mössbauer recognized that the converse of the recoil-free emission process, recoil-free resonant absorption, constitutes an ideal detector, one that is precisely tuned to the energy of the γ ray. Recoil-free resonant absorption makes it possible to compare with unprecedented precision the nuclear transition energy in a

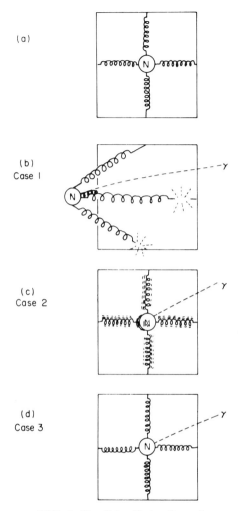

FIG. 1. Possible effects of recoil.

source with that in an absorber. While the measurement of the energies of such γ rays could be carried out in pre-Mössbauer days at best with a precision of 0.1%, the discovery of the Mössbauer effect has raised the precision by a factor of 10^9.

D. Mössbauer Spectroscopy

Chemical methods of spectroscopy are usually based upon resonance experiments where the resonance condition is observed by systematically perturbing a system and then noting the influence of the variation on the measurable parameters. In chemical applications of the Mössbauer effect the source and absorber nuclei are usually in different chemical states. The Doppler effect is used to bring the system into resonance. The relative motion of the source and absorber shifts the positions of their nuclear energy levels until they coincide. The relative velocity required is a measure of how mismatched the source and absorber are.

The experimental factors which enter into the design of a suitable spectrometer include the method chosen to Doppler-modulate the energy of the γ ray; detection of the γ ray of interest against the background of other radiation; collection, storage, and handling of the data; the geometry of the experiment; the need for cryostats; and the states of the sample and the host material for the source nuclei. A full discussion of these factors is beyond the scope of this review and can be found in several excellent books and articles.[2-6]

E. Parameters of Mössbauer Spectra

The parameters of the Mössbauer experiment are briefly summarized below with reference to the six parts of Fig. 2.

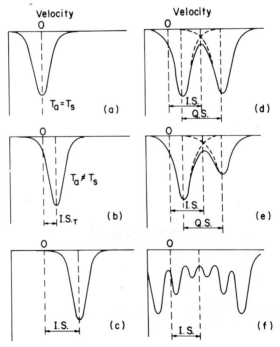

FIG. 2. Parameters of Mössbauer spectra. See text for discussion of parts.

(a) Source = absorber: The peak area is proportional to the recoil-free fraction f' of the absorber where

$$f' = \exp - (4\pi^2 \langle \bar{r}^2 \rangle / \lambda^2) \tag{1}$$

and $\langle \bar{r}^2 \rangle$ is the mean square of the amplitude of atomic displacement and λ is the wavelength of the scattered radiation.

(b) Source = absorber but the temperatures of the two solids are different: The temperature shift of the resonance line, δ_T, also called the second-order Doppler shift, is given by

$$\delta_T = E_R \langle \bar{v}^2 \rangle / 2c^2 \tag{2}$$

where $\langle \bar{v}^2 \rangle$ is the mean square velocity of the atom in vibrational motion and c is the speed of light.

(c) Source $\neq$ absorber: The isomer shift of the resonance line (I.S.) is given by

$$\text{I.S.} = \tfrac{4}{5}\pi Z e^2 R^2 (\Delta R/R)[|\psi(0)|_a^2 - |\psi(0)|_s^2] \tag{3}$$

where the nucleus is of atomic number Z and radius R and ΔR is the change in nuclear radius in decay from the excited isomeric state.

(d) Source $\neq$ absorber and the absorber nucleus experiences an electric field gradient, eq: The eigenvalues E_Q of the quadrupolar interaction Hamiltonian are given by

$$E_Q = [e^2 q Q / 4I(2I - 1)][3m^2 - I(I + 1)](1 + \tfrac{1}{3}\eta^2)^{1/2} \tag{4}$$

TABLE II

Spectroscopic Energies

Spectroscopy	Typical Energy of Radiation (cm^{-1})	Typical Energy Difference Detected (eV)
X ray: inner electronic states	10,000,000	10^3
Photoelectron: electronic states	—	10
Visible–ultraviolet: valence electronic states	10,000	1
Infrared–Raman: vibrational states	1,000	10^{-1}
Microwave: rotational states	10	10^{-1}
Electron spin resonance: electron spin states in a magnetic field	0.1	10^{-4}
Nuclear magnetic resonance: nuclear spin states in a magnetic field	0.001	10^{-7}
Mössbauer: nuclear states	100,000,000	10^{-8}

where I is the nuclear spin quantum number, m is the magnetic quantum number, eQ is the nuclear quadrupole moment, and η is the electric field gradient asymmetry parameter. For transitions involving an $I = \frac{3}{2}$ excited state and an $I = \frac{1}{2}$ ground state (e.g., ^{119m}Sn and ^{57}Fe), the quadrupole splitting is simply the difference of the eigenvalues of the $I = \frac{3}{2}$ nucleus evaluated for $m = \frac{3}{2}$ and $m = \frac{1}{2}$:

$$\text{Q.S.} = |E_Q(m = \tfrac{3}{2}) - E_Q(m = \tfrac{1}{2})|$$

$$\text{Q.S.} = \left|\frac{e^2qQ}{2}\left(1 + \frac{\eta^2}{3}\right)^{1/2}\right| \tag{5}$$

Nuclei with $I < 1$ have $eQ = 0$ and, therefore, exhibit no such interaction.

(e) Source $\neq$ absorber, the absorber nucleus experiences an electric field gradient, and the anisotropy of the recoil-free fraction, f', gives rise to asymmetries in the line intensities of doublet spectra, even in polycrystalline samples [the Goldanskii–Karyagin effect].

(f) Source $\neq$ absorber and the absorber nucleus experiences a magnetic field, H: The eigenvalues of the magnetic interaction Hamiltonian are given by

$$\Delta = -\mu H m/I \tag{6}$$

where μ is the magnetic moment of the nucleus and m and I are the magnetic and spin quantum numbers. Magnetic interactions are not generally of importance in organometallic systems and will not be stressed here.

While a detailed discussion of all the parameters which can be derived from Mössbauer spectroscopy and their significance is beyond the scope of this review, there follows a brief exposition of each with particular relation to the Mössbauer spectroscopy of organometallic compounds.

1. The Isomer Shift

As in the more familiar NMR spectra, Mössbauer spectra show a shift and frequently a splitting. The splittings in the Mössbauer case arise for totally different reasons than the NMR spin–spin couplings, and these will be discussed in the next section. Both shifts, however, arise from rather similar causes, but the physics of the Mössbauer shift is somewhat more straightforward. The term "isomer shift" (I.S. or δ) has been almost universally adopted.

Experimentally, the I.S. is the displacement of the resonance from zero relative velocity as shown in Fig. 2(c). The Mössbauer experiment always involves a pair of compounds, the emitter (source), and the absorber. When these are identical, the I.S. is zero. When two different chemical compounds are used, the I.S. represents the mismatch of the two sets of nuclear energy levels, expressed in Doppler shift energy units (millimeters per second). The precise value of the energy level separation in each compound is never measured directly, however. Relative motion of the source–absorber pair toward one another (taken as positive velocity), for example, adds energy to the emitted γ ray to bring an absorber nucleus with a larger level spacing into resonance. Thus a series of absorbers can be compared with a single defined chemical state (the source) or, by the use of data derived from a given source, with one another.

This mismatch of the two sets of nuclear levels arises from the interaction of the two positively charged nuclei with their electronic environments, and it may be considered as an electric monopole interaction.

The relationship between the I.S. and the electric field density at the nucleus is given by

$$\text{I.S.} = \tfrac{4}{5}\pi Z e^2 R^2 (\Delta R/R)[\rho_a(0) - \rho_s(0)] \tag{7}$$

where the absorber and source nuclei are of atomic number Z and radius R and experience total electron densities $\rho_a(0)$ and $\rho_s(0)$, respectively. The effect is called the isomer shift because it depends on the difference in the nuclear radii of the ground and excited isomeric states, ΔR. From this equation we see that the I.S. for any nuclide is the product of a constant term $[(4\pi/5)Ze^2]$, a nuclear term $[R^2 \Delta R/R]$, and an extranuclear, electronic term $[\rho_a(0) - \rho_s(0)]$.

Chemists replace $\rho(0)$, the total electron density at the nucleus, by $|\Psi(0)|^2$, the square of the total wavefunction for the atom evaluated at the nucleus:

$$\text{I.S.} = \text{const. } R^2(\Delta R/R)[|\Psi(0)|_a^2 - |\Psi(0)|_s^2] \tag{8}$$

In general, of the types of electrons in the atom, only the s-electron density does not vanish at the nucleus $[|\psi_{ns}(0)|^2 \neq 0]$ and we can further modify the expression to*

$$\text{I.S.} = \text{const. } R^2(\Delta R/R)[|\psi_{ns}(0)|_a^2 - |\psi_{ns}(0)|_s^2] \tag{9}$$

Thus the I.S. is directly sensitive to changes in s-electron density and, thereby, indirectly dependent upon the other electrons in the atom. To relate the I.S. to $|\psi_{ns}(0)|^2$ requires knowledge of the sign and magnitude of the fractional change in the nuclear charge radius on excitation, $\Delta R/R$. The sign of the change in the charge radius of ^{57}Fe, for example, is negative; of ^{119}Sn is positive.

---

* This is strictly true only for a point nucleus, while Eq. (7) is developed for nuclei having finite radii. Thus other types of electrons will be found within the nucleus as well, but their direct effect is expected to be small.

2. The Quadrupole Splitting

We will consider three types of multipole interactions of a nucleus with its environment in an atom: the electric monopole arising from the electric charge on the nucleus, eZ; the magnetic dipole arising from the magnetic dipole moment of the nucleus, μ; and the electric quadrupole arising from the electric quadrupole moment of the nucleus, eQ. The isomer shift discussed in the last section arises from the electric monopole interaction. The magnetic dipole interaction or nuclear Zeeman effect will be discussed in the next section. In this section we will briefly describe the interaction of the nuclear quadrupole with gradients in the extranuclear electric field.

A charge (pole) migrates in a field toward the opposite pole; a dipole rotates in a field to achieve the lowest energy orientation ($+ \rightarrow -$). A dipole migrates in the presence of a field gradient; a quadrupole rotates in a field gradient to achieve the lowest energy orientation. In addition to its stable equilibrium orientation, the dipole may be found in a higher-energy, unstable equilibrium orientation if its moment is exactly aligned with the external electric field. This ambiguity of equilibrium states in the classical sense may be also interpreted in quantum mechanical terms. Correspondingly, consideration of the quadrupole in the presence of an electric field gradient leads to similar equilibrium state ambiguities of various energies; in quantum mechanical terms one speaks of the removal of the degeneracy of the quadrupolar energy levels.

Any nuclear state with a spin greater than $\frac{1}{2}$ has a quadrupole moment, eQ. The energy level diagrams of ^{57}Fe and ^{119}Sn are similar and are represented in Fig. 3. The excited spin $\frac{3}{2}$ state in each nucleus has a quadrupole moment, and this state can be split into two by the interaction with an asymmetric electric field produced by the placement of the extranuclear charges. The presence of a nonzero field gradient at the nucleus is determined primarily by the symmetry of the distribution of electrons about the nucleus, which is in turn determined by the symmetry of the bonding of the tin atom in question. Cubic symmetry (tetrahedral or octahedral arrangement of identical attached atoms) will result in a zero field gradient, whereas quadrupolar splitting of spectral lines will be expected for lower symmetries. The magnitude of ΔE_Q is directly observed as the quadrupole splitting (Q.S.) which is given for the case of these two nuclei as

$$\Delta E_Q = E_{3/2} - E_{1/2} = \tfrac{1}{2}e^2Q(1 + \tfrac{1}{3}\eta^2)^{1/2} \tag{10}$$

where $E_{3/2}$ and $E_{1/2}$ are the energies associated with those values of the magnetic quantum number m as in Fig. 5, e is the electronic charge, and η, called the asymmetry parameter, is related to the magnitudes of the three principal moments of the electrostatic field gradient which are the second

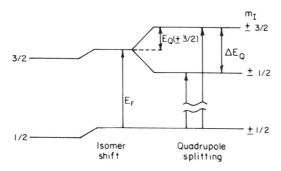

FIG. 3. The origin of quadrupole splitting.

derivatives of the potential with respect to each coordinate, $\partial^2 V/\partial z^2$, $\partial^2 V/\partial y^2$, and $\partial^2 V/\partial x^2$, and is given by

$$\eta \equiv \frac{(\partial^2 V/\partial x^2) - (\partial^2 V/\partial y^2)}{\partial^2 V/\partial z^2} \tag{11}$$

When the atom is at a lattice site which experiences an axially symmetric field, i.e., one in which $(\partial^2 V/\partial x^2) = (\partial^2 V/\partial y^2)$, $\eta = 0$, Eq. (12) reduces to

$$\Delta E_Q = \pm \tfrac{1}{2} e Q V_{zz} \tag{12}$$

where $V_{zz} = \partial^2 V/\partial z^2$ is the largest component of the diagonalized electric field gradient (efg) tensor. The product eQV_{zz} is the quadrupole coupling constant. In this manner the magnitude, but not the sign of the quantity eQV_{zz}, can be derived from a spectrum showing Q.S., even though for ^{57}Fe and ^{119}Sn the nucleus has a quadrupole moment only in the excited state. Nuclear quadrupole resonance (see this volume, Chapter 5) allows the determination of quadrupole interaction constants for nuclei which are quadrupolar in the ground state. Thus the two techniques complement one another.

The sign of ΔE_Q can be established by studying the relative intensities of the two lines of the doublet produced by a single-crystal absorber as a function of the angle of observation relative to known crystallographic axes or by the study of the relative intensities of the six lines of the spectrum produced in an applied magnetic field as discussed in Section I.E.3 below. Like the I.S., the Q.S. is produced as the product of a constant term $(e^2/2)$, a nuclear term (Q), and an extranuclear, atomic term $\{q[1 + (\eta^2/3)]^{1/2}\}$.

Two important sources of the electric field gradient in atoms are (a) charges on neighboring ions, and (b) the incompletely filled electron shells of the atom itself. Distant ions can contribute directly to the gradient if the symmetry of their array is lower than cubic. Furthermore, the efg from the distant charges can distort the electron distribution of the Mössbauer atom so as to create an efg of their own which usually serves to amplify the efg arising from the distant charges, a phenomenon known as antishielding. Antishielding is more important in ionic lattices, and we shall ignore it in discussions of molecular organometallic crystals. Dipole and higher-order moments of neighboring molecules are usually assumed to make only minor contributions to the gradient. The even more important efg due to electrons in partially filled, nonspherical orbitals is not a matter of concern for the main group elements such as tin, but play the dominant role for transition metals such as iron.

Our discussions will be simplified by the assumption that the efg at the tin nucleus is due only to the surrounding electrons and that the role played by the atoms directly bonded to the resonant tin atom is to distort this electronic environment from the spherical symmetry it otherwise has. Thus we expect no Q.S. when the tin atom occupies a site of tetrahedral or octahedral bonding symmetry.

3. The Magnetic Splitting

In Fig. 4 both the $+m$ and $-m$ states occur at the same energy in an asymmetric electric field. This degeneracy can be removed in a magnetic field. Such a field is experienced by the atoms in a ferromagnetic material, or by a nonmagnetic compound in an applied magnetic field. Under these conditions a nuclear Zeeman effect is observed, as depicted in Fig. 4 where each nuclear energy level is split into $(2m + 1)$ components. The selection rules which govern allowed transitions between these substates result in a six-line spectrum

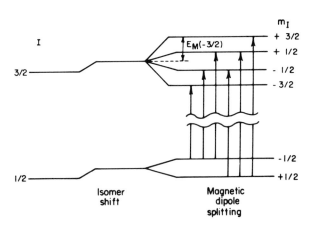

FIG. 4. The origin of magnetic splitting.

as in Fig. 2(f). The energies of these levels are given by Eq. (6). As in the case of the I.S. and Q.S., the energy depends both upon a nuclear factor μ and an extranuclear factor H. It is worth comparing this situation to that found in NMR spectroscopy. In NMR, transitions are found to occur between adjacent sublevels of the ground state, while in Mössbauer spectroscopy of nuclei in a magnetic field ("Magbauer") the observed transitions are between sublevels of the ground and excited nuclear states. The two techniques differ in energy by a factor of 10^{10}, but the definition of the Mössbauer γ ray allows these hyperfine interactions to be resolved.

The even spacing of the lines in Fig. 2(f) is modified if quadrupolar interactions are also present, and small quadrupolar perturbations can be detected in the magnetic experiment.[7]

4. The Magnitude of the Resonance Effect

The first three parameters considered above, I.S., Q.S., and nuclear Zeeman splitting, arise from the interaction of the nucleus with molecular and crystalline electric fields and internal and applied magnetic fields. The next parameters which will be considered depend upon the dynamics of motion of the emitting and absorbing nuclei.

The first of these dynamical parameters, the magnitude of the resonance effect, ε, is obtainable directly from the usual Mössbauer spectrum and is commonly reported as a percentage

$$\varepsilon = 100[(R_\infty - R_0)/R_\infty] \qquad (13)$$

where R_0 is the counting rate at the resonance maximum and R_∞ the counting rate at a Doppler velocity well removed from the resonance maximum. The measured magnitude (usually ~ 0–25%) is not a fundamental quantity, but depends upon a number of experimental factors such as the fraction of counts included due to non-recoil-free events, velocity-dependent solid angle effects, pulse shape-dependent counter dead-time effects, and the number of absorber nuclei in the optical path.

5. The Recoil-Free Fraction

The magnitude of the effect, ε, can be related under certain conditions to the recoil-free fraction f, but it should not be confused with it. The probability of achieving the Mössbauer effect depends upon the amplitude of thermal motion of the Mössbauer nucleus. When this motion is unbounded, the probability for the effect is zero. The fraction of recoil-free, or Mössbauer, events is dependent on the ratio of the mean square vibrational amplitude, $\langle \bar{r}^2 \rangle$, of the emitting or absorbing nucleus to the square of the wavelength λ

of the scattered radiation. This dependence can be understood on the basis of a classical explanation. Movement of the emitting atom over distances comparable to a wavelength during the emission process will destroy the phase coherence of the emitted wave, and parts of this wave will interfere destructively and be weakened at the natural frequency of the emitter.

The recoil-free fraction f is expressed as

$$f = \exp[-4\pi^2 \langle \bar{r}^2 \rangle / \lambda^2] \tag{14}$$

where $\langle \bar{r}^2 \rangle$ is taken in the direction of emission of the γ ray and averaged over the lifetime of the nuclear level involved in the γ-ray emission process. Unbounded motion, as in a liquid or gas, will cause the recoil-free fraction to vanish. It may be noted that Eq. (14) gives no indication that periodicity is required, and thus it is not surprising that the Mössbauer effect is observed in amorphous solids and frozen solutions. If the binding of the atom is anisotropic then the mean square amplitude of its motion will vary along the various crystallographic axes, and oriented single crystals will exhibit anisotropic recoil-free fractions.

A number of methods for evaluating the recoil-free fraction of a source, f, and an absorber, f', are available. Evaluation generally requires knowledge of the natural linewidth Γ and of the experimental source and absorber linewidths Γ_s and Γ_a. Unfortunately, obtaining reliable f' values is not simple or straightforward,[8-10] and many of the values quoted in the early literature show discrepancies and are unreliable.

6. Line Asymmetries in Doublet Spectra

Asymmetries in the intensities of component lines of doublet spectra of polycrystalline material were observed for organotin halides early on[11-18] and have been interpreted by Karyagin as due to the lattice-dynamic anisotropy in the recoil-free fraction f' of the absorber.[19] The phenomenon has been referred to as the Goldanskii–Karyagin effect.[21-22] That the anisotropy of the recoil-free fraction does not disappear even in a randomly oriented polycrystalline powder may seem at first surprising, but it comes about from the fact that the various hyperfine components themselves have intensities which are a function of the direction of emission relative to the electric field gradient axes and, therefore, provide means of taking spatially weighted averages of the recoil-free fraction.

7. The Linewidth

In the ideal transmission Mössbauer experiment, the observed linewidth Γ_{obs}, is simply twice the natural linewidth Γ, as given by the Heisenberg uncertainty principle

$$\Gamma = h/2\pi\tau \tag{15}$$

where τ is the mean lifetime of the excited state (e.g., 2.67×10^{-8} second for ^{119m}Sn) which gives the natural linewidth of 2.4×10^{-8} eV quoted in Table I or 0.31 mm/second in Doppler shift units. Thus the narrowest line observable would be 0.62 mm/second. In practice Γ_{obs} is always larger than 2Γ. Experimental factors responsible for line broadening include the finite thickness of the source and absorber, the presence of impurities containing the Mössbauer nucleus, thermal noise, the velocity resolution of the mover as influenced by vibration, solid state defects in the source or absorber, etc. Sections I.E.1–3 treated the electromagnetic effects on spectra giving rise to completely resolved hyperfine structure. However, in many experiments the hyperfine structure is not observed, and instead only broadened resonance lines are seen. In cases where atoms experience several different environments in the same material, the slightly shifted single lines will combine to produce an envelope with $\Gamma_{obs} > 2\Gamma$. Since line broadening may arise from several factors acting simultaneously, interpretation of its cause in any given case may be a very complicated and difficult problem.

8. Dependence of the Parameters on Temperature

It can be shown that the recoil-free fraction of a molecular crystal will always be smaller than that for a network polymeric lattice. In the former there are stronger bonds within the molecular unit and weaker bonds between units, while the unit itself has larger mass than in the latter case where all the bonds are strong. It can be further shown that the dependence of the recoil-free fraction on temperature is sharply increased in a molecular crystal, where the f' value can be very low at higher temperatures. This fact underlies the need to cryostat organometallic samples. At liquid nitrogen temperatures the f' value can be close to unity, even for molecular crystals. The f' value increases at any temperature if weak van der Waals interactions are replaced by hydrogen or donor–acceptor covalent bonds between discrete molecules in the crystal. The temperature dependence of the f' value diminishes at the same time.

The isomer shift is also temperature-dependent through a second-order Doppler shift. The temperature shift, $E_R\langle v^2\rangle/2c^2$ [see Eq. (2)], appears as a higher-order term in the treatment of resonance absorption. Since the temperature difference between the source and absorber is rarely greater than 225°K (ambient vs 77°K) and the total shift for this temperature difference is less than one linewidth, it is usual to ignore this correction. This discussion is based on the assumption that the s-electron density at the nucleus $|\psi_{ns}(0)|^2$ is independent of temperature. In principle, variation of $|\psi_{ns}(0)|^2$ with temperature is possible if there are strong fields in the solid which vary with temperature or if the hybridization of the bonding orbitals is a function of molecular vibrations or rotations.

For there to be a change of quadrupole splitting with temperature, it is necessary to have an electronic excited state which is close enough to the electronic ground state to be thermally populated at ambient temperatures. This effect is usually serious only for very heavy atoms. Electric field gradients might be expected to average out more readily at higher temperatures (as in NQR) and thereby diminish the Q.S.

The temperature dependence of the magnetic hyperfine interaction (Zeeman splitting) is, of course, closely related to the temperature dependence of the magnetism.

II. APPLICATIONS

The sections which follow discuss selected applications of Mössbauer spectroscopy to the solution of problems in the chemistry of organometallic compounds.

A. Composition

The well-known complex "Prussian blue" is formed by treating a solution of potassium ferrocyanide with an excess of iron(III):

$$Fe^{3+} + [Fe(CN)_6]^{4-} \longrightarrow Fe_4[Fe(CN)_6]_3 \qquad (16)$$

"Turnbull's blue," on the other hand, is formed by treating a solution of potassium ferricyanide with an excess of iron(II):

$$Fe^{2+} + [Fe(CN)_6]^{3-} \longrightarrow Fe_3[Fe(CN)_6]_2 \qquad (17)$$

A third, colloidal material, the so-called "soluble Prussian blue," is obtained by mixing solutions of either ferricyanide and iron(III) or ferrocyanide and iron(II) in 1:1 molar ratios:

$$\begin{matrix} Fe^{3+} + [Fe(CN)_6]^{4-} \searrow \\ & \{Fe[Fe(CN)_6]\}^{1-} \qquad (18) \\ Fe^{2+} + [Fe(CN)_6]^{3-} \nearrow \end{matrix}$$

The structures and colors of these three complexes have been matters of discussion for many years. The ^{57}Fe Mössbauer spectra of the first two materials are identical and that of the third is very similar.[22-26] Each shows

6. Mössbauer Spectra

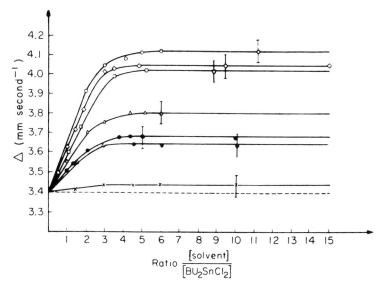

FIG. 5. The quadrupole splitting values measured at different molar ratios of solvated dibutyltin dichloride: ○, dimethyl sulfoxide (DMSO); ◇, dimethylformamide (DMF); □, hexamethyltriamidophosphate (HMPA); △, dimethoxyethane (DME); ●, tetrahydrofuran (THF); ⊙, diethoxyethane (DEE); ×, diethyl ether.

an overlapping doublet and singlet pattern arising from Fe^{3+} and $[Fe(CN)_6]^{4-}$, respectively, with the intensities in the approximately correct molar ratios. Their formulas can now be written as $Fe_x[Fe(CN)_6]_y$, with the cation in the oxidized form and the anion in the reduced form. Thus on mixing the solutions in the formation of "Turnbull's blue," the overall raction:

$$Fe^{2+} + [Fe(CN)_6]^{3-} \longrightarrow Fe^{3+} + [Fe(CN)_6]^{4-} \qquad (19)$$

which is consistent with the redox potentials of the ions must occur.

B. Solution Interactions

Solvents play an important role in determining the course and rate of organometallic reactions. Electron-donating aprotic solvents interact with metal atoms in organometallic compounds in solution. Figure 5 shows the magnitude of the Q.S. of dibutyltin dichloride measured at different molar ratios of various common solvents. The coordination of solvent molecules apparently increases the electric field gradient at the tin nucleus from that found in the neat solid dibutyltin dichloride. From Fig. 5 it is possible to

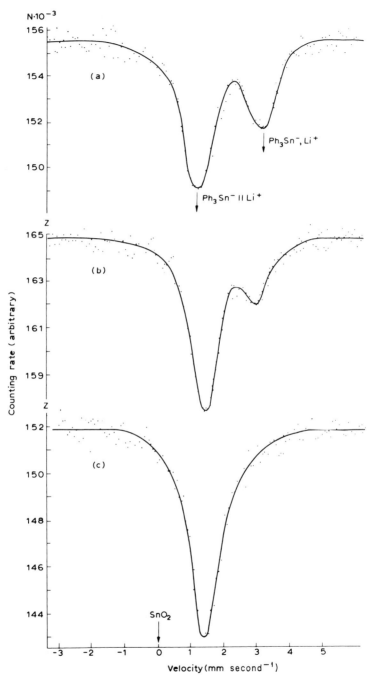

FIG. 6. Spectra of solutions of $(C_6H_5)_3Sn^-Li^+$ (a) in THF; (b) in a mixture of THF + DMSO at a volume ratio DMSO:THF 0.07; (c) in a mixture of THF + DMSO at a volume ratio DMSO:THF = 3.40.

rank the solvents according to their polarizing effect on the tin field gradient as DMSO > DMF > HMPA > DME > THF > DEE > diethyl ether. The break in the curves occurs at a mole ratio of 2–3:1. Further dilution causes no further changes in the Q.S. value.[27] In another study the Q.S. was found to increase generally with the dipole moment of the solvent to $\mu = 1.6$ D, and decrease somewhat at $\mu = 3$ D. The dependence on concentration is like that shown in Fig. 5.[28] Solvent-dependent spectra have also been recorded for triphenyltin lithium solutions. The I.S. values seen in Fig. 6 are said to correspond to the two types of ion pairs, $(C_6H_5)_3Sn^-Li^+$, and $(C_6H_5)_3Sn^-//Li^+$ where the ions are separated by solvent molecules. Their relative concentrations are seen to vary with the solvating ability of the medium.[29] The interaction of triethyltin vinyl with butyllithium in benzene solution[30] and the formation of complexes of $(CH_3)_3SnCH_2Cl$ with various polar aprotic solvents[31] have been studied by ^{119m}Sn Mössbauer techniques. The interaction of iodine with various organic solvents has been investigated by ^{129}I Mössbauer studies of frozen solutions.[32,33]

C. Kinetics

Nondestructive analysis of the solid phase can be carried out without interrupting the course of a reaction. Tin-119m Mössbauer spectroscopy has been used to study the diorganotin compounds $[R_2Sn]_n$ which contain tin–tin bonds (see Section I). These materials undergo slow oxidation on exposure to air; for example

$$[(n\text{-}C_4H_9)_2Sn]_n + \tfrac{1}{2}O_2 \longrightarrow [(n\text{-}C_4H_9)_2SnO]_n \qquad (20)$$

and the process has been followed by ^{119m}Sn Mössbauer spectroscopy as shown in Fig. 7.[34] The tin(II)-substituted carboranes constitute a genuine example of an organotin(II) species, and the spectra can be used to follow the slow oxidation which proceeds when an ethereal solution is exposed to air.[35] The hydrolysis of the five-membered tin(II) heterocycle o-phenylenedioxytin(II) has been followed by Mössbauer techniques which reveal that oxidation accompanies the hydrolysis.[36] The radiation-induced exchange between tetraphenyltin and tin(IV) iodide to produce component spectral lines due to $(C_6H_5)_nSnI_{4-n}$ has also been studied.[37] More recently the action of n-butyllithium on triethylvinyltin has been studied for various mole ratios.[38] The reaction of $FeCl_2$ with triazole derivatives has been studied by ^{57}Fe Mössbauer spectroscopy[39] as has the reduction of π-cyclopentadienyliron tricarbonyl iodide where the products were elucidated.[40]

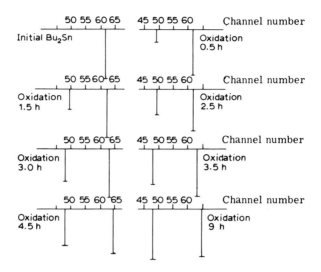

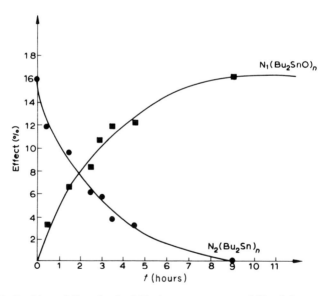

FIG. 7. Position of lines in the Mössbauer spectrum of $(Bu_2Sn)_n$ as a function of the time of its oxidation by air at room temperature. The value of the effect is proportional to the length of the line. The velocity value of the channel is 0.124 mm/second. Zero velocity relative to SnO_2 corresponds to channel No. 49.5.

D. Surface Studies

Mössbauer spectroscopy because of its sensitivity to the strength and angular distribution of binding and to the density of electrons at the nucleus can be used to study the nature and range of molecular forces in the surface zone. In experiments using refractory light element oxide substrates, the adsorbant will be transparent to γ rays, and the usual transmission geometry can be employed. Opaque adsorbants such as heavy metals can be studied by the use of back-scattering techniques now being intensively developed.[41] Results from studies of this type should aid in the understanding of adsorption, corrosion, and catalytic phenomena. The application of the Mössbauer effect to the study of surface phenomena has been reviewed.[42,43] In the area of organometallic chemistry the adsorption of tetramethyltin on γ-aluminum oxide or silica gel has been found to give rise to new resonances due to Sn–O bond formation, especially at elevated temperatures, while the spectrum of tetramethyltin adsorbed on activated carbon under similar conditions with temperatures up to 300°C is unchanged. Atmospheric oxygen was shown not to participate in these processes. The resonance of adsorbed liquid tetramethyltin can be observed at room temperature.[44,45] Iron(II) and iron(III) have been likewise studied absorbed on cotton,[45] on zeolites,[47–49] and on organic ion exchange resins.[50,51] Gold cyanide complexes have been studied in contact with magnesium oxide and alumina substrates by ^{197}Au Mössbauer spectroscopy.[52] The lattice dynamics of krypton in organic clathrates as studied by the Mössbauer effect in ^{83}Kr shows frequencies not elucidated by infrared measurements.[53]

E. Polymerization

The recoil-free fraction f' discussed in Section I.E.5 is related in the lattice dynamics of the solid to the displacement of the absorber Mössbauer atom from its equilibrium position. The value of f' should then also be a function of the binding force constants holding the absorber atom in the lattice, being larger the more rigid the lattice. Examples of compounds with high recoil-free fractions include many which are hard, refractory, or highly associated. The spectra of these compounds can often be recorded at room temperature. Organometallic compounds generally form molecular crystals, i.e., consisting of molecules which retain their identity when assembled into a solid held together by weak van der Waals and dipole–dipole forces. Two kinds of interactions are important in such solids—the binding of the Mössbauer atom in the molecule and the binding of the molecule in the solid. The force constant which enters into the magnitude of f' will in general depend upon contributions from both intra- and intermolecular forces. The temperature dependence of f' is determined primarily by intermolecular motions and the dependence is less severe in cases where intermolecular forces are strong.

Strong association, such as that found in dimethyltin diformate,[54] dioctyltin oxide,[55] and the associated heterocycle o-phenylenedioxytin(II),[36] gives rise to larger f' values. The organotin compounds for which Mössbauer spectra are obtained easily at room temperature have one feature in common: the tin atom is held in a polymeric lattice.[56] Compounds with higher values of f' such as bis(toluene-3,4-dithiolato)tin are known to form a network lattice with strong intermolecular bonds.[57,58] It is interesting that bis(1,2-ethanedithiolato)tin, in which the organic chelating group lacks the requirement for planarity, shows a much weaker resonance at room temperature than its toluene analogue,[59] as do the tin–oxygen heterocycles where additional substitution either on the organic group[36] or on the tin atom[60] apparently disrupts somewhat the original structure of the solid with its intermolecular binding. The data for the various dialkyltin oxides where bulkier alkyl groups are substituted can be understood in the same terms.[61,62]

Figure 8 shows the ^{57}Fe spectra of ferrocene (1) and two products of the polymerization of ferrocene, a soluble copolymer (2) and an insoluble polyferrocene. The spectrum of the soluble copolymer (2) is similar to that of ferrocene itself, a doublet with a strongly temperature-dependent f' value. The insoluble material has a spectrum which seems to be the result of the superposition of two doublets, the outer of which is similar to that of ferrocene itself. The weaker temperature dependence of the f' value for the inner doublet suggests that this feature can be assigned to bridged ferrocene groups. From the ^{57}Fe Mössbauer data it was possible to estimate the degree of cross-linking in the insoluble polymer.[63,64] The quantitative determination of the degree of cross-linking has been used in a study of ferrocene copolymers to evaluate new oxidizing agents for the transformation of ferrocene to ferrocinium. The degree of oxidation found from the ^{57}Fe Mössbauer data is much lower than that obtained from standard chemical redox methods, and it was concluded that the oxidation involves organic side chains and proceeds to some extent through polymer degradation. Only the appearance of ferrocinium iron would be detected by the Mössbauer experiment.[65]

As noted in Section I.E.6, the anisotropy of lattice binding can be detected in the Mössbauer spectrum as an asymmetry in the intensities of the component lines of doublet patterns. The Goldanskii–Karyagin effect can be detected even in randomly oriented polycrystalline powders. The strongly anisotropic recoil-free fraction most often arises from fastening the Mössbauer nucleus into polymer chains or planes. The one- or two-dimensional intermolecular binding is one of the typical examples of strongly anisotropic interactions in, for example, organotin chemistry where such linear or network association is now recognized as a common feature.[57] In addition, those

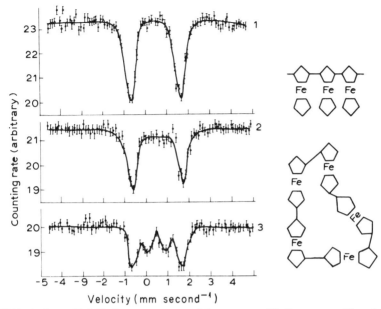

FIG. 8. Iron-57 Mössbauer spectra of ferrocenes: (*1*) ferrocene, (*2*) soluble polyferrocene, (*3*) insoluble polyferrocene.

atoms which when attached to tin produce the largest quadrupole splittings are just those which would enhance further association at tin. For example, the crystal structures of trimethyltin fluoride and hydroxide show that these compounds are associated into one-dimensional polymeric chains through fluorine and hydroxyl bridging to tin.[66,67] Likewise, the tin atoms in trimethyltin cyanide are pentacoordinated by bridging C≡N groups to form infinite chains along the crystal c axis. The X-ray anisotropic thermal parameters indicate that the mean square displacement of the tin atom is least along the chain axis, and the eccentricity of the vibrational ellipsoid is maximized in an intermediate range of temperature.[68] These studies have been extended to two-dimensional, polymeric dioctyltin oxide.[69,70] At high and low temperatures the thermal motion tends more toward a vibrational sphere, and the Goldanskii–Karyagin effect may vanish. In addition, the ratio of line intensities may reverse, as observed for triphenyltin hydroxide and trimethyltin cyanide, as a function of temperature.[70]

F. Radiation Damage

Chemical changes affecting the Mössbauer atom in organometallic compounds exposed to radiation can be detected by Mössbauer spectroscopy.

Often the details of the chemical change can be elucidated from the spectral changes. For example, irradiation of di-n-butyltin sulfate with ^{60}Co γ rays produces a material whose spectrum contains a new resonance in the tin(II) region which can be assigned to $SnSO_4$, indicating that fission of the tin–carbon bonds has occurred. Other spectral features have been assigned to (n-C_4H_9)$_2$SnSO$_3$. Irradiation in air results in oxidation to tin(IV) oxide.[71] Radiation damage in the polymer [—(n-C_4H_9)$_2$Sn(OCOCCH$_3$CH$_2$—)]$_n$ has also been followed by Mössbauer spectroscopy.[71] Irradiation of tetraphenyltin and tin(IV) iodide mixtures with 1.6-MeV electrons gives rise to a spectrum in which components due to $(C_6H_5)_n SnI_{4-m}$ can be identified.[72] The stabilization of carbon-chain polymers to oxidation by use of organotin compounds has been studied using dibutyltin dimaleate in polyethylene; the radiation damage from ^{60}Co γ rays can be followed by Mössbauer spectroscopy.[73]

Oxalate salts, because of their sensitivity to radiation, have potential for use as radiation dosimeters, and this possible application has motivated studies of the chemical effects of photolysis and radiolysis of these materials. Mössbauer spectroscopy is an ideal technique for following the course of the radiation-induced chemical changes, and tin(II) oxalate has been studied in this way.[74] After the iron(III) oxalate $Fe_2(C_2O_4)_3 \cdot nH_2O$ ($n = 5$) is irradiated with γ rays from a ^{60}Co source (2×10^8 R) or with the light output of a mercury vapor lamp its Mössbauer spectrum consists of four well-resolved resonances, two of which arise from the unchanged material, while the two peaks of lower intensity correspond to the spectral features of iron(II) oxalate. It was found that both γ radiolysis and photolysis yield qualitatively similar results. Quantitative estimates of dosage are apparently possible for such systems.[75] More recent studies reveal that n, γ reactions in iron(II) oxalate produce oxidation.[76–79]

Before closing this section it is necessary to mention the possible chemical consequences of the production of the decay of the Mössbauer nuclide itself. The tin Mössbauer source nuclide ^{119m}Sn is produced from ^{118}Sn by a process involving the capture of a thermal neutron. The nuclear process converts one isotope of tin to another. However, the ^{119m}Sn isomer can also be produced from the 38-hour half-life isotope, ^{119}Sb. Here one begins with a fifth group atom in a typical fifth-group chemical environment, so that the tin atom produced by the transmutation finds itself initially in a very strange situation. Indeed, iron Mössbauer γ rays arise from ^{57}Co decay, iodine γ's from ^{127}Te and ^{129}Te decay, xenon γ's from ^{129}I decay, etc. In each case the consequences of the chemical changes necessary to accommodate the newly formed atoms are of interest. Antimony γ rays arise from the decay of ^{121m}Sn, in turn produced from the thermal neutron irradiation of ^{121}Sn. In this case it is of interest to discover the final state of the antimony atom in its fourth-group environment.

The Mössbauer effect itself is ideally suited for the identification of the nonequilibrium "anomalous" charge states created by the processes discussed above. The case of ^{57}Fe Mössbauer spectroscopy is particularly favorable since characteristic I.S., Q.S., and magnetic hyperfine splittings have been established for many of the electronic configurations of iron. For example, such after-effects have been reported in the compounds cobalt(III) acetylacetonate[80] and cobalticinium tetraphenylborate.[81] In the latter case the decay produces a ferricinium ion spectrum which is accompanied by a component which differs significantly from that of ferrocene itself. Evidence for the production of new chemical entities is also found in the hydrated oxalates. The oxalate ion exhibits an important reducing effect in the (n, γ) hot atom chemistry of cobalt and iron compounds.[82] It has recently been shown that the decay of ^{119m}Sn in $K_6Sn_2(C_2O_4)_7 \cdot 4H_2O$ produces an anomalous charge state of tin,[83] as does the electron capture decay of ^{151}Gd^{3+} incorporated in $Er_2(C_2O_4)_3 \cdot 10H_2O$ produce the anomalous charge state ^{151}Er^{2+}.[83a] These anomalous charge states are difficult to detect in compounds of high electrical conductivity due to their extremely short lifetimes.

G. Spin States of Iron

Inorganic, high-spin compounds of iron with six identical ligands can be separated into the ferric, with electron configuration $3d^54s^0$ ($^6S_{5/2}$), and the ferrous, with $3d^64s^0(^5D)$. The ferric high-spin compounds have a spherically symmetric half-filled shell which would give no contribution to the electric field gradient. Typically, small quadrupole splittings are observed which are primarily due to the location of cations and anions in the crystal lattice. The major contribution to the field gradient in the ionic ferrous compounds comes from the extra electron, which gives rise to large quadrupole splittings.

For six-coordinated organometallic iron compounds of low spin the situation is reversed. In this case the iron(III) compounds show a larger Q.S. than analogous iron(II) compounds since in covalent ferrous compounds the effective atomic number of iron becomes 36, while in octahedral ferric compounds the d-orbital vacancy makes a contribution to the electric field gradient giving rise to a small Q.S., as in the FeA_6^{-3} compounds. Ligand substitution gives rise to larger Q.S. values, again with Fe(III) > Fe(II) for analogous compounds.

The interesting five-coordinated bis(N,N-dialkyldithiocarbamato)iron(III) halides contain a trivalent iron atom whose ground electronic term is an orbital singlet and spin quartet. The quartet ground term results from the low symmetry (C_{2v}) local environment of the iron. Several complexes of the general formula $Fe(S_2CNR_2)_2X$ have been studied by Mössbauer spectroscopy at temperatures down to 1.2°K where the bromide gives a simple doublet, while the chlorides give multiline patterns. The doublet lines of the bromide derivative split further into a triplet–doublet pattern when subjected to an external magnetic field. This behavior is indicative of the combined effect of a large, randomly oriented magnetic hyperfine field, together with a quadrupole interaction. In the chlorides, on the other hand, the R = CH_3 compound is paramagnetic with a well-defined magnetic hyperfine splitting. Line broadening is observed due to relaxation among electronic levels. For the R = C_2H_5 derivative, the spectrum is characteristic of ferromagnetic order at the low temperatures employed. When R = i-C_3H_7 the compound is paramagnetic. In this study the Mössbauer parameters were used in conjunction with data for X-ray crystallography, magnetic susceptibility, and single crystal electron paramagnetic resonance studies.[84]

Several complex iron-bearing biological materials have been studied to determine the number and type of iron spin environments. For example, it appears that the seven iron atoms in the electron-transport protein ferredoxin are distributed in two nonequivalent, but closely similar environments in a 2:5 ratio with all the iron low-spin iron(III),[85] not in the Fe^{2+}, Fe^{3+}, or in Fe(VI) states as had been suggested. Hemoglobin has been shown to undergo a change in spin state when it takes up oxygen to become oxyhemoglobin.[86] The change from high-spin to low-spin ferrous can be monitored by use of the Mössbauer spectra.

H. The Sign of the Quadrupole Coupling Constant

In the usual Mössbauer experiment conducted in the absence of external fields the magnitude, but not the sign of the quadrupole coupling constant, e^2qQ, is obtained from the Q.S. It is of interest to determine the sign of e^2qQ since theoretical descriptions of the bonding yield predictions of this sign, and it is frequently the case that different descriptions yield opposite predictions.

A detailed study of the sign of the quadrupole coupling constant in ferrocene was carried out by recording the ^{57}Fe Mössbauer spectrum in the presence of the 40-kG field produced by a superconducting magnet at 4.2°K. The normal doublet spectrum of ferrocene shown in Fig. 9 was further split under these conditions into a lower-energy triplet and a higher-energy doublet resulting from the removal of the 3d-orbital degeneracy. That the triplet is at lower energy indicates that e^2qQ is positive,[87] in agreement with the molecular orbital calculations of Dahl and Balhausen.[88] A crystal field treatment yields a negative sign of e^2qQ.

Reports of the sign of the quadrupole interactions in tin(IV)[89-92] and tin(II)[93] compounds have recently appeared. The signs in trimethyltin fluoride and hydroxide are the same and indicate that π bonding is of little importance in influencing either the magnitude or the sign of the Q.S.[90] Calculated

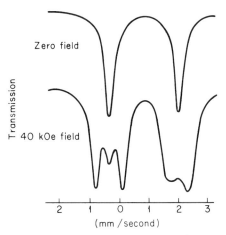

FIG. 9. Schematic effect of splitting the quadrupolar spectrum of ferrocene by a 40-kG magnetic field.

spectra for several situations in which quadrupole splitting and applied magnetic field at tin are varied have been published.[94]

I. Oxidation State

For iron and tin compounds the magnitude of the I.S. has been found to be related to the oxidation state of the element. For tin the separation of the I.S. values in compounds formally classified as Sn(IV) and Sn(II) about the value of white tin has been applied with benefit to several types of organotin systems. In iron compounds of both Fe(II) and Fe(III) the spin of the 3d electrons can be predominantly paired (low spin) or unpaired (high spin). Knowledge of the spin state of the iron allows assignment of the oxidation state on the basis of the magnitude of the I.S. only if the iron is high spin since for low spin compounds the I.S. values for the two oxidation states overlap.

Genuine examples of organic derivatives of tin(II) are very few.[95] The so-called diorganotin compounds, known for over a century as examples of tin(II) species, all have I.S. values below white tin.[96] Their structure shows that polymerization has taken place to yield tin(IV) species[95]:

$$n\mathrm{R_2Sn} \longrightarrow -[-\underset{\underset{R}{|}}{\overset{\overset{R}{|}}{\mathrm{Sn}}}-]_n- \qquad (21)$$

Larger substituent groups might slow down the rate of polymerization, as in the tin(II)-substituted carboranes.[35] The chemistry of the curious, angular sandwich compound $(C_5H_5)_2Sn^{97-99}$ has recently been investigated.[100] Its Mössbauer spectrum contains a small doublet with an I.S. greater than that of white tin. Polymerization or oxidation converts the tin to tin(IV)[100]:

$$n(C_5H_5)_2Sn \longrightarrow -[(C_5H_5)_2Sn]_n- \qquad (22)$$

$$(C_5H_5)_2Sn \xrightarrow{O_2} -[-(C_5H_5)_2SnO-]_n- \qquad (23)$$
$$[(C_5H_5)_2Sn]_n \xrightarrow{O_2}$$

as does reaction with Grignard reagents[101]:

$$2(C_5H_5)_2Sn + 2RMgX \longrightarrow [R(C_5H_5)_2SnMgX]_2 \qquad (24)$$

or disproportionation with mercury(II) chloride[102]:

$$(C_5H_5)_2Sn + HgCl_2 \longrightarrow Hg + (C_5H_5)_2SnCl_2 \qquad (25)$$

or its action as a carbeneoid species with dimethylacetylenedicarboxylate.

$$2(C_5H_5)_2Sn + 2RC{\equiv}CR \longrightarrow \begin{array}{c} H_5C_5 \quad C_5H_5 \\ R\diagdown Sn \diagup R \\ \\ R \diagup Sn \diagdown R \\ C_5H_5 \quad C_5H_5 \end{array}$$

However, in other reactions the tin(II) species utilizes its lone pair electrons to act as a Lewis base:

$$(C_5H_5)_2Sn + BF_3 \xrightarrow{(Ref.\ 103)} (C_5H_5)_2Sn \longrightarrow BF_3 \qquad (26)$$

$$2(C_5H_5)_2Sn + 2FeCl_3 \xrightarrow{(Ref.\ 102)} [(C_5H_5)_2SnFeCl_3]_2 \qquad (27)$$

or undergoes displacement reactions in which the cyclopentadienyl groups are transferred[102]:

$$(C_5H_5)_2Sn + 2(CH_3)_3GeCl \longrightarrow SnCl_2 + 2(CH_3)_3GeC_5H_5 \qquad (28)$$

$$(C_5H_5)_2Sn + FeCl_2 \longrightarrow SnCl_2 + (C_5H_5)_2Fe \qquad (29)$$

or insertion into the C_5H_5–Sn system takes place:

$$(C_5H_5)_2Sn + 2SO_2 \longrightarrow Sn(O\overset{\overset{O}{\|}}{S}C_5H_5)_2 \qquad (30)$$

preserving the tin as tin(II).

In contrast to Eq. (27) above, the reaction of tin(II) halides with transition metal halides has been used to produce a vast number of new compounds variously classified as containing tin(II) donor groups or tin(IV)–metal bonds.[95,104,105] The I.S. values for these compounds indicate that these compounds are formed via insertion of the tin(II) species into the metal–halogen bond, and all can be classified as tin(IV).[106] Apparently π interactions are not important in these systems.[107]

Tin porphyrin derivatives have been studied as model systems in an attempt to elucidate the mechanism of metal incorporation into the porphyrin ring. The incorporation of tin starting with tin(II) materials results in a final product containing a single tin(IV) atom in a sandwich arrangement with two porphyrin rings[108] similar to the analogous phthalocyanine derivative, Pc_2Sn, whose structure[109] and Mössbauer spectrum[110-112] are known. Several other tin(IV) porphyrin compounds where ligands are attached in a *trans* arrangement at tin have been studied,[112] and the synthesis of a genuine tin(II) porphyrin[113] has recently been confirmed by Mössbauer techniques.[114]

J. Applications to Biological Systems[115]

The Mössbauer spectrum of a living thing, the hydrogeneous bacterium *B. Hydrogenomonas Z-1*, has been recorded.[116] The bacteria were grown in an iron(III) citrate solution enriched in ^{57}Fe. The spectrum contains components due to both iron valence states, a doublet arising from Fe(II) and a singlet from the Fe(III) present in the cytochrome system of these bacteria, despite only $^{57}Fe(III)$ having been introduced into the medium.

The presence of iron in biological material opens the possibility of using ^{57}Fe Mössbauer spectroscopy to elucidate the formal oxidation state of the iron, the strength of binding of iron to neighboring atoms, the amplitude of thermal motion in various directions in bound iron and the metabolism of iron in organisms where ^{57}Fe can be used as a tracer atom. A great difficulty arises because of the miniscule concentration of iron atoms in biological materials of very high molecular weight. Enrichment in the Mössbauer isotope is almost always necessary, but this can be carried out by the species itself *in vivo*. In this way, by injecting rats with $^{57}FeCl_3$ the iron-57 content of their hemoglobin was raised to 6% in one week. The effect of changing various ligands bound to the iron atoms in the blood could then be studied. The results are shown in Fig. 10. Spectra (a) and (b) are characteristic of oxyhaemoglobin, whereas in (c), (d), and (e) other ligands are attached to the haemoglobin molecule. The similarity of spectra (d) and (e) was interpreted to suggest that the carbon dioxide was bound to another part of the molecule.[117]

A change in the oxidation state of iron during the fixation of nitrogen by azobacteria has recently been observed[118] by monitoring the Mössbauer spectra.

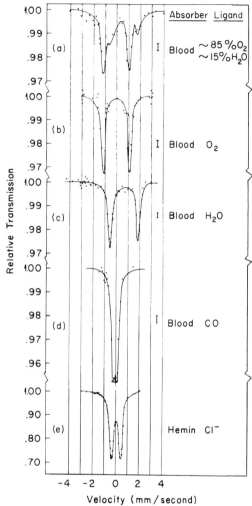

FIG. 10. Mössbauer spectra obtained by using a room-temperature ^{57}Co in platinum source and the following absorbers at 5°K. (a) Venous rat red cells; (b) human red cells saturated with O_2 gas; (c) human red cells saturated with N_2 gas; (d) human red cells saturated with CO gas; (e) crystalline hemin.

K. Structural Determinations

Spectroscopic methods are the usual choice of organometallic chemists seeking to elucidate structural formulas as these techniques are generally fast, give high precision data, and are easily automated. Such savings of work and time, however, are often accompanied by substantial ambiguities in the interpretation of the acquired data, so that a combination of various spectroscopic studies is required, almost without exception, for adequate structural characterization. Mössbauer spectroscopy is an exception to these generalizations only in that its data acquisition time is rather long.

The spectral parameters of primary interest in the application of Mössbauer spectroscopy to the determination of molecular structure are the isomer shift (I.S.), the quadrupole splitting (Q.S.), the recoil-free fraction (f') and for ^{119m}Sn spectroscopy, the quotient of the quadrupole splitting divided by the isomer shift [i.e., ρ = Q.S./I.S.; I.S. relative to ^{119m}SnO$_2$ or Ba^{119m}SnO$_3$[119]]. The full-width at half-maximum absorption (Γ) is primarily a nuclear characteristic and is of little use in this context although, in principle, it could be perturbed by chemically induced variations in nuclear relaxation times. The presence of several similar chemical sites for the Mössbauer nucleus may lead, however, to an observed line broadening if the differences between the isomer shifts of these sites are small compared to Γ. Motional narrowing has recently been used to explain a large decrease in the linewidths of the ^{57}Fe spectrum of tris(acetylacetonato)iron(III) [Fe(acac)$_3$] upon its exposure to substantial doses of ^{60}Co γ radiation.[120] This decrease in Γ from 1.86 to ≈ 0.65 mm/second is reportedly due to the partial conversion of the γ-irradiated Fe(acac)$_3$ to [Fe(acac)$_3$]$^-$ with the resultant extra electron providing an efficient relaxation mechanism through its ability to jump between adjacent complexes.

The magnitude of the recoil-free fraction of organotin compounds has been found not to correlate in any simple way with such properties as melting or boiling points, ligand mass, coordination number, nor with molecular weight.[121] The temperature dependence of f', on the other hand, is generally suppressed for polymeric compounds and variable temperature studies may therefore allow an estimation of the degree of polymerization. Dimethyltin difluoride, for example, exhibits a pronounced ^{119m}Sn Mössbauer doublet resonance at ambient temperatures (Fig. 11), and its structure is known to consist of a polymeric network of (CH$_3$)$_2$SnF$_4$ octahedra joined through bridging fluorine atoms[122]; dimethyltin dichloride, on the other hand, exhibits extremely weak ^{119m}Sn resonance under similar conditions in accord with the loosely bound polymeric structures surmised from the results of both X-ray diffraction[123,124] and ^{35}Cl nuclear quadrupole resonance[124-126] investigations. Additional factors, such as increased coordination about the tin atom or the presence of strongly ionic lattices, may also diminish the temperature dependence of f' and their potential contributions should always be considered.

The number of resolvable resonances may be used to establish the number of differing chemical sites of the Mössbauer nuclei, and the relative intensity (ε) of each such resonance may be used to infer the distribution of the Mössbauer

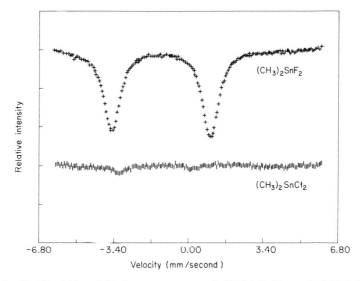

FIG. 11. Tin-119m Mössbauer spectra of $(CH_3)_2SnF_2$ and $(CH_3)_2SnCl_2$ recorded at ambient temperature with a $Ba^{119m}SnO_3$ source.

nuclei among these sites. The failure to account appropriately for hyperfine splittings may lead to an incorrect determination of the number of sites, but the recognition of such interactions is often possible through the application of strong external magnetic fields or, more commonly, by intuitive means. The exact distribution among the sites is not directly available from the observed resonance intensities since ε, a bulk property, is a function of both the number of nuclei and the chemical forces present at the site. Variations of the latter type alone may cause a significant change in the observed resonance intensities (e.g., by a factor of 2).

The isomer shift is a useful parameter for structural determinations insofar as it may indicate the formal oxidation state of the Mössbauer atom. The relative overall nuclear electron density also reflects, to some extent, the coordination number about the Mössbauer nucleus; for example, the isomer shifts for both the organotin(II) and -tin(IV) compounds are found generally to decrease upon complexation.[1] Similarities in the isomer shifts (along with other data) have recently been used to infer structural similarities among the stannous halide complexes of the form $SnX_2 \cdot L$ (X = Cl, Br; L = piperazine, 2-morpholine, 2-piperidine) which are considered to be polymeric with the second ligand molecule in the 1:2 morpholine and piperidine complexes at least weakly bonded to the central tin atom.[127] A comparison of the isomer shifts of trimethyltin sulfate, selenate, and chromate with those of trimethyltin cyanide, isothiocyanate, acetate, and assorted haloacetates has likewise been used to infer five-coordinate polymeric structures for the first three compounds.[128] Such comparisons can be ambiguous, though, and care must be taken so as not to overinterpret the data.

Perhaps the most useful parameter for drawing structural conclusions from the Mössbauer experiment is the quadrupole splitting. The electric field gradient present at the nuclear site under investigation is in molecules partly due to the spatial distribution of the surrounding electrons and should, therefore, reflect the bonding geometry. The precise electronic charge distribution actually varies both in space and with time, but the time dependency is expected to average over the duration of the Mössbauer event and may thus be disregarded. The calculation of the gradient may be attempted by either classical or quantum methods. In the second approach, the expectation value of the electric field gradient operator ($\mathcal{H}_q$) is evaluated over the total molecular wavefunction (ψ_m)

$$(V_{zz})_{\text{electronic}} = \sum_{\text{electrons}} \int \psi_m(\mathcal{H}_q)\psi_m^* \, d\tau \tag{31}$$

where $\mathcal{H}_q = (3\cos^2\theta - 1)/r^3$, θ and r represent spherical coordinates of the electron, and τ represents all space. The summation is carried out over all the electrons since $\mathcal{H}_q$ is a one-electron operator. The nuclear contributions (from all of the nuclei in the molecule) to V_{zz} must then also be included, but they are more readily obtainable by classical methods. The procedure seems straightforward, but difficulties immediately arise in approximating the required molecular wavefunctions. Sternheimer shielding or antishielding parameters[129] along with appropriate screening constants are invariably needed to describe the complex molecular electron distributions with still manageable functions. Calculations of this sort are, in practice, quite imprecise; an excellent summary of approximate methods of evaluating the electric field gradient has recently become available.[130]

In addition to the intramolecular contributions, one must also consider the extramolecular contributions to the electric field gradient, particularly in the case of ionic compounds. Usually little is known about the relative magnitudes of these interactions; conventional lattice summations are sometimes carried out on the basis of point-charge approximations with partial success. Such summations are exceedingly difficult to perform with precision because of the need to include both permanent dipole moment and variable molecular polarizability terms. The extramolecular contribution may again be significantly altered by distortions induced in the molecular wavefunctions described by Sternheimer antishielding factors.

Recent attempts at interpretation of the quadrupole splitting in terms of molecular structure have generally assumed that the gradient is principally determined by the immediate electronic environment of the Mössbauer nucleus and, almost by necessity, have ignored any variations of the Sternheimer factors with changes in structure. It should be noted here that most organometallic compounds do not form ionic lattices. These assumptions have led to simple models in which the atoms directly bonded to the Mössbauer atom are approximated by point charges of characteristic magnitude (Q_i) and distance (r_i) from the Mössbauer nucleus.[131-135,173] On the basis of these models the quadrupole splitting of the essential octahedral *trans-* and *cis-* A_2MB_4 isomers are predicted to be in the ratio of 2:1, respectively. Such *trans* geometry in R_2SnB_4 complexes (where R = organic group) leads to splittings of ≈4 mm/second, whereas the *cis* geometry leads to a splitting of ≈2 mm/second. This rule of thumb has been used, for example, to indicate a *trans* arrangement of hydrocarbon groups in di-*n*-butyl(dipicolinato)tin(IV) and divinyl(dipicolinato)tin(IV) and a *cis* configuration in diphenyl(dipicolinato)tin(IV),[136] where the observed quadrupole splittings were 4.35, 4.02, and 1.94 mm/second, respectively. Some representative results[134] of the point-charge model approach for various structures are given in Figs. 12-17 where the parameter P_i is defined by $P_i \equiv Q_i/r_i^3$.

Whether the ^{119m}Sn quadrupole splitting is primarily induced by ligand electronegativity differences[137-140] or by ligand-tin ($p \rightarrow d)\pi$ interaction differences[119,142,143] has been the subject of lively controversy. On the one hand, there is the apparent (although thus far limited) success of the point-charge model for predicting the relative magnitudes of the quadrupole splitting for various geometries, but on the other hand, there is the observa-

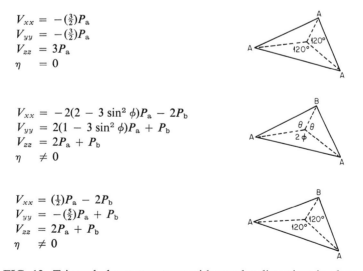

FIG. 12. Trigonal planar structures with angular distortions in plane.

$V_{xx} = 2(2 - 3\sin^2\theta)P_a$
$V_{yy} = 2(2 - 3\sin^2\theta)P_a$
$V_{zz} = 4(1 - 3\cos^2\theta)P_a$
$\eta = 0$

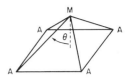

$V_{11} = -6(1 - \sin^2\phi)P_a - 2P_b$
$V_{22} = 3(1 - 2\sin^2\phi)P_a + P_b$
$V_{33} = 3P_a + P_b$
$\eta \neq 0$

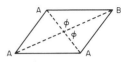

$V_{11} = 2P_a + 2(1 - 3\sin^2\theta)P_a$
$V_{22} = 2(1 - 3\sin^2\phi)P_a + 2P_b$
$V_{33} = 2(1 - 3\cos^2\phi)P_a + 2(1 - 3\cos^2\theta)P_b$
$\eta \neq 0$

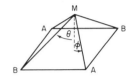

$V_{xx} = -2(2 - 3\sin^2\theta)P_a - 2(2 - 3\sin^2\phi)P_b$
$V_{yy} = 2(1 - 3\sin^2\theta)P_a + 2(1 - 3\sin^2\phi)P_b$
$V_{zz} = 2P_a + 2P_b$
$\eta \neq 0$

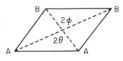

FIG. 13. Distorted square planar structures.

$V_{xx} = (5 - 6\sin^2\theta)P_a$
$V_{yy} = V_{xx}$
$V_{zz} = -2(5 - 6\sin^2\theta)P_a$
$\eta = 0$

$V_{xx} = 2(2 - 3\sin^2\theta)P_a + P_b$
$V_{yy} = V_{xx}$
$V_{zz} = -4(2 - 3\sin^2\theta)P_a - 2P_b$
$\eta = 0$

$V_{xx} = 2(1 - 3\sin^2\theta)P_a + 3P_b$
$V_{yy} = 2P_a + 3(1 - 2\sin^2\phi)P_b$
$V_{zz} = 2(1 - 3\cos^2\theta)P_a - 6\cos^2\phi P_b$
$\eta \neq 0$

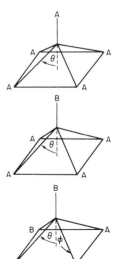

FIG. 14. Various distorted five-coordinated structures.

$$V_{xx} = 2P_a^{ax} - 6\cos^2\phi P_a^{eq}$$
$$V_{yy} = 2P_a^{ax} + 3(1 - 2\sin^2\phi)P_a^{eq}$$
$$V_{zz} = -(4P_a^{ax} - 3P_a^{eq})$$
$$\eta \neq 0$$

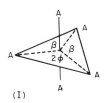

(I)

$$V_{xx} = P_a^{ax} - 6\cos^2\phi P_a^{eq} + P_b^{ax}$$
$$V_{yy} = P_a^{ax} + 3(1 - 2\sin^2\phi)P_a^{eq} + P_b^{ax}$$
$$V_{zz} = -2P_a^{ax} + 3P_a^{eq} - 2P_b^{ax}$$
$$\eta \neq 0$$

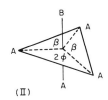

(II)

$$V_{11} = 2P_a^{ax} + 2(1 - 3\cos^2\phi)P_a^{eq} - 2P_b^{eq}$$
$$V_{22} = 2P_a^{ax} + 2(1 - 3\sin^2\phi)P_a^{eq} + P_b^{eq}$$
$$V_{33} = -4P_a^{ax} + 2P_a^{eq} + P_b^{eq}$$
$$\eta \neq 0$$

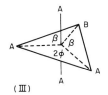

(III)

$$V_{xx} = -6\cos^2\phi P_a^{eq} + 2P_b^{ax}$$
$$V_{yy} = 3(1 - 2\sin^2\phi)P_a^{eq} + 2P_b^{ax}$$
$$V_{zz} = 3P_a^{eq} - 4P_b^{ax}$$
$$\eta \neq 0$$

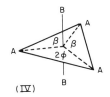

(IV)

$$V_{11} = P_a^{ax} + 2(1 - 3\cos^2\phi)P_a^{eq} + P_b^{ax} - 2P_b^{eq}$$
$$V_{22} = P_a^{ax} + 2(1 - 3\sin^2\phi)P_a^{eq} + P_b^{ax} + P_b^{eq}$$
$$V_{33} = -2P_a^{ax} + 2P_a^{eq} - 2P_b^{ax} + P_b^{eq}$$
$$\eta \neq 0$$

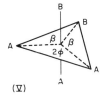

(V)

$$V_{11} = 2P_a^{ax} - 2P_a^{eq} + 2(1 - 3\cos^2\phi)P_b^{eq}$$
$$V_{22} = 2P_a^{ax} + P_a^{eq} + 2(1 - 3\sin^2\phi)P_b^{eq}$$
$$V_{33} = -4P_a^{ax} + P_a^{eq} + 2P_b^{eq}$$
$$\eta \neq 0$$

(VI)

FIG. 15. Trigonal bipyramids with angular distortions in equatorial plane.

$V_{xx} = 6(1 - \sin^2 \theta)P_a$
$V_{yy} = V_{xx}$
$V_{zz} = -12 \cos^2 \theta P_a$
$\eta = 0$

$V_{xx} = (5 - 6\sin^2 \theta)P_a + P_b$
$V_{yy} = V_{xx}$
$V_{zz} = 2(1 - 6\cos^2 \theta)P_a - 2P_b$
$\eta = 0$

$V_{xx} = 2(2 - 3\sin^2 \theta)P_a + 2P_b$
$V_{yy} = V_{xx}$
$V_{zz} = 4(1 - 3\cos^2 \theta)P_a - 4P_b$
$\eta = 0$

FIG. 16. Octahedra with metal above or in plane.

$V_{xx} = (1 - 6\cos^2 \phi)P_a - P_b$
$V_{yy} = P_a - P_b$
$V_{zz} = -2(1 - 3\cos^2 \phi)P_a + 2P_b$
$\eta \neq 0$

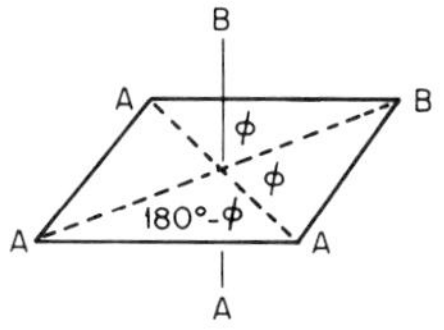

$V_{xx} = -6\cos^2 \phi P_a$
$V_{yy} = 3(1 - 2\sin^2 \phi)P_a + 3P_b$
$V_{zz} = 3(P_a - P_b)$
$\eta \neq 0$

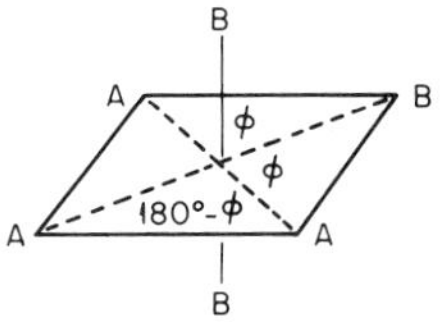

$V_{11} = 3(1 - 2\cos^2 \theta)P_a + 3(1 - 2\cos^2 \phi)P_b$
$V_{22} = 3(1 - 2\sin^2 \theta)P_a + 3(1 - 2\sin^2 \phi)P_b$
$V_{33} = 0$
$\eta \neq 0$

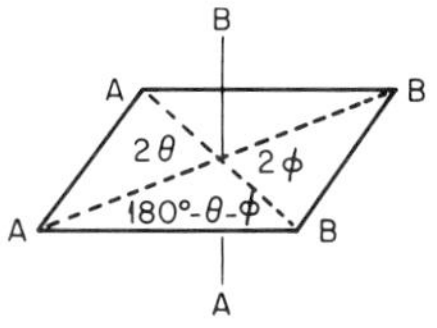

FIG. 17. Octahedra with angular distortion in plane.

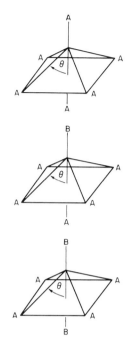

tion that a well-resolved quadrupole splitting is almost never present in ^{119m}Sn spectroscopy except when the tin atom is bonded simultaneously to ligands both with and without lone pair electrons. This latter observation has become widely formalized as "Greenwood's Rule".[142,143] The controversy promises to continue, although it is becoming evident that neither explanation gives a sufficient and necessary condition for a resolvable splitting.[1]

In ^{57}Fe spectroscopy, the 3d-orbital populations of the iron atom will depend upon whether it is high or low spin, tetrahedral or octahedral, or ferrous or ferric. The relative contributions of the 3d orbitals to the electric field gradient are easily obtained through the use of the electric field gradient operator and the appropriate hydrogenic wavefunctions. The results are summarized in Fig. 18.

One may then, in principle, calculate a quadrupole splitting for any compound for which the electronic configuration of the Mössbauer atom is known. The reverse procedure—characterization of the electronic configuration from the spectral splitting—is more difficult as a particular value of the electric field gradient is not unique to any one configuration. In addition, the results shown in Fig. 18 disregard a large number of additional factors known to contribute to the magnitude of the gradient.

Early in the history of ^{119m}Sn spectroscopy it was pointed out that a "synthetic" parameter ρ, defined by

$$\rho \equiv Q.S./I.S.$$

seemed to correlate to a large extent with coordination number about the tin atom.[21] Compounds with $\rho < 1.8$ usually contain four-coordinated tin atoms while in those with $\rho > 2.1$ tin was five- or six-coordinated. No theoretical significance is attached to this parameter, though, and the observed correlation evidently reflects the independent observations that the quadrupole splitting generally increases and that the isomer shift generally decreases with increasing coordination number. Although not strictly true, (and although an undefined gap is left in the range $1.8 \leqslant \rho \leqslant 2.1$), this rule of thumb is

	V_{xx}	V_{yy}	V_{zz}
$3d_{z^2}$	$-2/7$	$-2/7$	$+4/7$
$3d_{x^2-y^2}$	$+2/7$	$+2/7$	$-4/7$
$3d_{xy}$	$+2/7$	$+2/7$	$-4/7$
$3d_{xz}$	$+2/7$	$-4/7$	$+2/7$
$3d_{yz}$	$-4/7$	$+2/7$	$+2/7$

FIG. 18. Relative contributions of the 3d orbitals to the electric field gradient.

6. Mössbauer Spectra 273

often invoked in support of higher coordination in organotin(IV) chemistry. The isomer shift used in the calculation of ρ is relative to SnO_2 or $BaSnO_3$.

Additional specific applications of Mössbauer spectroscopy now follow to demonstrate the particular problems in structural chemistry to which this technique has been applied. Along with the more common spectroscopic methods, the Mössbauer experiment may often be used to eliminate some structural possibilities; it does not, except in few instances, give sufficient information to characterize unambiguously the actual molecular configuration. The Mössbauer technique is hampered by the same basic interpretative difficulties which have severely limited the usefulness of pure nuclear quadrupole resonance spectroscopy in this application.

The elucidation of the structure of triiron dodecacarbonyl $[Fe_3(CO)_{12}]$ presents one of the most interesting examples,[144] clearly illustrating typical problems encountered by the Mössbauer spectroscopist. The two linear geometries of Fig. 19 were initially proposed on the basis of freezing point depression studies in iron pentacarbonyl as the solvent, which had found the material to be a trimer of $Fe(CO)_4$.[145] Additional support for these proposals was generated by data gathered through infrared spectroscopic,[146,147] kinetic,[148] and preliminary X-ray diffraction[149,150] experiments. Although the diamagnetism of $Fe_3(CO)_{12}$[151-153] was initially thought to be inconsistent with a so-called 3-3-3-3 configuration,[146] subsequent molecular orbital calculations showed that such a ground state could be allowed for this model.[154] An X-ray diffraction study published in 1957 claimed, however, that such a linear model could not account for the observed data[155] and went on to suggest a structure with an equilateral triangle of iron atoms involving some sort of iron–iron bridging through the carbonyls.[156] The

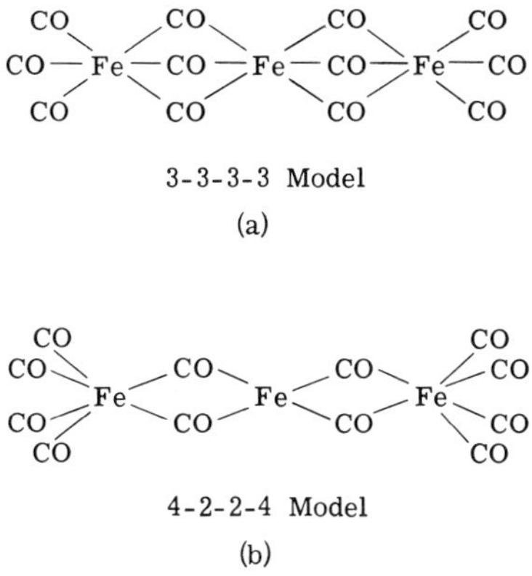

3-3-3-3 Model

(a)

4-2-2-4 Model

(b)

FIG. 19. Two linear structures initially proposed for $Fe_3(CO)_{12}$.

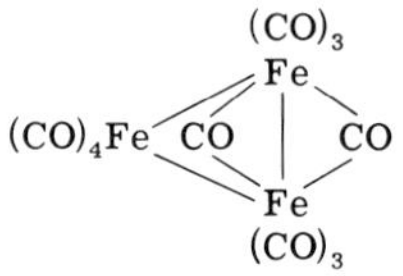

```
        (CO)₃
        /Fe\
(CO)₄Fe—CO  |  CO
        \Fe/
        (CO)₃
```

FIG. 20. The trimeric-type structure of $Fe_3(CO)_{12}$.

^{57}Fe Mössbauer spectrum of $Fe_3(CO)_{12}$, on the other hand, exhibits three distinct resonances of equal intensity which were assigned as a singlet surrounded by the two wings of a doublet with twice as much of the iron in the site experiencing an electric field gradient.[157-164] The apparent presence of two distinct, chemically different iron sites, one with cubic symmetry, led the investigators to conclude that only a linear model (such as the 4-2-2-4 or the 3-3-3-3 model) would be consistent with the data. Triangular models were discarded since it was not expected that any site in such a structure would possess sufficient symmetry to give rise to a singlet resonance. Subsequently, both Mössbauer and infrared data for the related $[HFe_3(CO)_{11}]^-$ anion suggested an analogous unsymmetrical structure.[165,166] The known structure of $Fe_3(CO)_{11}P(C_6H_5)_3$[167,168] also lent support to this suggestion. The validity of this model (Fig. 20) was subsequently confirmed by a detailed single crystal X-ray investigation of triiron dodecacarbonyl.[144] The apparent lack of an electric field gradient at the unique iron site has recently been clarified by two studies in which small quadrupole splittings (≈ 0.15 mm/second) were resolved at both ambient and liquid nitrogen temperatures.[169,170] The reason for such a small gradient at this site is as yet unclear. The large doublet to the pseudooctahedral $Fe_2Fe(CO)_4$ group. Recent Mössbauer studies of the isoelectronic, anionic, polynuclear iron carbonyls of the form of the isoelectronic, anionic, polynuclear iron carbonyls of the form $[(C_2H_5)_4N][MFe_2(CO)_{12}]^-$, where M = Mn, Tc, or Re, indicate analogous structures with the unique iron atom of $Fe_3(CO)_{12}$ substituted in these compounds by the second metal and apparently confirm the previous assignments of the Mössbauer resonances in $Fe_3(CO)_{12}$.[169] The iron in these compounds gives rise to a splitting of ≈ 0.90 mm/second, in contrast to the value 0.43 mm/second observed for $Mn_2Fe(CO)_{14}$ which is known to consist of a planar iron tetracarbonyl inserted between two axial manganese pentacarbonyls.[171] The ^{57}Fe spectrum of $Na_2[Fe(CO)_4]$, on the other hand, is composed of a single absorption, in accord with a tetrahedral anion, while $Fe(CO)_5$ shows a distinct doublet, as would be expected from a trigonal bipyramidal configuration about the iron atom.[159-162,164]

Mössbauer spectroscopy has also played an important role in the structural investigation of ferricinium ferrichloride $[(C_5H_5)_2Fe_2Cl_4]$ for which two structures had been proposed,[172] the ion pair, $[(C_5H_5)_2Fe^{III}]^+[Fe^{III}Cl_4]^-$, and the nonionic dimer,

$$\begin{array}{ccc} C_5H_5 & Cl & C_5H_5 \\ \diagdown & \diagup \diagdown & \diagup \\ & Fe^{III} \quad Fe^{III} & \\ \diagup & \diagdown \diagup \diagdown & \\ Cl & Cl & Cl \end{array}$$

The ^{57}Fe spectrum was found to be a singlet, in better agreement with the latter structure since the ionic form would be expected to exhibit the absorptions of two distinct iron sites. Again the failure to resolve a quadrupole splitting could be explained on the basis of its small magnitude. However, ferricinium ferrichloride prepared from ^{57}Fe-enriched iron(II) chloride and unenriched ferrocene gave only unenriched ferrocene on decomposition, showing the ionic structure to be the only possible formulation.[174] In order for a singlet Mössbauer resonance to arise from such a structure, the cation and anion must not only possess identical isomer shifts, but must also fail to experience any significant electric field gradient. More recently two resonances have been resolved from the previous single absorption.[175]

These difficulties are not unique to ^{57}Fe spectroscopy. The crystal structure of the addition complex of dimethyltin dichloride and 2,2′,2″-terpyridyl shows the compound properly to be formulated as

$$[(CH_3)_2SnCl \cdot \text{terpyridyl}]^+[(CH_3)_2SnCl_3]^-.^{176,177}$$

The cation contains the tin atom in a six-coordinated octahedral configuration (2,2′,2″-terpyridyl has three coordination sites), whereas the anion contains the tin atom in a five-coordinated trigonal bipyramidal configuration. Both ions are distorted from the perfect geometries. Although the ^{119m}Sn Mössbauer spectrum was discussed initially in terms of a simple doublet,[178] it can be decomposed by computer techniques into two distinct, but overlapping doublets[179] as shown in Fig. 21. Both tin sites exhibit surprisingly similar values of isomer shift and quadrupole splitting; the calculated intensities differ, however, by a factor of 2, indicating the variability of recoil-free fractions with chemical sites since the two sites are equally populated. The structure of the presumably analogous

$$[(C_6H_5)_2SnCl \cdot \text{terpyridyl}]^+[(C_6H_5)_2SnCl_3]^-$$

has not been investigated, but its ^{119m}Sn Mössbauer spectrum is similar.

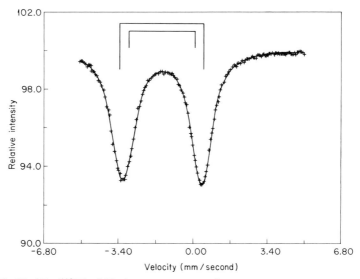

FIG. 21. The ^{119m}Sn Mössbauer spectrum of

$$[(CH_3)_2SnCl \cdot terpy]^+ [(CH_3)_2SnCl_3]^-$$

recorded at 80°K with the isomer shift relative to Ba^{119m}SnO$_3$ at room temperature.

The resonances of the cation and anion again overlap almost completely, and in this case, the observed intensities for the two sites are equal. A somewhat different complex results from n-butyltin trichloride with 2,2′,2″-terpyridyl; the complex is formulated as $[n\text{-}C_4H_9SnCl_2 \cdot terpyridyl]_2^+ \cdot [n\text{-}C_4H_9SnCl_5]^{2-}$ in which both ions involve six coordination about the tin atom. Again a simple doublet is obtained which can be computer fitted to a pair of doublet lines. The calculated intensities of the two sites are again almost equal in this case. The particular assignment of each doublet has proved to be elusive. Isolation of either the cation or anion through the use of non-tin-containing gegen-ions results in spectra whose parameters are sufficiently different to indicate the presence of varying extramolecular contributions to the electric field gradients.[178,179] The ^{119m}Sn Mössbauer spectra are thus of limited help in the determination of the structures of these complexes.

Complexes of iron(III) with bidentate organosulfur ligands are generally five coordinated in the solid state,[180-182] although it has recently been suggested that six coordination is achieved in solution through the coordination of available solvent molecules.[183] Several complexes containing bis(N,N-diethyldithiocarbamato)-(I), bis(maleonitriledithiolato)-(II), and bis(toluenedithiolato)-(III) groups have now been examined in both the solid and frozen solution phases by ^{57}Fe spectroscopy.[184] The quadrupole splittings decrease from 2.65–3.10 in the solid state to 0.70–0.85 mm/seconds in ≈ 0.03 M

$$\left[\begin{array}{c} S \\ S \end{array}\!\!\!C\!-\!N\!\!\!\begin{array}{c} C_2H_5 \\ C_2H_5 \end{array}\right]^{-} \quad \left[\begin{array}{c} S \\ S \end{array}\!\!\!\begin{array}{c} C \\ \| \\ C \end{array}\!\!\!\begin{array}{c} CN \\ CN \end{array}\right]^{2-} \quad \left[\begin{array}{c} S \\ S \end{array}\!\!\!\bigcirc\!\!\!\begin{array}{c} CH_3 \\ \end{array}\right]^{2-}$$

(I) (II) (III)

dimethylformamide. This dramatic decrease in the magnitude of the splitting is consistent with the higher symmetry expected for the presumably six-coordinated complex in solution, and the interpretation is corroborated by the similarly small splittings (0.52–0.66 mm/second) found for tris(N,N'-diethyldithiocarbamato)iron(III), the structure of which involves six-coordinated iron.[185] The isomer shifts were found not to change significantly from their solid state values of 0.6–0.7 mm/second.[185]

The magnitude of the ρ parameter has been used to rule out some of the possible structures for the metal–carbonyl complexes of tin-containing arene ligands of the form $(CH_3)_n Sn(C_6H_5)_{4-n} \cdot m M(CO)_3$ ($n = 2, 3; m = 1, 2$; M = Cr, Mo).[186,187] Since all of these complexes gave ρ ≪ 1.8, it was concluded that these compounds are four-coordinated at tin, thus eliminating the possibility of intramolecular "end bridging" through a carbonyl group which might, for example, in $(CH_3)_2Sn(C_6H_5)_2[Cr(CO)_3]_2$, place the tin atom in a *trans*-$Sn(CH_3)_2(C_6H_5)_2O_2$ octahedral environment or through $(d \to d)$–π bonding from the transition metal. The small splittings (≈ 0.6–0.9 mm/second) observed in these compounds are thought to arise from imbalances in the σ framework around the tin atom.

Both the isomer shifts and the experimental line widths of the ^{57}Fe resonances in a series of iron phosphonate compounds have played a role in a proposal for their structure.[188] The isomer shifts of tris(methylmethylphosphonato)iron(III), tris(ethylethylphosphonato)iron(III), and tris(*i*-propylmethylphosphonato)iron(III) were found to be approximately 20% more positive than that of tris(acetylacetonate)iron(III),[189] which serves as a possible structural model.[190] This large difference could not be rationalized on the basis of absorber temperature variations, and it was concluded that $(d \to d)$–π interactions between phosphorus and iron might be responsible for the observed effect.

The linewidths recorded for the Mössbauer resonances in these compounds were thought too large to be attributable to finite absorber thickness effects, as is also the case for tris(acetylacetonato)iron(III).[191] Apparently spin–spin relaxation collapses the expected hyperfine splitting.[192] The spin–spin interaction depends critically upon the metal–metal distance. On this basis, the internuclear Fe–Fe distances in the phosphonato compounds must be smaller than that in the model compound, as the linewidth is greater there.

The systematic investigation of series of analogous compounds may yield information of structural significance when the spectral data are compared with the predictions of simple theoretical arguments. Such an approach has been recently used in a ^{119m}Sn Mössbauer study of the trialkyltin halides and pseudohalides, R_3SnX (where R = alkyl; X = F, Cl, Br, I, OH), and the di-n-butyltin halides, $(C_4H_9)_2SnX_2$ (where X = F, Cl, Br, I).[194] The theoretical discussion is based upon an extension of the Townes and Dailey method,[195] but it should be noted that this procedure involves many assumptions when applied to electronic distributions of this complexity.

Complexes of the type SnX_4B_2 {where X = Cl, Br, and I and where B = R_3PO(R = C_2H_5, C_4H_9, C_6H_5), $(C_6H_5)_3AsO$, $(CH_3)_2SO$, $(C_6H_5)_3As$, R_3P(R = C_4H_9, C_6H_5, C_8H_{17}), $C_6H_5(CH_3)_2P$, and B_2 = o-$[(C_6H_5)_2P]C_6H_4$, o-$(CH_3)_2N(C_6H_4)P(C_6H_5)_2$, and $[o$-$(CH_3)_2N(C_6H_4)]_2P(C_6H_5)$} have been studied by ^{119m}Sn Mössbauer and far-infrared techniques. Cis and trans isomers have been distinguished by the number of ν(Sn—X) and δ(Sn—X) bands they exhibit; those assigned trans on this basis are found to exhibit Mössbauer quadrupole splittings of $\approx$1 mm/second. Point-charge calculations predict that the Q.S. for cis isomers should be one half that for trans, and the Q.S. for these isomers is barely resolvable as expected. In one case, $SnCl_4 \cdot 2P(C_6H_5)_3$, the infrared and Mössbauer evidence leads to opposite stereochemical assignments, but dipole moment data in solution suggest a mixture of cis and trans isomers for this complex.[196]

Additional structural application of Mössbauer spectroscopy may be encountered in previous sections of this review, particularly in the sections devoted to polymerization and biological applications. Organometallic examples with nuclei other than ^{119m}Sn and ^{57}Fe have been extremely limited. It has been reported, though, that the ^{197}Au [a $\frac{1}{2}$(MI)$\frac{1}{2}$ transition] spectrum of $KAu(CN)_2$ shows a well-resolved doublet in agreement with the expected nonzero electric field gradient for a linear sp-bonded $[Au(CN)_2]^-$ ion.[193]

References

[1] J. J. Zuckerman, Advan. Organometal. Chem. 9, 21 (1970).
[2] "Chemical Applications of Mössbauer Spectroscopy" (V. I. Goldanskii and R. H. Herber, eds.). Academic Press, New York, 1968.
[3] I. J. Gruverman, "Mössbauer Effect Methodology," Vol. 1 (1965), Vol. 2 (1966), Vol. 3 (1967), Vol. 4 (1968), Vol. 5 (1970). Plenum, New York.
[4] "Mössbauer Effect Data Index 1958–1965" (A. H. Muir, K. J. Ando, and H. M. Coogan, eds.). Wiley (Interscience), New York, 1966.
[5] "The Mössbauer Effect and Its Application in Chemistry" (C. W. Seidel,

ed.). Advances in Chemistry Series, Vol. 68. American Chemical Society, Washington, 1967.
[6] G. K. Wertheim, "The Mössbauer Effect: Principles and Applications." Academic Press, New York, 1964.
[7] B. A. Goodman and N. N. Greenwood, *Chem. Commun.* p. 1105 (1969).
[8] C. Hohenemser, *Phys. Rev.* **139**, A185 (1965).
[9] H. A. Stöckler and H. Sano, *Nucl. Instr. Methods* **44**, 103 (1966).
[10] H. A. Stöckler, H. Sano, and R. H. Herber, *J. Chem. Phys.* **47**, 1567 (1967).
[11] A. Yu. Aleksandrov, N. N. Delyagin, K. P. Mitrofanov, L. S. Polak, and V. S. Shpinel, *Dokl. Akad. Nauk SSSR* **148**, 126 (1963).
[12] V. A. Bryukhanov, V. I. Goldanskii, N. N. Delyagin, L. A. Korytko, E. F. Makarov, I. P. Suzdalev, and V. S. Shpinel, *Sov. Phys.—JETP* **16**, 321 (1963).
[13] V. I. Goldanskii, G. M. Gorodinsky, S. V. Karyagin, L. A. Korytko, L. M. Krizhansky, E. F. Makarov, I. P. Suzdalev, and V. V. Khrapov, *Proc. Acad. Sci. USSR, Phys. Chem. Sect.* **147**, 766 (1962).
[14] V. I. Goldanskii, E. F. Makarov, and V. V. Khrapov, *Sov. Phys.—JETP* **17**, 508 (1963).
[15] V. I. Goldanskii, E. F. Makarov, and V. V. Khrapov, *Phys. Lett.* **3**, 344 (1963).
[16] V. I. Goldanskii, E. F. Makarov, R. A. Stukan, V. A. Trukhtanov, and V. V. Khrapov, *Proc. Acad. Sci. USSR* **151**, 598 (1963).
[17] V. A. Komissarova, A. A. Sorokin, and V. S. Shpinel, *Sov. Phys.—JETP* **23**, 800 (1966).
[18] V. S. Shpinel, A. Yu. Aleksandrov, G. K. Ryansnyi, and O. Yu. Okhlobystin, *Sov. Phys.—JETP* **21**, 47 (1965).
[19] S. V. Karyagin, *Dokl. Akad. Nauk SSSR* **148**, 110 (1963).
[20] H. A. Stöckler and H. Sano, *Phys. Rev.* **165**, 406 (1968).
[21] R. L. Collins and J. C. Travis, *Mössbauer Effect Methodol.* **3**, 123 (1967).
[22] L. M. Epstein, *J. Chem. Phys.* **36**, 2731 (1962).
[23] J. F. Duncan and P. W. R. Wigley, *J. Chem. Soc.* p. 1120 (1963).
[24] W. Kerler, W. Neuwirth, E. Fluck, P. Kuhn, and B. Zimmermann, *Z. Physik.* **173**, 321 (1963).
[25] K. Otto and A. Ito, *Rev. Mod. Phys.* **36**, 459 (1964).
[26] A. Ito, M. Suenaga, and K. Ono, *J. Chem. Phys.* **48**, 3597 (1968).
[27] V. I. Goldanskii, O. Yu. Okhlobystin, V. Ya. Rochev, and V. V. Khrapov, *J. Organometal. Chem.* **4**, 160 (1965).
[28] A. Yu. Aleksandrov, Y. A. G. Dorfman, O. L. Lependina, K. P. Mitrofanov, M. V. Plotnikova, L. S. Polak, A. Ya. Temkin, and V. S. Shpinel, *Russ. J. Phys. Chem. (Eng. Transl.)* **38**, 1185 (1964).
[29] K. A. Bilevich, V. I. Goldanskii, V. Ya. Rochev, and V. V. Khrapov, *Izv. Akad. Nauk SSSR, Ser. Khim.* p. 1705 (1969).
[30] A. Yu. Aleksandrov, V. I. Goldanskii, L. A. Korytko, V. A. Maltsev, and N. A. Plate, *Vysokomol. Soed.* **10**, 109 (1968).
[31] V. V. Khrapov, V. I. Goldanskii, A. K. Prokofiev, V. Ya. Rochev, and R. G. Kostyanovskii, *Izv. Akad. Nauk SSSR, Ser. Khim.* p. 1261 (1968).
[32] S. Bukshpan and T. Sonnino, *J. Chem. Phys.* **48**, 4442 (1968).
[33] S. Bukshan, C. Goldstein, and T. Sonnino, *J. Chem. Phys.* **49**, 5477 (1968).
[34] V. I. Goldanskii, V. Ya. Rochev, and V. V. Khrapov, *Dokl. Akad. Nauk SSSR* **156**, 909 (1964).

[35] A. Yu. Aleksandrov, V. I. Bregadze, V. I. Goldanskii, L. I. Zakharkin, O. Yu. Okhlobystin, and V. V. Khrapov, *Dokl. Akad. Nauk SSSR* **165**, 593 (1965).
[36] A. J. Bearden, H. S. Marsh, and J. J. Zuckerman, *Inorg. Chem.* **5**, 1260 (1966).
[37] V. I. Goldanskii, G. M. Gorodinsky, S. V. Karyagin, L. A. Korytko, L. M. Krizhansky, E. F. Makarov, I. P. Suzdalev, and V. V. Khrapov, *Proc. Acad. Sci. USSR, Phys. Chem. Sect. (Eng. Transl.)*, **147**, 776 (1962).
[38] A. Yu. Aleksandrov, V. I. Goldanskii, L. A. Korytko, V. A. Malizev, and N. A. Plate, *Vysokomol. Soed., Ser. B* **10**, 209 (1968).
[39] R. A. Stukan, V. I. Goldanskii, E. F. Makarov, B. V. Borshagovskii, N. S. Kochetkova, M. I. Rybinskaya, and A. N. Nesmeyanov, *Dokl. Akad. Nauk SSSR* **170**, 354 (1966).
[40] B. V. Borsghagovskii, V. I. Goldanskii, S. P. Gubin, L. I. Denisovich, and R. A. Stukan, *Teor. Eksperim. Khim. Khim.*, **5**, 372 (1969).
[41] R. L. Collins, in "Mössbauer Effect Methodology" (I. J. Gruverman, ed.), Vol. 4, p. 129. Plenum, New York, 1968.
[42] M. C. Hobson, Jr., *J. Electrochem. Soc.* **115**, 175C (1968).
[43] M. J. D. Low, in "The Solid–Gas Interface" (E. A. Flood, ed.), Vol. 2. p. 947. Marcel Dekker, New York, 1967.
[44] A. N. Karasev, L. S. Polak, E. B. Shlikhter, and V. S. Shpinel, *Kinetics Catalysis (USSR) (Eng. Transl.)* **6**, 630 (1965).
[45] A. N. Karasev, L. S. Polak, E. B. Shlikhter, and V. S. Shpinel, *Russ. J. Phys. Chem. (Eng. Transl.)* **39**, 1670 (1965).
[46] M. Ianakieva, J. P. Quiles, M. Chene, T. Christov, R. Chevalier, and M. Belakhasky, *C. R. Acad. Sci.* **267C**, 1013 (1968).
[47] W. N. Delgass, R. L. Garten, and M. Boudart, *J. Chem. Phys.* **50**, 4603 (1969).
[48] W. N. Delgass, R. L. Garten, and M. Boudart, *J. Phys. Chem.* **73**, 2970 (1969).
[49] J. Morice and L. V. C. Rees, *Trans. Faraday Soc.* **64**, 1388 (1968).
[50] V. I. Goldanskii, I. P. Suzdalev, and A. S. Plachinda, *Dokl. Akad. Nauk SSSR* **185**, 629 (1969).
[51] A. Johansson, *J. Inorg. Nucl. Chem.* **31**, 3273 (1969).
[52] W. N. Delgass, M. Boudart, and G. Parravano, *J. Phys. Chem.* **72**, 3563 (1968).
[53] Y. Hazony and S. L. Ruby, *J. Chem. Phys.* **49**, 1478 (1968).
[54] R. H. Herber and H. A. Stöckler, in "Applications of the Mössbauer Effect in Chemistry and Solid-State Physics," *Tech. Rept. Ser. No. 50*, p. 110. I.A.E.A., Vienna, 1966.
[55] H. A. Stöckler and H. Sano, *J. Chem. Phys.* **50**, 3813 (1969).
[56] R. C. Poller, J. N. R. Ruddick, B. Taylor, and D. L. B. Roley, *J. Organometal. Chem.* **24**, 341 (1970).
[57] R. Okawara and M. Wada, *Advan. Organometal. Chem.* **5**, 137 (1967).
[58] R. Poller, *Proc. Chem. Soc.* p. 312 (1963).
[59] L. M. Epstein and D. K. Straub, *Inorg. Chem.* **4**, 1551 (1965).
[60] J. J. Zuckerman, unpublished results.
[61] A. Yu. Aleksandrov, K. P. Mitrofanov, O. Yu. Okhlobystin, L. S. Polak, and V. S. Shpinel, *Proc. Acad. Sci. USSR, Phys. Chem. Sect. (Eng. Transl.)* **153**, 974 (1963).
[62] V. I. Goldanskii, E. F. Makarov, R. A. Stukan, V. A. Trukhtanov, and V. V. Khrapov, *Proc. Acad. Sci. USSR, Phys. Chem. Sect. (Eng. Transl.)* **151**, 598 (1963).

[63] V. F. Belov, T. P. Vishnyakova, V. I. Goldanskii, E. F. Makarov, Ya. M. Paushkin, T. A. Sokolinskaya, R. A. Strukan, and V. A. Trukhtanov, *Dokl. Acad. Nauk SSSR* **159**, 831 (1964).

[64] V. I. Goldanskii, E. F. Makarov, R. A. Stukan, Ya. M. Paushkin, T. P. Vishnyakova, and T. A. Sokolinskaya, in "Semiconductive Copolymers with Conjugated Bonds," p. 128. TSNIITE Neftekhim., Moscow, 1966.

[65] L. A. Aliev, T. P. Vishnyakova, Ya. M. Paushkin, A. A. Pendin, T. A. Sokolinskaya, and R. A. Stukan, *Izv. Akad. Nauk SSSR*, 306 (1970).

[66] H. C. Clark, R. J. O'Brien, and J. Trotter, *Proc. Chem. Soc.* p. 85 (1964); *J. Chem. Soc.* p. 2332 (1964).

[67] E. O. Schlemper and D. Britton, *Inorg. Chem.* **5**, 507 (1966).

[68] H. A. Stöckler and H. Sano, *J. Chem. Phys.* **50**, 3813 (1968).

[69] H. A. Stöckler and H. Sano, *Mössbauer Effect Methodol.* **5**, 3 (1970).

[70] H. A. Stöckler and H. Sano, *Phys. Lett.* **25A**, 550 (1967).

[71] A. Yu. Aleksandrov, N. N. Delyagin, K. P. Mitrofanov, L. S. Polak, and V. S. Shpinel, *Soviet Phys.—JETP (Eng. Transl.)* **16**, 1467 (1963).

[72] V. I. Goldanskii, G. M. Gorodinsky, S. V. Karyagin, L. A. Korytko, L. M. Krizhansky, E. F. Makarov, I. P. Suzdalev, and V. V. Khrapov, *Proc. Acad. Sci. USSR, Phys. Chem. Sect. (Eng. Transl.)* **147**, 766 (1962).

[73] A. Yu. Aleksandrov, V. I. Bregadze, V. I. Goldanskii, L. I. Zakharkin, O. Yu. Okhlobystin, and V. V. Khrapov, *Dokl. Akad. Nauk SSSR* **165**, 593 (1965).

[74] A. M. Babeshkin, A. Bekker, E. N. Eframov, and A. N. Nesmeyanov, *Vestn. Mosk. Univ. Khim.* **24**, 78 (1969).

[75] N. Saito, H. Sano, T. Tominaga, and F. Ambe, "Application of the Mössbauer Effect to Chemistry and Solid State Physics," *Tech. Rept. Ser. No. 50*, p. 251. I.A.E.A., Vienna, 1966.

[76] J. Fenger and K. E. Siekierska, *Radiochim. Acta* **10**, 172 (1968).

[77] P. Gutlich, S. Odar, B. W. Fitzsimmons, and N. E. Erickson, *Radiochim. Acta* **10**, 147 (1968).

[78] G. K. Wertheim and D. N. E. Buchanan, *Chem. Phys. Lett.* **3**, 87 (1969).

[79] Y. D. Perfilev, L. A. Kulikov, A. M. Babeshkin, and A. N. Nesmeyanov, *Vestn. Mosk. Univ. Khim.* **24**, 112 (1969).

[80] G. K. Wertheim, W. R. Kingston, and R. H. Herber, *J. Chem. Phys.* **37**, 687 (1962).

[81] G. K. Wertheim, W. R. Kingston, and R. H. Herber, *J. Chem. Phys.* **38**, 2106 (1963).

[82] N. Saito, H. Sano, and T. Tominaga, *Chem. Ind.* p. 1622 (1964).

[83] H. Sano and M. Kano, *Chem. Commun.* p. 601 (1969).

[83a] P. Glentworth, A. L. Nichols, N. R. Large, and R. J. Bullock, *Chem. Commun.* p. 206 (1971).

[84] H. H. Wickman and A. M. Trozzolo, *Symp. Faraday Soc.* **1**, 21 (1967).

[85] D. C. Blomstrom, E. Knight, Jr., W. D. Phillips, and J. F. Weiher, *Proc. Nat. Acad. Sci. U.S.* **51**, 1085 (1964).

[86] G. Lang, *J. Appl. Phys.* **38**, 915 (1967).

[87] R. L. Collins, *J. Chem. Phys.* **42**, 1072 (1965).

[88] J. P. Dahl and C. F. Balhausen, *Mat. Fys. Medd. Dan. Vid Selsk.* **33**, 5 (1961).

[89] B. A. Goodman and N. N. Greenwood, *Chem. Commun.* p. 1105 (1969).

[90] N. E. Erickson, M. Cefola, and E. O. Kazimir, *Abstr. 157th ACS Nat. Meet.* Minneapolis, Minn., April, 1969.

[91] N. E. Erickson, *Chem. Commun.* p. 1349 (1970).
[92] B. W. Fitzsimmons, *J. Chem. Soc. A* p. 3235 (1970).
[93] T. C. Gibb, B. A. Goodman, and N. N. Greenwood, *Chem. Commun.* p. 774 (1970).
[94] T. C. Gibb, *J. Chem. Soc. A* p. 2503 (1970).
[95] J. D. Donaldson, *Prog. Inorg. Chem.* **8**, 287 (1966).
[96] V. I. Goldanskii, V. Ya. Rochev, and V. V. Khrapov, *Proc. Acad. Sci. USSR, Phys. Chem. Sect. (Eng. Transl.)* **156**, 571 (1964).
[97] E. O. Fischer and H. Grubert, *Z. Naturforsch.* **11b**, 423 (1956).
[98] L. D. Dave and G. Wilkinson, *J. Chem. Soc.* p. 3684 (1959).
[99] A. Almenningen, A. Haaland, and T. Motzfeldt, *J. Organometal. Chem.* **7**, 97 (1967).
[100] P. G. Harrison and J. J. Zuckerman, *J. Amer. Chem. Soc.* **91**, 6885 (1969).
[101] P. G. Harrison, J. G. Noltes, and J. J. Zuckerman, *J. Organometal. Chem.* in press.
[102] P. G. Harrison and J. J. Zuckerman, unpublished results.
[103] P. G. Harrison and J. J. Zuckerman, *J. Amer. Chem. Soc.* **92**, 2577 (1970).
[104] J. F. Young, *Advan. Inorg. Chem. Radiochem.* **11**, 91 (1968).
[105] V. I. Baranovskii, V. P. Sergeev, and B. E. Dzevitskii, *Dokl. Akad. Nauk SSSR* **184**, 632 (1969).
[106] D. E. Fenton and J. J. Zuckerman, *Inorg. Chem.* **8**, 1771 (1969).
[107] D. E. Fenton and J. J. Zuckerman, *J. Amer. Chem. Soc.* **90**, 6226 (1968).
[108] B. F. Burnham and J. J. Zuckerman, *J. Amer. Chem. Soc.* **92**, 1547 (1970).
[109] W. E. Bennett, *Abstr. 152nd Nat. Meet. A.C.S.* New York City, 1966.
[110] H. A. Stöckler and R. H. Herber, *J. Chem. Phys.* **45**, 1182 (1966).
[111] H. A. Stöckler and R. H. Herber, *J. Chem. Phys.* **46**, 2020 (1967).
[112] M. O'Rourke and C. Curran, *J. Amer. Chem. Soc.* **92**, 1501 (1970).
[113] D. G. Whitten, private communication, 1970.
[114] B. Y. K. Ho and J. J. Zuckerman, unpublished results.
[115] Yu. Sh. Moshkovskii, in "Chemical Applications of Mössbauer Spectroscopy" (V. I. Goldanskii and R. H. Herber, eds.), p. 524. Academic, New York, 1968.
[116] Yu. Sh. Moshkovskii, E. F. Makarov, G. A. Zavarzin, N. Ya. Vedenina, S. S. Mardanyan, and V. I. Goldanskii *Biofiz.* **11**, 357 (1966).
[117] U. Gonser, R. W. Grant, and J. Kregzde, *Science* **143**, 680 (1964).
[118] Yu. Sh. Moskovskii, I. D. Ivanov, R. A. Stukan, G. I. Matchanov, S. S. Mardanyan, Yu. M. Belov, and V. I. Goldanskii, *Dokl. Akad. Nauk SSSR* **174**, 215 (1967).
[119] R. H. Herber, H. A. Stöckler, and W. T. Reichle, *J. Chem. Phys.* **42**, 2447 (1965).
[120] G. M. Bancroft, K. G. Dharmawardena, and A. J. Stone, *Chem. Commun.* p. 6 (1971).
[121] See Ref. 1, p. 76ff.
[122] E. O. Schlemper and Walter C. Hamilton, *Inorg. Chem.* **5**, 995 (1966).
[123] A. G. Davies, H. J. Milledge, D. C. Puxley, and P. J. Smith, *J. Chem. Soc. A* p. 2862 (1970).
[124] J. D. Graybeal and A. Berta, *U.S. Nat. Bur. Std. Spec. Publ.* **301** (1969).
[125] E. D. Swiger and D. Graybeal, *J. Am. Chem. Soc.* **87**, 1464 (1965).
[126] J. Green and J. D. Graybeal, *J. Am. Chem. Soc.* **89**, 4305 (1967).
[127] J. D. Donaldson and D. G. Nicholson, *Inorg. Nucl. Chem. Lett.* **6**, 151 (1970).

[128] B. F. E. Ford, J. R. Sams, R. G. Goel, and D. R. Ridley, *J. Inorg. Nucl. Chem.* **33**, 23 (1971).
[129] R. M. Sternheimer, *Phys. Rev.* **80**, 102 (1950); **84**, 244 (1951); **86**, 316 (1952); **95**, 736 (1954); **105**, 158 (1957); **130**, 1423 (1963); **132**, 1637 (1963); **146**, 140 (1966).
[130] E. A. C. Lucken, "Nuclear Quadrupole Coupling Constants." Academic, New York, 1969.
[131] R. V. Parish and R. H. Platt, *Inorg. Chim. Acta* **4**, 65 (1970).
[132] B. W. Fitzsimmons, N. J. Seeley, and A. W. Smith, *Chem. Commun.* p. 390 (1968).
[133] B. W. Fitzsimmons, N. J. Seeley, and A. W. Smith, *J. Chem. Soc. A* p. 143 (1969).
[134] N. W. G. Debye and J. J. Zuckerman, in "Developments in Applied Spectroscopy," Vol. 8. Plenum, New York, 1970.
[135] M. G. Clark, *Mol. Phys.* **20**, 257 (1971).
[136] V. Naik and C. Curran, *Inorg. Chem.* **10**, 1017 (1971).
[137] R. V. Parish and R. H. Platt, *J. Chem. Soc. A* p. 2145 (1969).
[138] M. Cordey-Hayes, R. D. Peacock, and M. Vucelic, *J. Inorg. Nucl. Chem.* **29**, 1177 (1967).
[139] R. V. Parish and R. H. Platt, *Chem. Commun.* p. 1118 (1968).
[140] T. Chivers and J. R. Sams, *Chem. Commun.* p. 249 (1969).
[141] N. N. Greenwood and J. N. R. Ruddick, *J. Chem. Soc. A* p. 1679 (1967).
[142] T. C. Gibb and N. N. Greenwood, *J. Chem. Soc. A* p. 43 (1966).
[143] N. N. Greenwood, P. G. Perkins, and D. H. Wall, *Symp. Faraday Soc.* **1**, 51 (1967).
[144] L. F. Dahl, *J. Amer. Chem. Soc.* **91**, 1351 (1969).
[145] W. Hieber and E. Becker, *Chem. Ber.* **63B**, 1405 (1930).
[146] R. K. Sheline, *J. Amer. Chem. Soc.* **73**, 1615 (1951).
[147] J. W. Cable and R. K. Sheline, *Chem. Rev.* **56**, 1 (1956).
[148] D. F. Keeley and R. F. Johnson, *J. Inorg. Nucl. Chem.* **11**, 33 (1959).
[149] R. Brill, *Z. Krist.* **77**, 36 (1931).
[150] O. S. Mills, *Chem. Ind.* p. 73 (1957).
[151] H. Freudlich and E. J. Cuy, *Chem. Ber.* **56B**, 2264 (1923).
[152] S. Berkman and H. Zocher, *Z. Physik. Chem.* **124**, 318 (1926).
[153] H. G. Cutforth and P. W. Selwood, *J. Amer. Chem. Soc.* **65**, 2414 (1943).
[154] D. A. Brown, *J. Inorg. Nucl. Chem.* **11**, 33 (1959).
[155] L. F. Dahl and R. E. Rundle, *J. Chem. Phys.* **26**, 1751 (1957).
[156] L. F. Dahl and R. E. Rundle, *J. Chem. Phys.* **27**, 323 (1957).
[157] R. H. Herber, R. B. King, and G. K. Wertheim, *Inorg. Chem.* **3**, 101 (1964).
[158] R. H. Herber, W. R. Kingston, and G. K. Wertheim, *Inorg. Chem.* **2**, 153 (1963).
[159] E. Fluck, W. Kerler, and W. Neuwirth, *Angew. Chem.* **75**, 461 (1963).
[160] E. Fluck, W. Kerler, and W. Neuwirth, *Angew. Chem. Int. Ed. Engl.* **2**, 277 (1963).
[161] E. Fluck, W. Kerler, and W. Neuwirth, *Kagaku No Ryoiki* **19**, 195 (1965).
[162] W. Kerler, W. Neuwirth, E. Fluck, P. Kuhn, and B. Zimmerman, *Z. Physik* **173**, 321 (1963).
[163] W. Kerler, W. Neuwirth, and E. Fluck, *Z. Physik* **175**, 200 (1963).
[164] M. Kalvius, V. Zahn, P. Kienle, and H. Eicher, *Z. Naturforsch.* **17a**, 494 (1962).
[165] L. F. Dahl and J. F. Blount, *Inorg. Chem.* **4**, 1373 (1965).

[166] N. E. Erickson and A. W. Fairhall, *Inorg. Chem.* **4**, 1320 (1965).
[167] D. J. Dahm and R. A. Jacobson, *Chem. Commun.* p. 496 (1966).
[168] D. J. Dahm and R. A. Jacobson, *J. Amer. Chem. Soc.* **90**, 5106 (1968).
[169] W. Lindauer, H. W. Spiess, and R. K. Sheline, *Inorg. Chem.* **9**, 1694 (1970).
[170] K. Farmery, M. Kilner, R. Greatrex, and N. N. Greenwood, *J. Chem. Soc. A* p. 2339 (1969).
[171] P. A. Agron, R. D. Ellison, and H. A. Levy, *Acta Crystallogr.* **23**, 1079 (1967).
[172] R. A. Stukan and L. P. Yurieva, *Dokl. Akad. Nauk SSSR* **167**, 1311 (1966).
[173] N. W. G. Debye, Ph.D. thesis, Cornell University, 1970.
[174] R. A. Stukan, thesis, Acad. Sci. USSR Inst. Chem. Phys., Moscow, 1965.
[175] V. I. Goldanskii, private communication to E. Fluck, quoted in Ref. 2.
[176] F. W. B. Einstein and B. R. Penfold, *Chem. Commun.* p. 780 (1966).
[177] F. W. B. Einstein and B. R. Penfold, *J. Chem. Soc. A* p. 3019 (1968).
[178] N. W. G. Debye, E. Rosenberg, and J. J. Zuckerman, *J. Am. Chem. Soc.* **90**, 3234 (1968).
[179] N. W. G. Debye, unpublished results, 1970.
[180] B. F. Hoskins, R. L. Martin, and A. H. White, *Nature (London)* **211**, 627 (1966).
[181] W. C. Hamilton and R. Spratley, *Acta Crystallogr. A* **21**, 142 (1966).
[182] A. Davison, N. Edelstein, R. H. Holm, and A. H. Maki, *Inorg. Chem.* **3**, 814 (1964).
[183] C. Furlani, *Coord. Chem. Rev.* **3**, 141 (1968).
[184] J. L. K. F. DeVries, J. M. Trooster, and E. DeBoer, *Inorg. Chem.* **10**, 81 (1971).
[185] A. H. White, R. Roper, E. Kokot, H. Waterman, and R. L. Martin, *Austr. J. Chem.* **17**, 294 (1964).
[186] P. G. Harrison, J. J. Zuckerman, T. V. Long, II, T. P. Poeth, and B. R. Willeford, *Inorg. Nucl. Chem. Lett.* **6**, 627 (1970).
[187] Thomas P. Poeth, P. G. Harrison, T. V. Long, II, B. R. Willeford, and J. J. Zuckerman, *Inorg. Chem.* **10**, 522 (1971).
[188] J. J. Kokolas, D. N. Kramer, A. A. Temperley, and R. Levin, *Spectrosc. Lett.* **2**, 283 (1969).
[189] R. H. Herber, *Mössbauer Effect Methodol.* **1**, 7 (1965).
[190] J. J. Kokolas, D. N. Kramer, F. Block, and R. Levin, *Spectrosc. Lett.* to be published, noted in Ref. 188.
[191] G. K. Wertheim, W. R. Kingston, and R. H. Herber, *J. Chem. Phys.* **37**, 687 (1962).
[192] J. W. G. Wignall, *J. Chem. Phys.* **44**, 2462 (1966).
[193] D. A. Shirley in Ref. 2, p. 507.
[194] V. Kotkhekar and V. S. Shpinel, *Zh. Strukt. Khim.* **10**, 37 (1969).
[195] C. H. Townes and B. P. Dailey, *J. Chem. Phys.* **17**, 782 (1949).
[196] P. G. Harrison, B. C. Lane, and J. J. Zuckerman, *Inorg. Chem.*, **11**, 1537 (1972).

Automated Chemical Structure Analysis Systems 7

SHIN-ICHI SASAKI

I.	Introduction.	285
II.	Structure Determination of Alkanes with Monomethyl Side Chain	286
III.	Structure Determination of Oligopeptides	289
IV.	Structure Determination of Aliphatic Ketones	293
V.	Structure Determination of Aliphatic Ethers	299
VI.	Construction of Molecular Structure from Suitable Fragments	304
	A. Data Analysis of NMR Spectrum	306
	B. Example of Designation of "Components" by Computer-Aided Spectrometry	309
	C. Method of Building Up Molecular Structure from Designated "Components"	315
	D. Results	317
	References	320

I. INTRODUCTION

A vast number of mass, NMR, and IR spectra are available at present time. This large number of spectra allows the development of computerized programs suitable for identifying compounds of which spectra already have been recorded. Magnetic tape generally has been used to store these spectral data, and retrieval programs have been used for the identification.[1,2] However, it is obvious that the number of organic compounds is almost infinitely great, and it is impossible to expect that the spectra for all of them will ever be stored on magnetic tape. Therefore, it will not be easy to identify certain organic compounds by the sorting method alone. Thus the electronic computer is now being applied to the determination of whole structure of organic molecules—including compounds with unknown structure—from a higher level of data analysis by making it act as an artificial intelligence. The following is taken from the most important and interesting recent papers concerning automated structure determination of organic compounds.

II. STRUCTURE DETERMINATION OF ALKANES WITH MONOMETHYL SIDE CHAIN

Pettersson and Ryhage describe a computer method for identification of aliphatic saturated normal and monomethyl substituted hydrocarbons from C_6 to C_{30}.[3] The relative intensity (greater than or equal to 0.1%) of each m/e value for each unknown compound was punched on cards and used as input data. The class of compound should first be decided from the spectral data. For this purpose a rectangular array method[4] was adopted. In the rectangular array the intensities of every 14 mass numbers are summed, i.e., the intensities of 14, 28, 42, 56, ... as well as the intensities of 15, 29, 43, 57, ..., etc. The magnitudes of the 14 sums (S_1–S_{14}) give a relatively good indication of the class of compound. As shown in Table I, for the saturated hydrocarbon the S values listed in decreasing order of magnitude are: S_2, S_{14}, S_1, S_3, S_{13}, and S_{12}, whereas for the unsaturated hydrocarbon the following sequence is obtained: S_{14}, S_2, S_1, S_{12}, S_{13}, and S_3. Although it is difficult to distinguish between normal fatty acids and their methyl esters by this method, the saturated hydrocarbons can at least be distinguished from the other class of compounds. After this procedure, the mass spectral data are treated by the subroutine for straight chain and monomethyl substituted hydrocarbons.

Mass spectra of saturated straight chain hydrocarbons are very easy to interpret since they all follow the same fragmentation pattern; for example, m/e 43, 57, or 71 appears as base peak and the intensities of m/e 85, 99, 113, ..., etc. decrease continuously until the $(M - 15)$ peak is reached for hydrocarbons with more than five carbon atoms. To determine the position of the methyl group of methyl branched hydrocarbons, Eq. (1) was used and

TABLE I

Rectangular Array of n-Nonacosane (1), 2-Dodecene (2), n-Heptylic Acid (3), and Methyl Octadecanoate (4)[a]

Compound	S_1	S_2	S_3	S_4	S_5	S_6	S_7	S_8	S_9	S_{10}	S_{11}	S_{12}	S_{13}	S_{14}
1	112	607	37	2	1	1	1	2	1	3	2	24	25	209
2	181	227	8	1	0	0	0	1	0	3	3	52	39	321
3	42	67	4	110	113	15	1	1	1	1	1	11	7	68
4	29	86	10	182	165	26	2	1	0	6	2	18	11	94

[a] B. Pettersson and R. Ryhage, *Anal. Chem.* **39**, 790 (1967).

the ratios of peak intensities were calculated:

$$Q1(n) = I(29 + n \times 14)/I(43 + n \times 14) \tag{1}$$

where I = intensity and n = values greater than or equal to 1 until the peak $57 + n \times 14$ is missing in the mass spectrum.

As an example, the $Q1$ values calculated for 4-methyloctane based on the spectrum shown in Fig. 1 are given below:

$$Q1(1) = I(43)/I(57) = 3.9$$
$$Q1(2) = I(57)/I(71) = 0.9$$
$$Q1(3) = I(71)/I(85) = 1.0$$
$$Q1(4) = I(85)/I(99) = 20.1$$
$$Q1(5) = I(99)/I(113) = 1.8$$
$$Q1(6) = I(113)/I(127) \text{ (missing)}$$

The peak $57 + n \times 14$, which is missing, is in this case $57 + (5 \times 14) = 127$; n is accordingly equal to 5. $Q1$ is usually greater than or equal to 1 for saturated straight chain hydrocarbons unless the denominator is the base peak. In this case the subroutine disregards the $Q1$ value, which can be only $Q1(1)$ or $Q1(2)$ and continues. $Q1$ values less than 1 are also disregarded when the dividend and divisor in the $Q1$ ratio are of about same intensity and intensities are less than 1%. The latter $Q1$ values are set equal to 1 and counted as 1 in the calculation of the number of methyl groups. Thus the computer prints: "Methyl- or alkyl-substituted hydrocarbon" and the position at which the irregularity of the spectrum pattern occurs is stored ($m = n$), 4-methyloctane would consequently give $m = 2$. The last peak $43 + n \times 14$, which is found in the spectra is (M − 15), (M − 29), or (M − 1). If the intensities of $43 + n \times 14 + 15$, $43 + n \times 14 + 29$, $43 + n \times 14 + 1$ are below 0.1%, the computer prints "Molecular ion not found, but calculated. Check the molecular weight." The program for calculation of real molecular weights was also developed by the authors. The following results were obtained by this technique.

Compound Tested	Computer Printout
$CH_3(CH_2)_{12}CH_3$	Straight chain hydrocarbon. Molecular weight, 198

$CH_3(CH_2)_8CH(CH_2)_9\text{---}CH_3$
  |
  CH_3
	Me- or alkyl-substituted hydrocarbon. Molecular weight not found, but calculated. Check the molecular weight. M. W. 296, Me- group in position 10

Mass–charge ratio (m/e)	Relative intensities for ionizing voltage of 70 V	Mass–charge ratio (m/e)	Relative intensities for ionizing voltage of 70 V
27	20.4	69	2.62
28	2.97	70	20.9
29	21.5	71	27.5
30	0.49	72	1.48
37	0.03	73	0.03
38	0.35	77	0.11
39	9.64	78	0.01
40	1.66	79	0.07
41	25.7	80	0.01
42	9.51	81	0.04
43	100.	82	0.02
44	3.31	83	0.20
45	0.04	84	18.2
50	0.17	85	28.2
51	0.51	86	1.80
52	0.22	87	0.05
53	1.75	97	0.03
54	0.74	98	3.06
55	11.8	99	1.42
56	7.96	100	0.10
57	25.4	111	0.01
58	1.09	112	0.44
59	0.02	113	0.82
62	0.02	114	0.08
63	0.08	127	0.02
64	0.01	128	2.46
65	0.21	129	0.24
66	0.07		
67	0.39		
68	0.14		

FIG. 1. Mass spectrum of 4-methyloctane (cited from A. P. I., No. 247).

The following three methods were combined to determine if the compound has alkyl group(s) or more than one methyl group.

(1) The ratios $Q1(n)$ [see Eq. (1)] are calculated up to the (M − 15) peak. If some $Q1$ values are greater than 50, this is very significant for low molecular weight hydrocarbons ($m/e < 200$) with alkyl group(s) or more than one methyl group. These high $Q1$ values appear when the straight chain has a quarternary carbon atom with two methyl or alkyl groups or if the methyl substituents are in an α or β position to each other. High $Q1$ values also

appear for monomethyl substituted hydrocarbons with the methyl group in position 2, because in this case the (M − 43) is of rather high and the (M − 29) is of low intensity.

(2) Low molecular weight hydrocarbons with alkyl group(s) or more than one methyl group sometimes have more than two $Q1$ values less than 1 and high molecular weight hydrocarbons ($m/e > 200$) more than three $Q1$ values less than 1.

(3) The hydrocarbons with alkyl group(s) or more than one methyl group have the ratio between the second most intense peak less than 0.5. This is particularly the case for low molecular weight compounds where several ion fragments have the same intensity.

The following result was obtained by this method:

Compound Tested	Computer Printout
$CH_3CH(CH_2)_2{-}CH(CH_2)_6CH_3$ $\quad\;\;\vert\qquad\qquad\;\;\vert$ $\quad CH_3\qquad\quad\;\; CH_3$	Me- or alkyl substituted hydrocarbon M. W. 198 Probably not a monomethyl substituted hydrocarbon

The authors state that only about 2% of the answers were wrong for the mass spectra of known methyl-substituted hydrocarbons checked. Twenty per cent sorted out as uncalculated, and the remainder gave correct answers.

III. STRUCTURE DETERMINATION OF OLIGOPEPTIDES

Papers on the determination of the amino acid sequence in oligopeptides were published by Biemann's[5] and McLafferty's groups[6] at the same time. In general, the sequence of amino acids in these peptides is determined by a stepwise degradation procedure which is monitored by total hydrolysis with amino acid analysis by column chromatography. This procedure is time consuming and requires relatively large samples. Recently, another determination method using mass spectrometry has been presented.[7] This clarified the fact that the fragmentation patterns of the peptides exhibit cleavage of the chain involving two main pathways:

A YNHCHR$_1$CO ┼─ NHCHR$_2$CO ┼─── NHCHR$_n$CO ┼─ OR
　　　　　　A$_1$　　　　　　A$_2$　　　　　　A$_n$

B YNHCHR$_1$ ┼─ CONHCHR$_2$ ┼─ CO ── NHCHR$_n$ ┼─ COOR
　　　　　B$_1$　　　　B$_2$　　　　　　B$_n$

Here Y and R are protecting groups of terminal amino and carboxy groups, respectively, to obtain sufficient volatility and stability (Table II).

Amino acid sequence is elucidated utilizing a computer program of the following general format based on the recognition that barring rearrangements, the structure of a linear molecule is determined unequivocally by using the possible fragments which contain only one end of the chain. The initial step is the identification of all peaks corresponding in mass to peaks possible from fragmentation schemes A and B. The spectrum is first checked for a peak corresponding to the sum of the exact masses of the N-derivative moiety and the glycine unit, $NHCH_2CO$, for example, (see Table III)

$$46.03722(CD_3CO) + 57.02146 = 103.05868$$

The search is repeated using combinations of the N derivative with each of the other possible amino acids. Fragment ions corresponding to scheme B are checked by subtracting the exact mass of CO (27.99491 amu) from each of these combinations. To identify the next amino acid unit of the chain, the process is repeated, with the addition of mass (value from Table III) of the newly identified amino acid unit to the sum of exact masses described above. This process is repeated until the search for an additional chain member finds no suitable match for either A or B. A check is now made for the molecular ion by adding the exact mass of the ester group OR to the mass of the identified chain A_n. A fit then establishes the molecular size, and identification is complete.

TABLE II

Ion Source Temperature (°C) at which the Mass Spectrum was Obtained[a]

Peptide	Temp.
N-Trideuterioacetyl-Ala-Leu-Ala-Val-Val-Val methyl ester	210
N-Trifluoroacetyl-Leu-Gly-Phe methyl ester	150
N-Acetyl-Pro-Gly-Phe-Gly methyl ester	190
N-Trifluoroacetyl-His-Pro-Tyr methyl ester	200
N-Trifluoroacetyl-His-Met-(β-O-methyl-Asp) methyl ester	190
N-Trifluoroacetyl-Pro-His-Leu methyl ester	215
N-Trideuterioacetyl-Pro-Phe-His-Leu-Leu methyl ester	235
N-Carbobenzoxy-Val-(O-t-butyl-Glu) methyl ester	140
N-Carbobenzoxy-Ile-(S-benzyl-Cys)-Ser methyl ester	190
N-Carbobenzoxy-Val-Gly-Ala-Leu-Ala methyl ester	200

[a] M. Senn, R. Venkataraghavan, and F. W. McLafferty, *J. Amer. Chem. Soc.* **88**, 5593 (1966).

TABLE III

Exact Masses of N Derivatives of Amino Acids[a]

	N derivatives	Mass	
	CH₃CO	43.01839	
	CD₃CO	46.03722	
	CF₃CO	96.99012	

Amino Acid	Mass	Amino Acid	Mass
Gly	57.02146	Asp-O-CH₃	129.04259
Ala	71.03711	Met	131.04048
Ser	87.03203	His	137.05891
Pro	97.05276	Glu-O-CH₃	143.05824
Val	99.06841	Phe	147.06841
Thr	101.04768	Tyr	133.06322
Leu	113.08406	Try	186.07931

[a] M. Senn, R. Venkataraghavan, and F. W. McLafferty, *J. Amer. Chem. Soc.* **88**, 5594 (1966).

The scheme for the search is schematically illustrated by Biemann in Fig. 2, using the N-carbobenzoxy peptide ester Cbz-Ser-Phe-Leu-OMe as a concrete example (i.e., Y = C₆H₅CH₂OCO, R = CH₃). To the initial mass 163.0633 (Y—NH—CH) is added the mass of the first amino acid side chain in the list (the mass of H, 1.0078, for glycine), and the resulting sum (the amino ion for N-terminal gylcine) is compared with the masses of experimentally found compound ions. If one of these agrees with that sum within a certain limit, then glycine is a possibility for the N-terminal amino acid. This is further tested by adding the mass of CO and searching the data for a fit with this sum. If either one of these are found, the search continues for the next amino acid by adding the mass of C₂H₂NO₂ and repeating the addition of the side-chain masses and CO (the second and third row of arrows in Fig. 2 represent the same amino acids indicated in the first row). If neither the amino nor the aminoacyl fragment corresponding to N-terminal glycine is found in the data, the next amino acid (alanine) is tested in the first segment, and so forth.

Output from computer interpretation of the mass spectrum of CD₃CO-Ala-Leu-Ala-Val-Val-Val-OCH₃ is shown in Fig. 3.

The results expressed in the papers contributed from M.I.T. and Purdue University may be summarized as follows: The method developed by these groups was successfully applied to di- through hexa-peptides containing

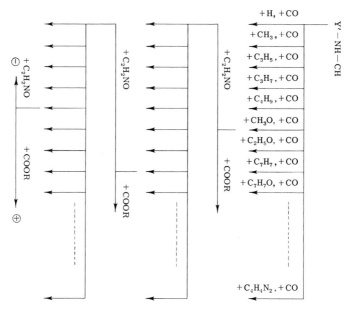

FIG. 2. Schematic representation of the basic outline of the program logic of the sequence search. [From K. Biemann, C. Cone, B. R. Webster, and G. P. Arsenault, *J. Amer. Chem. Soc.* **88**, 5601 (1966).]

1 Amino acid sequence analysis
Sample number 66-2
N-Trideuteroacetylmethyl ester
The following are the sequence found in this peptide

Sequence identity	Found	Calc	Error
A-1 ALA	89.079700	89.079410	−0.29
B-2 LEU	117.073170	117.074321	1.15
A-2 LEU	202.164000	202.163470	−0.53
B-2 LEU	230.158401	230.158380	−0.02
A-3 ALA	273.201201	273.200500	−0.70
B-3 ALA	301.196100	301.195491	−0.61
A-4 VAL	372.267500	372.268990	1.49
B-4 VAL	400.264310	400.263900	−0.41
A-5 VAL	471.336600	471.337400	0.80
B-5 VAL	499.333301	499.332310	−0.99
B-6 VAL	598.402000	598.400721	−1.28
Mol-ion	629.421000	629.419110	−1.89

Amino acid sequence in this peptide is
∗(D)AC-ALA ∗LEU ∗ALA ∗VAL ∗VAL ∗VAL∗

FIG. 3. Output from computer interpretation of mass spectrum of a peptide. [From K. Biemann, C. Cone, B. R. Webster, and G. P. Arsenault, *J. Amer. Chem. Soc.* **88**, 5602 (1966).]

glycine, alanine, serine, proline, valine, threonine, leucine, asparagine, α-aminoadipic acid, lysine, glutamic acid, methionine, histidine, phenylalanine, tyrosine, and δ-benzyl-cysteine.

IV. STRUCTURE DETERMINATION OF ALIPHATIC KETONES

An approach to computer interpretation of the mass spectra of aliphatic ketones was presented by Djerassi and his co-workers.[8] This approach is based on the capability of a computer program (DENDRAL[9]) to manipulate

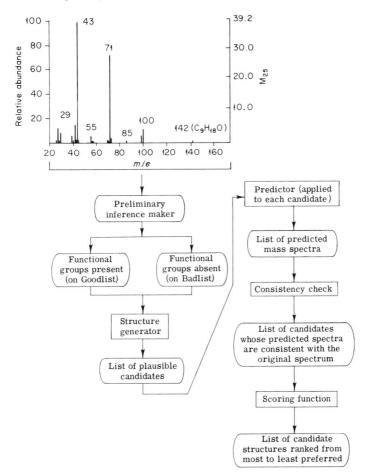

FIG. 4. Mass spectrum of unknown aliphatic ketone and the conceptualization of heuristic DENDRAL. [From A. M. Duffield, A. V. Robertson, C. Djerassi, B. G. Buchanan, G. L. Sutherland, E. A. Feigenbaum, and J. Lederberg, *J. Amer. Chem. Soc.* **91**, 2979 (1969).]

structural representations of organic molecules and their functional groups and to generate rigorously exhaustive and irredundant lists of structures including the candidates for a given problem.

The basic approach they have used in the computer interpretation of mass spectra is represented diagramatically in Fig. 4. The DENDRAL program can enumerate all the possible acyclic chemical structures of a given empirical formula.

The program is fed a low-resolution mass spectrum of an unknown compound and the empirical composition of the molecular ion. However, in the case of the example shown in Fig. 4, nine possible empirical formulas containing $C_9H_{18}O$ are calculated for the molecular weight of 142.[10] Therefore, as long as the system is using a low-resolution mass spectrum, the determination of empirical formulas is a problem to be solved by other analytical approaches. If the molecular formula is decided as $C_9H_{18}O$, then 1936 possible acyclic structures (aldehydes, ketones, unsaturated alcohols and ethers) can be computed for this composition (Table IV).[11]

A theory of mass spectral fragmentation processes is then employed and stored in the Preliminary Inference Maker to decide what functional groups are present within the molecular structure of the unknown. For instance, ketone mass spectra can be recognized because of their simple fragmentation modes (Scheme 1).[12] Cleavage adjacent to oxygen, followed by

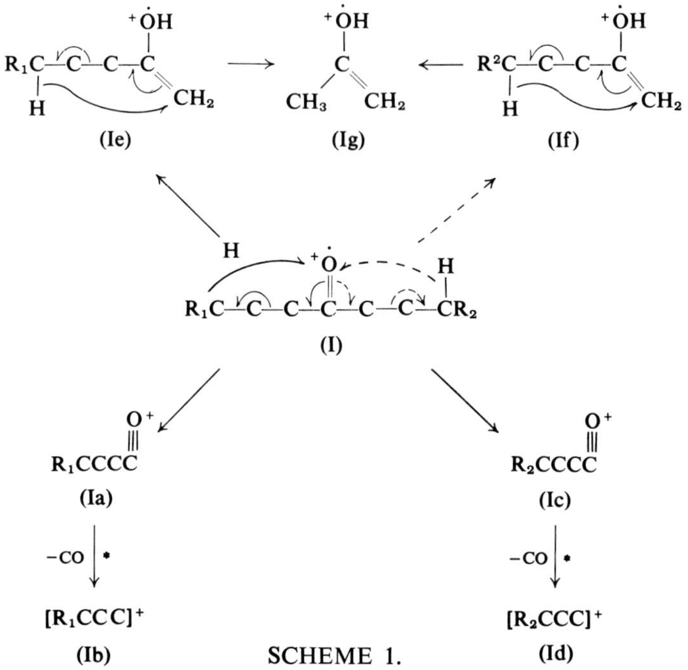

SCHEME 1.

by the KUMT by Kudo[b])

Section	Comp	\multicolumn{12}{c}{Number of carbon atoms}

Section	Comp	1	2	3	4	5	6	7	8	9	10	11	12
A	C_nH_{2n+2}	1	1	1	1	3	5	9	18	35	75	159	355
	C_nH_{2n}		1	2	5	15	25	56	139	338	852	2877	
	C_nH_{2n-2}		1	2	4	9	23	55	152	375	1048		
	C_nH_{2n-4}			0	2	6	21	59	195	563	1823		
	C_nH_{2n-6}				1	4	15	45	182	629	2270	8057	
	C_nH_{2n-8}					0	5	21	110	511	2113		
	C_nH_{2n-10}						1	8	45	262	1304		
	C_nH_{2n-12}							0	9	77	532		
	C_nH_{2n-14}								1	13	135		
	C_nH_{2n-16}								0	0	17		
	C_nH_{2n-18}										1		
B	$C_nH_{2n+2}O$	1	2	3	7	14	32	72	171	405	989	2460	6123
	$C_nH_{2n}O$	1	3	9	21	74	211	596	1684	4145	13406		
	$C_nH_{2n-2}O$		1	4	15	47	156	492	1544				
	$C_nH_{2n-4}O$		0	2	7	32	566	2687					
C	$C_nH_{2n+2}O_2$	0	2	6	18	48	133	357	990	2688			
	$C_nH_{2n}O_2$	1	3	8	32	110	380	1233	4030				
D	$C_nH_{2n+3}N$	1	2	4	8	17	39	89	211	507	1238	3057	
	$C_nH_{2n+1}N$	1	2	5	14	40	111	304	845	2322			
E	$C_nH_{2n+4}N_2$	1	5	11	34	84	235	623	1724				
	$C_nH_{2n+2}N_2$	2	8	24	78	241	751	2334					
F	$C_nH_{2n+3}NO$	2	6	18	50	137	365	995	2727				
	$C_nH_{2n+1}NO$	2	9	31	105	350	1116	3574					
G	$C_nH_{2n+3}NO_2$	2	9	43	160	533	1756	5617					
	$C_nH_{2n+1}NO_2$	3	17	83	362	1430							
H	$C_nH_{2n+3}NO_3$	0	7	56	288	1313							
	$C_nH_{2n+1}NO_3$	3	17	130	751	3740							

[a] J. Lederberg, G. L. Sutherland, B. G. Buchanan, E. A. Feigenbaum, A. V. Robertson, A. M. Duffield, and C. Djerassi, *J. Amer. Chem. Soc.* **91**, 2975 (1969). [b] S. Sasaki, Y. Kudo, S. Ochiai, and H. Abe, *Mikrochim. Acta*, **1971**, 726.

decarbonylation of the product, yields two ion fragmentation pathways (I) → (Ia) → (Ib) and (I) → (Ic) → (Id) depicted in Scheme 1. A simple mathematical relationship then exists between the masses of ions (Ia) and (Ic), and the molecular ion M, viz., (Ia) + (Ic) = M + 28. To distinguish ketones from aldehydes, alcohols, and ethers the Preliminary Inference Maker searches for two significant peaks in the spectrum which satisfy this relationship. The program must also recognize the decarbonylation products (Ib) and (Id) in the next step. Thus it is ascertained that 82 of the 1936 possible isomers of $C_9H_{18}O$ are ketones.

The program now gives the inferred functional group highest priority, by placing it on Goodlist. The ions having masses 58, 72, 86, 86, and 58 corresponding to McLafferty rearrangement products are searched (Scheme 2).[13]

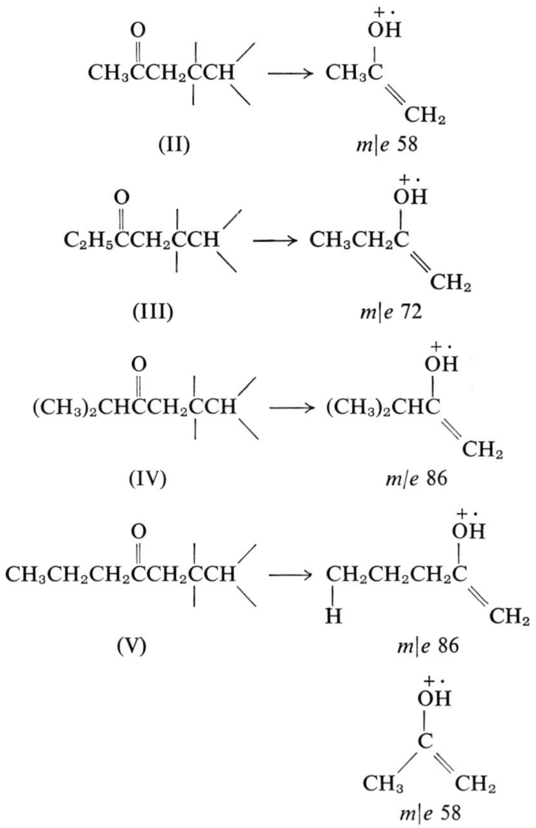

$$CH_3\overset{O}{\underset{}{\overset{\|}{C}}}CH_2\overset{}{\underset{}{C}}CH\diagup\diagdown \longrightarrow CH_3\overset{\overset{+\cdot}{OH}}{\underset{}{C}}\diagdown_{CH_2}$$

(II) m/e 58

$$C_2H_5\overset{O}{\underset{}{\overset{\|}{C}}}CH_2\overset{}{\underset{}{C}}CH\diagup\diagdown \longrightarrow CH_3CH_2\overset{\overset{+\cdot}{OH}}{\underset{}{C}}\diagdown_{CH_2}$$

(III) m/e 72

$$(CH_3)_2CH\overset{O}{\underset{}{\overset{\|}{C}}}CH_2\overset{}{\underset{}{C}}CH\diagup\diagdown \longrightarrow (CH_3)_2CH\overset{\overset{+\cdot}{OH}}{\underset{}{C}}\diagdown_{CH_2}$$

(IV) m/e 86

$$CH_3CH_2CH_2\overset{O}{\underset{}{\overset{\|}{C}}}CH_2\overset{}{\underset{H}{C}}CH\diagup\diagdown \longrightarrow CH_2CH_2CH_2\overset{\overset{+\cdot}{OH}}{\underset{}{C}}\diagdown_{CH_2}$$

(V) m/e 86

$$\overset{\overset{+\cdot}{OH}}{\underset{CH_3 \diagdown CH_2}{C}}$$

m/e 58

SCHEME 2. Rearrangement of aliphatic ketones. [From A. M. Duffield, A. V. Robertson C. Djerassi, B. G. Buchanan, G. L. Sutherland, E. A. Feigenbaum, and J. Lederberg, *J. Amer. Chem. Soc.* **91**, 2979 (1969).]

Any substructure thus found is placed on Goodlist and only those chemical structures which contain this unit are generated by the Structure Generator. In the case of the sample shown in Fig. 4, the program fails to recognize the mass numbers corresponding to the McLafferty rearrangement products. Thus all possible acyclic ketones of the proper composition will be produced by the Structure Generator, excluding those which contain the subgraphs represented by (II)–(V) in Scheme 2. The following eight candidate structures were produced by the Structure Generator:

(1) 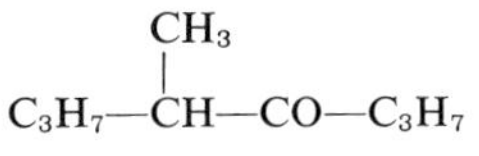 $C_3H_7-\underset{\underset{CH_3}{|}}{CH}-CO-C_3H_7$

(2) $\underset{\underset{CH_3}{/}}{\overset{CH_3}{\diagdown}}CH-\underset{\underset{CH_3}{|}}{CH}COC_3H_7$

(3) $C_2H_5\underset{\underset{C_2H_5}{|}}{CH}-CO-C_3H_7$

(4) $C_2H_5-\underset{\underset{CH_3}{|}}{\overset{\overset{CH_3}{|}}{C}}-CO-C_3H_7$

(5) $C_3H_7CHCOCH\underset{CH_3}{\overset{CH_3}{\diagup}}$ with CH_3 on left C

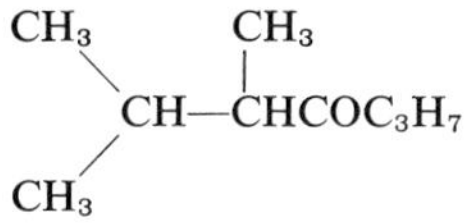

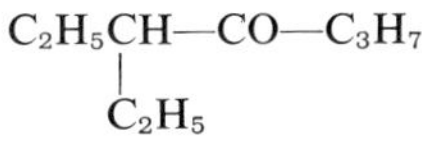

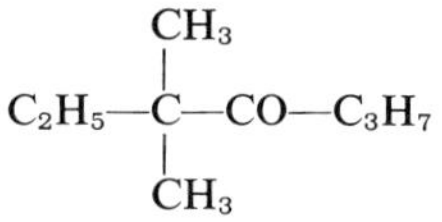

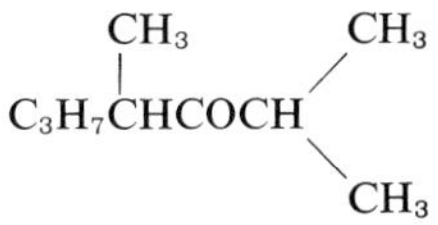

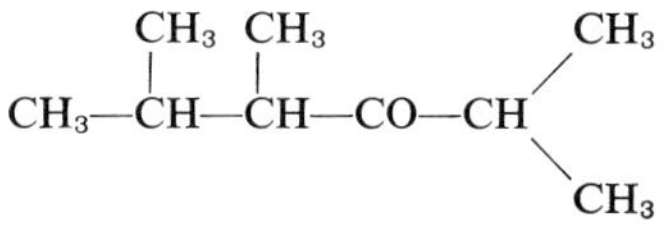

(6) $CH_3-\underset{\underset{}{}}{\overset{\overset{CH_3}{|}}{CH}}-\underset{\underset{}{}}{\overset{\overset{CH_3}{|}}{CH}}-CO-CH\underset{CH_3}{\overset{CH_3}{\diagup}}$

(7) $\underset{C_2H_5}{\overset{C_2H_5}{\diagdown}}CHCO-CH\underset{CH_3}{\overset{CH_3}{\diagup}}$

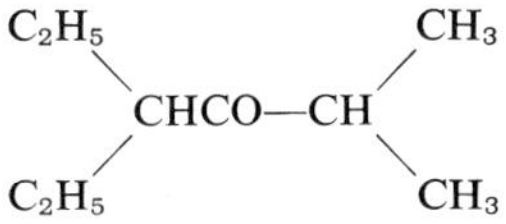

(8) 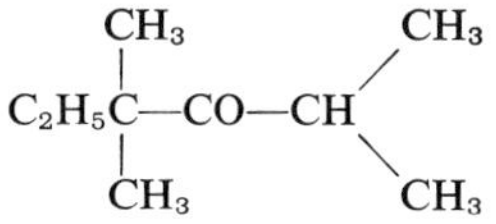 $C_2H_5\underset{\underset{CH_3}{|}}{\overset{\overset{CH_3}{|}}{C}}-CO-CH\underset{CH_3}{\overset{CH_3}{\diagup}}$

Each of these structures is successively scrutinized by the Predictor which deduces a hypothetical mass spectrum for each candidate structure. Below are the lists of predicted mass-intensity pairs for these structures as obtained by the computer:

Candidate 1 (43, 44) (71, 100) (72, 13) (99, 55) (100, 13) (142, 9)
         2 (43, 44) (71, 100) (72, 4) (99, 55) (100, 2) (114, 2) (142, 9)
         3 (43, 44) (71, 100) (86, 4) (99, 55) (114, 4) (142, 9)
         4 (43, 42) (71, 100) (86, 4) (99, 57) (114, 4) (142, 9)
         5 (43, 50) (71, 100) (99, 50) (100, 15) (142, 10)
         6 (43, 50) (71, 100) (99, 50) (100, 2) (142, 10)
         7 (43, 50) (71, 100) (99, 50) (114, 2) (142, 10)
         8 (43, 48) (71, 100) (99, 51) (114, 2) (142, 10)

The program either rejects or accepts structures on the basis of a comparison of the observed (Fig. 4) and predicted mass spectra. Candidates 5 and 6 show no anomalous predicted peaks when compared to Fig. 4, and thus they remain viable candidates.

TABLE V

Heuristic DENDRAL's Interpretation of Mass Spectra of Some Aliphatic Ketones[a]

Compound	No. of aliphatic Isomers	No. of aliphatic Ketones	No. of candidates from Structure generator	No. of candidates from Consistency check	Ranking of candidates
2-Butanone	11	1	1	1	1st, 2-butanone
3-Pentanone	14	3	1	1	1st, 3-pentanone
3-Hexanone	91	6	1	1	1st, 3-hexanone
2-Methylhexan-3-one	254	15	1	1	1st, 2-methylhexan-3-one
3-Heptanone	254	15	2	2	Tie for 1st, 3-heptanone and 3-one
3-Octanone	698	33	4	4	1st, 3-octanone
4-Octanone	698	33	2	1	1st, 4-octanone
2,4-Dimethylhexan-3-one	698	33	4	3	Tie for 1st, 2,4-dimethylhexan-3-one and 2,2-dimethylhexan-3-one
6-Methylheptan-3-one	698	33	4	4	1st, 3-octanone; tied for 2nd, 6-methylheptan-3-one, 5-methylheptan-3-one, and 5,5-dimethylhexan-3-one
3-Nonanone	1936	82	7	7	1st, 3-nonanone
2-Methyloctan-3-one	1936	82	4	3	Consistency check eliminated correct structure because no McLafferty +1 peak was present in original mass spectrum
4-Nonanone	1936	82	4	0	Consistency check eliminated all candidates since no peak was present at m/e 114 (McLafferty rearrangement) in original mass spectrum

[a] A. M. Duffield, A. V. Robertson, C. Djerassi, B. G. Buchanan, G. L. Sutherland, E. A. Feigenbaum, and J. Lederberg, *J. Amer. Chem. Soc.* **91**, 2981 (1969).

The "admissible" molecules are then arranged in an ordered list by the Scoring Function. However, in the present case, the Scoring Function was unable to distinguish between two structures and ranked them both as equally plausible.

Table V summarizes the program's interpretations of mass spectra for several aliphatic ketones.

V. STRUCTURE DETERMINATION OF ALIPHATIC ETHERS

The low-resolution mass spectra of aliphatic ethers are also interpreted by the use of DENDRAL algorithm.[14] The computation process is almost the same as that for the aliphatic ketones, except that the NMR subroutine was newly added as shown in Scheme 3. After inserting the mass spectrum (Fig. 5) and molecular formula of the unknown to the program, the type of ether is designated along the rules for ether identification (Scheme 4) stored in the Preliminary Inference Maker. When the spectrum shown in Fig. 5 is fed to program, the types of ether 2 and ether 4A were designated. Three structural formulas, $n\text{-}C_4H_9OC_2H_5$, $iso\text{-}C_3H_7\text{-}CH_2OC_2H_5$, and $t\text{-}C_4H_9OC_2H_5$ are computed as plausible candidates by Structure Generator.

The Predictor attempts to decide which candidate is the most plausible just as in the case of the aliphatic ketones, and the Scoring Function ranks $t\text{-}C_4H_9OC_2H_5$ as its first preference. However, as is frequently found, mass spectrometry alone is insufficient to separate the correct structure from three or four other dialkyl ethers. Thus an NMR subroutine was incorporated to predict NMR spectra of aliphatic ethers into the final stage of the heuristic DENDRAL. The NMR data necessary for the prediction of chemical shift are stored as a correlation table taken from "Infrared Absorption Spectroscopy" by Nakanishi.[15] The spectra for three candidates are computed by the program as shown at the top of page 301. Comparison of the NMR spectrum (1.09, 3H, T)(1.13, 9H, S)(3.32, 2H, Q) of the unknown compound, $C_6H_{14}O$, in Fig. 5 with the above-mentioned predicted spectra strongly suggests the candidate 3 is the most plausible.

Table VI records other examples in which DENDRAL examined known spectra as "unknown," utilizing the mass spectral information above or combining it with NMR data.

As demonstrated in Table VI, the use of this program either resulted in one or two candidates, if the number of theoretical possibilities is relatively low (less than 200), or at least in a drastic reduction of the number of possible structures (e.g., 10 candidates out of 989 possibilities) using mass spectral input alone. If NMR data are also employed, then a further reduction, usually leading to a single structure, is possible.

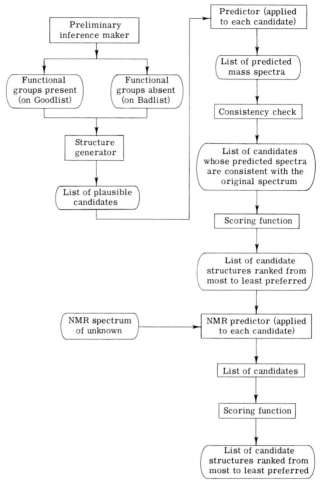

SCHEME 3. Conceptualization of heuristic DENDRAL. [From G. Schroll, A. M. Duffield, C. Djerassi, B. G. Buchanan, G. L. Sutherland, E. A. Feigenbaum, and J. Lederberg, *J. Amer. Chem. Soc.* **91**, 7440 (1969).]

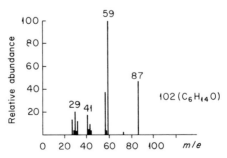

FIG. 5. Mass spectrum of unknown aliphatic ether.

	δ Value	Number of Hydrogens	Multiplicity
Candidate 1	0.90	3	T
	1.30	3	T
	1.40	2	M
	1.90	2	M
	3.40	2	T
	3.40	2	Q
Candidate 2	0.90	6	D
	1.30	3	T
	2.00	1	M
	3.40	2	D
	3.40	2	Q
Candidate 3	1.30	9	S
	1.30	3	T
	3.40	2	Q

As already mentioned, the structures of alkanes, oligopeptides, and aliphatic ketones can be determined automatically by the interpretation of their mass spectra. A computer interpretation of mass spectra of these compounds is rather easy in view of the available fundamental knowledge of their mass spectrometric fragmentation modes.[12,13] Aliphatic ethers, on the

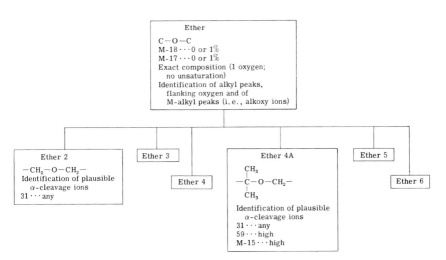

SCHEME 4. Rules for ether identification. (Rules for ether 3, 4, 5, and 6 are omitted.) [From G. Schroll, A. M. Duffield, C. Djerassi, B. G. Buchanan, G. L. Sutherland, E. A. Feigenbaum, and J. Lederberg, *J. Amer. Chem. Soc.* **91**, 7441 (1969).]

TABLE VI

Heuristic DENDRAL Interpretation of the Mass Spectra of Some Aliphatic Ethers[a]

Compound	Number of aliphatic Isomers	aliphatic Ethers	Number of candidates from Structure generator	Number of candidates from Consistency check	Ranking of candidates
1. C–O–C(C)(C)C	14	6	2	2	Correct structure ranked below ethyl n-propyl
2. C–C–O–C(C)(C)C	14	6	4	4	Correct structure ranked first
3. C–C–O–C–C–C–C (with C branch)	32	15	2	2	Correct structure tied with ethyl isobutyl
4. C–C–O–C(C)(C)–C–C	32	15	2	2	Correct structure tied with ethyl n-butyl
5. C–C(C)–O–C(C)–C	32	15	6	6	Correct structure tied with n-propyl isopropyl
6. C–C–O–C(C)–C–C	32	15	3	3	Correct structure ranked first
7. C–C–C–O–C–C–C	32	15	1	1	Correct structure ranked first

#	Structure				Result
8.	C-C(C)(C)-O-C(C)(C)	32	15	10	Correct structure ranked first
9.	C-C-C-C-O-C-C-C-C (with C branch)	72	33	2	Correct structure tied with n-propyl isobutyl
10.	C-C(C)-O-C(C)-C	72	33	1	Correct structure ranked first
11.	C-C-C-C-O-C-C-C-C-C (with C branch)	171	82	3	Correct structure tied with n-butyl isobutyl and diisobutyl
12.	C-C-C(C)-O-C(C)-C	171	82	15	Di-t-butyl ranked first; Correct structure tied for second with isopropyl isoamyl
13.	C-C-O-C-C-C-C-C-C-C	405	194	13	Correct structure tied with 12 other ethyl ethers
14.	C-C-C-C-O-C-C-C-C-C	405	194	8	Correct structure tied with 7 other (C_4)—O—(C_5) ethers
15.	C-C-C-C-C-O-C-C-C-C-C	989	482	10	Correct structure tied with 9 others (C_5)—O—(C_5) ethers
16.	C-C(C)(C)-C-O-C-C-C(C)(C)	989	482	10	Correct structure ranked first

[a] G. Schroll, A. M. Duffield, C. Djerassi, B. G. Buchanan, G. L. Sutherland, E. A. Feigenbaum, and J. Lederberg, *J. Amer. Chem. Soc.* **91**, 7444 (1969).

other hand, showing ambiguity in their mass spectral patterns, were identified by the incorporation of NMR subroutine into the final stage of the program in order to obtain unequivocal answers. Structure determination of compounds such as aliphatic alcohols giving more complex spectra seems to be impossible by the use of mass spectrometry alone.*

VI. CONSTRUCTION OF MOLECULAR STRUCTURE FROM SUITABLE FRAGMENTS

A computation method different from the above-described techniques has been developed by Sasaki and co-workers.[16] It is a rather difficult problem to predict either mass or NMR spectrum of organic compounds except aliphatic ketones, ethers, etc. Sasaki's method assumes that spectral information designates all possible partial structures with which to construct organic molecules. The most probable molecular structure(s) are built up from the designated partial structures. At the outset it is necessary to introduce the work of Munk and his co-workers concerned with the construction of molecular structures from fragments afforded by chemical and physicochemical evidence.[17] The structural implications of the data presented thus far are summarized in the partial structures shown below for N-acetylactinobolamine ($C_9H_{15}NO_3$). The search for the structure of actinobolamine was facilitated by an

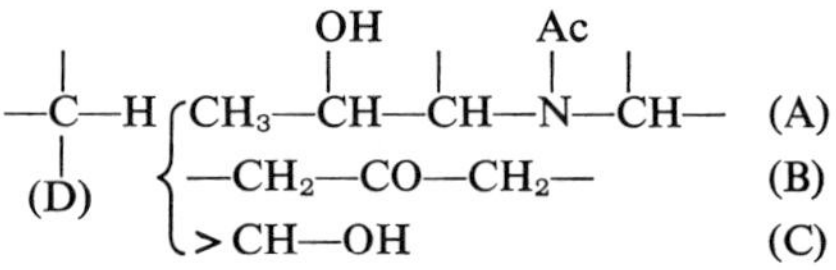

$$
\begin{array}{l}
\mid\text{OH}\text{Ac} \\
\mid\mid\mid\mid \\
-\text{C}-\text{H}\ \ \lceil\text{CH}_3-\text{CH}-\text{CH}-\text{N}-\text{CH}- \quad \text{(A)} \\
\mid\{-\text{CH}_2-\text{CO}-\text{CH}_2- \text{(B)} \\
(\text{D})\lfloor>\text{CH}-\text{OH}\text{(C)}
\end{array}
$$

examination of all structures consistent with the partial structure shown. For this purpose, the four fragments (A–D) of partial structure were fed into a computer programmed to generate all the possible ways of joining the fragments together (i.e., "structures") consistent with the following requirements: (1) fragment C must join directly to A (required by periodate oxidation studies) and (2) structures with multiple bond linkages are excluded. An examination of the six structures (II)–(VII) generated by the computer reveals only one expression, namely (II), consistent with the chemical and spectral properties of actinobolamine.

Sasaki's group has reported on attempts at the automated structure elucidation of organic compounds having fewer than 15 carbons, fewer than one oxygen, and fewer than one site of unsaturation.[18] Experiments which

---

* Private discussion with Dr. A. M. Duffield, Stanford University, March 1970.

were carried out on the degree of accuracy of structure determination of the *various* compounds, chiefly by using NMR, and IR* and UV* as supplementary means have been reported.[16] Since only known compounds were used as examples in the present research, the work was carried out on the premise that the molecular formulas are previously known. The method used by them for determining the structure is summarized as follows: First, atoms, partial structures, and skeletons (these are given as the general term of "component") which are necessary and sufficient enough to construct the molecular structure are chosen beforehand, and there are over 150 kinds of "components" at present (Table VII). Next, how many of which "component" which can be present for a certain sample compound are obtained from the spectral information supplied by the analysis of spectra of NMR, IR, and UV. Then, after the careful comparison of molecular formula with NMR spectrum, "components" which are neither in excess nor deficiency

---

* The data reduction of IR and UV spectra is omitted in this text.

TABLE VII

List of Components

Code No.	Component
1–6	t-Butyl[a]
7–12	gem-diMe
13–28	Me
29–34	Ac
35–40	MeO
41–46	Et
47–53	iso-Pr
54–60	Me—CH—
61–64, 93, 99	—CH—
65–85, 98	—CH$_2$—
86–91	Ph
94, 95, 132, 133	Ar
96, 134	Cyclopropenyl
97, 135	Ketene type
100, 131	—C≡C—
101–107	—CHO
108–113	—COOH
114–119	—OOCH
92, 120–122	—OH
123–129	—O—
130, 136, 137	Carbon
138–158	—CO—

[a] The first line expresses six kinds of t-butyls different in the property and environment, such as t-BuO, t-Bu—Ar—, t-Bu—CO—, t-Bu—C=C, t-Bu—C≡C—, and t-Bu—C—. This applies corresponding to another cases.

for forming one molecule are collected together and these are designated as one "set." The number of "sets" will be small if the designation of "components" from spectral information is accurate; otherwise, the number will be quite large. Finally the molecular structure(s) is built up using "components" in each "set" as material. The flow diagram of the above-mentioned process is shown in Scheme 5.

A. Data Analysis of NMR Spectrum

The positions, peak heights, and area intensities of all NMR signals are computed on line and the digitalyzed spectrum is thus analyzed according to the flow diagram shown in Scheme 6.

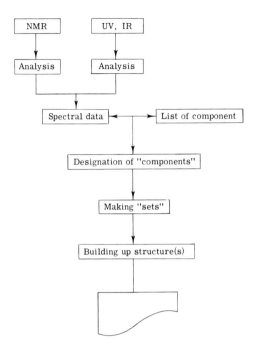

SCHEME 5. Flow diagram of automated chemical structure analysis.

If oxygen is contained in the molecular formula and the information from the IR spectrum supports the presence of OH, it is first necessary to redetermine the spectrum at different temperature. Only the —OH signal position should be temperature dependent. After the —OH signal is assigned, the relative positions of all adjacent signals are measured, and the spectrum is divided and classified into groups of 20 Hz or larger. Next, the area intensity of each subdivided group is calculated, and hydrogen(s) is apportioned to each group in accordance with the molecular formula. If the number of hydrogen(s) of each group can be decided in this manner, a set of signals which takes the pattern which can be recognized easily such as A_2X_3, AX_3, and AX_6 is consolidated and those which are not included in this set become subjects for consideration. If necessary, the latter groups are divided again and hydrogen(s) is allotted to each group. Finally the most probable partial structures are computed as NMR spectral information by comparison with the chemical shift table (Table VIII) prepared by using past literature and data books as references. In order to explain the above method concretely, a model system such as that in Fig. 6 are included. Now, if a distance which exceeds 20 Hz is present between A–B only, this spectrum is divided into two groups of G_1 and G_2. If the signals marked with x are

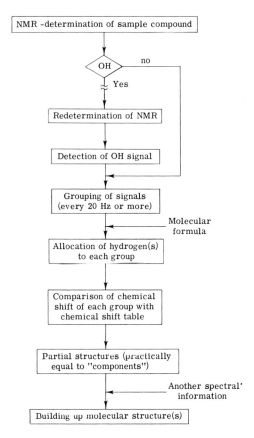

SCHEME 6. Flow diagram of NMR data processing.

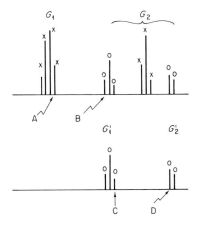

FIG. 6. Dividing of NMR spectrum.

assigned to the A_2X_3 pattern on the basis of the number of hydrogen(s) apportioned from the results of area calculations and measurement of signal intervals, these are excluded from consideration, and only those which have been indicated by the o mark are taken into consideration. In case the distance between C–D is over 20 Hz, G_2 is divided into G_1' and G_2', then finally all of these groups are collated with the chemical shift table and those which indicate an A_2X_3 pattern are assigned as having either (CO)—C_2H_5 or (O)—C_2H_5 structures. All possible partial structures which G_1' and G_2' involve are also suggested by use of the chemical shift table. The partial structure which can be obtained from NMR information is in general a "component," as shown in Table VII.

B. Example of Designation of "Components" by Computer-Aided Spectrometry

This is an example for showing analysis of a spectrum and the kinds of "components" which are assigned when ethyl levulinate ($C_7H_{12}O_3$) is used as sample.

NMR spectrum of this compound was recorded as follows (Sweep Range = 540–30 Hz):

No.	Position	Area	Height	
1	256.8	181	32	Group 1
2	249.8	621	94	
3	242.6	745	102	
4	235.4	249	38	
5	168.2	270	19	Group 2
6	162.8	900	61	
7	158.0	947	87	
8	153.4	938	96	
9	148.8	664	60	
10	143.6	186	20	
11	129.2	2490	425	
12	127.6	119	24	
13	82.2	710	124	Group 3
14	75.0	1313	244	
15	67.8	649	105	
Total area = 10982				

TABLE VIII
NMR Chemical Shift (Ref. TMS)

Code	Structure	Chem. shift range (Hz)	Code	Structure	Chem. shift range (Hz)
1	t-Bu—O	84–60	24	—CH$_2$CO	90–48
2	—Ar	96–72	25	—CH$_2$—C=C	90–48
3	—CO	90–60	26	—CH$_2$C≡C	90–48
4	—C=C	90–54	27	—CH$_2$C	84–30
5	—C≡C	90–54	28	—CH—	90–30
6	—C	66–36	29	—C(=O)—O	150–108
7	CH$_3$\\C—O / CH$_3$	84–48	30	—C(=O)—Ar	150–108
8	—Ar	84–66	31	CH$_3$C(=O)—O—C	150–108
9	—CO	90–54	32	—C=C	150–108
10	—C≡C	90–48	33	—C≡C	150–108
11	—C≡C	90–48	34	—C	150–108
12	—C	84–36	35	CH$_3$O—O	210–186
13	CH$_3$C—O	90–48	36	—Ar	246–210
14	—Ar	108–60	37	—C(=O)	246–210
15	—CO	84–42	38	—C=C	240–210
16	—C=C	84–42	39	—C≡C	240–210
17	—C≡C	84–42	40	—C	210–186
18	—C	90–30	41	Et—O	84–54
19	CH$_3$—C≡C	132–108	42	—Ar	90–54
20	—C≡C	144–90			282–186
21	—Ar	168–120			222–144

7. Automated Analysis Systems

No.	Structure		No.	Structure	
44	—C≡C—	90–48, 162–102	65	CCH₂—O—	282–186
45	—C≡C—	90–48, 180–114			
46	—C—	84–30, 144–90	66	—Ar	222–144
47	iso-Pr—O—	84–54, 300–90	67	—C=	168–108
48	—O—C—	90–54, 420–276			
			68	O	162–102
49	—Ar, —C=O	90–54, 300–90	69	—C=C—	180–114
50		90–48, 300–90	70	—C≡C—	144–0
			71	—C—	300–252
51	—C=C—	90–48, 300–90	72		320–252
52	—C≡C—	90–48, 300–90	73		324–240
53	—C—	84–30, 300–90			
54	CH₃CH—O—	90–30, 300–90	74	O—CH₂—O—, —Ar	318–240
55	—OC—	90–30, 468–276	75	—C=O	312–228
			76	—C=O	252–210
56	—Ar	90–30, 300–90	77	ArCH₂—Ar	252–192
57	—C=O—	90–30, 300–90			
			78	—C=C	246–192
58	—C≡C—	90–30, 300–90	79	—C≡C—	246–192
59	—C≡C—	90–30, 300–90	80	—C—	240–162
60	—C—	90–30, 300–90			
61	CH	300–0	81	—C=CCH₂—C≡C—	240–150
62	CH—OC	420–276	82	—C=CCH₂—C≡C—	264–192
			83	—C≡CCH₂—C=C—	216–150
63	CH(OC)₂=O	468–390	84	—C≡CCH₂—C≡C—	264–192
			85	—C=O	264–192
64	CH(OC)₃=O	480–390	86	C₆H₅—O—	

TABLE VIII —(continued)

Code	Structure	Chem. shift range (Hz)	Code	Structure	Chem. shift range (Hz)
87	C₆H₅—Ar		107	CH₂—CHO	600–540
88	—C— ‖ O		108	—O—COOH	600–360
			109	—Ar—	600–360
89	C≡C		110	O=C—	600–360
90	C≡C		111	C=C—	600–360
91	—C—		112	C≡C—	600–360
92	Ar—OH		113	—C—	600–360
93	H of Ar	540–396	114	—O—O—	504–468
			115	ArO—	504–468
			116	C—O—	504–468

94	CH₂ Ar
	\ /
	  O
	  |
	  O
		378–330	117	C=C—O—
				‖
				O
					504–468

			118	C≡C—O—	504–468
95	—702055—		119	C—O—	504–468
96	H \ C=C	480–390	120	—C—OH	Whole region
			121	—O—OH	Whole region
97	O=C=C	480–228			
98	CH₂=	396–264			
99	H \ C=				
100	H—C≡C	480–228			
101	Ar—CHO	192–120			
102	—C— ‖ O	612–540			
103	C=C—	612–540			
104	C≡C—	612–540			
105	—C—	600–540			

When the distances of signal–signal were measured, 4–5 and 12–13 yield values over 20 Hz and consequently, the spectrum was divided into groups 1, 2, and 3. Then, 12 hydrogens were apportioned from the ratio of area of each group as follows: two to group 1, seven to group 2, and three to group 3. From the facts that groups 1 and 3 are composed of four and three signals, and contain two and three hydrogens, respectively, and that $J_{H\text{-}H}$ is 7.2 Hz, these two groups are assigned as typical A_2X_3 type patterns. Therefore, the presence of —O—CH_2CH_3 is strongly suggested by comparison with Table VIII. This partial structure corresponding to (41) of Table I is stored and used for the construction of molecular structure.

Next, the "components" which can be present for group 2 with seven hydrogens are explained. According to the NMR data processing program (Scheme 6), the presence of the following components can not be neglected for group 2 in view of the fact that the presence of —O—CH_2CH_3 (41) has been confirmed, the number of sites of unsaturation is two from the molecular formula, and the presence of a carbonyl group can be detected in the IR spectrum*: (1), (3), (4), (6), (7), (9), (10), (12), (13), (15), (16), (18), (20), (22), (24), (25), (27), (28), (29), (31), (32), (34), (37), (38), (40), (61), (62), (65), (67), (68), (70), (71), (73), (74), (80), (81), (97), (98), (99), (102), (105), (106), (107), (110), (113), (116), (117), and (119). In order to find which of these "components" are the most suitable for group 2, the spread of the chemical shift table (Table VIII) and each candidate is studied carefully with the number of hydrogens held at seven. At this stage, a study of all candidates is omitted to avoid complication, and only the possibility of the presence or absence of the following four candidates is studied here: (1) t-Bu—O, (34) CH_3—CO—C—, (67) C—CH_2—CO, and (70) —C—CH_2C—. The spread of the chemical shift of (1) was 84–60 Hz; thus the presence of this "component" in group 2 was neglected. The spread of the chemical shift of (34) was 108–150 Hz, and this was confirmed from the point of view of chemical shift. Also, an A_3 type pattern (that is, a signal having three protons) should appear in order for (34) to be present, and the signal (No. 11) of the higher magnetic field side of group 2 fulfills this condition. It is difficult to detect the peculiar pattern types of (67) and (70), even if hydrogens are attached to the carbon at left side of (67) and carbons at both sides of (70). Consequently, the only method in such cases is to decide on the presence or absence based only on chemical shift. Therefore, (67) and (70) are not discarded but remain as "components" for building up the structure.

Finally, the following "components" will remain for group 2 if each is examined carefully with the NMR analytical program in this manner: (20) CH_3—C≡C, (31) CH_3COCO, (32) CH_3COC≡C, (34) CH_3COC—,

* The use of the IR subroutine is not described in this text.

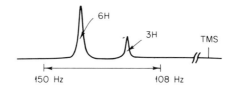

FIG. 7. Chemical shift range of CH_3C-C (34).
            $\parallel$
            O

(61) CH, (67) C—CH$_2$—CO, (68) C—CH$_2$—C=C, (70) C—CH$_2$—C, (81) CO—CH$_2$—C=C, and (83) C=C—CH$_2$—C=C.

When the types of "components" have been decided, useless computation can be avoided if the numbers of each "component" or at least the upper limit can be decided. The following operation is carried out for this purpose. First, only the chemical shifts are taken into consideration for study of methine and methylene groups. For example, (67) —C—CH$_2$—CO indicates spreading of chemical shift at 168–108 Hz, and consequently all areas of signals which appear in this range are summed. The value obtained by dividing this with the area of 1 proton × 2 is taken as the upper limit of the number of (67) present. According to this method, the upper limit of (67) is 3 in case of ethyl levulinate:

$$\frac{270 + 900 + 947 + 938 + 664 + 186 + 2490 + 119}{2 \times (10982/12)} \fallingdotseq 3$$

Also, the upper limit of the number of methyl groups is decided by taking into consideration the number of signals which appear in its range and the area of each signal. For instance, if two signals appear in the chemical shift range in which (34) CH$_3$CO—C should appear and if the left is composed of 6H and the right side 3H (Fig. 7), the upper limit of the presence of (34) becomes 1 even if the number of signals is two. However, in case of methyl which does not indicate a pattern which can be identified as easily, the upper limit is decided by calculation of the total area/3 × area of 1 proton. The upper limits of the number of above-designated "components" obtained by this method are given below, and this is the conclusion of NMR analysis for ethyl levulinate:

Comp.	(20)	(31)	(32)	(34)	(61)	(67)	(68)	(70)	(81)	(83)
Max.	1	1	1	1	7	3	3	3	2	1

C. Method of Building Up Molecular Structure from Designated "Components"

Sasaki's group developed the KUMT (= KUMITATE = build up) program for building up the "components" (designated by spectral information) into one molecule. This program is composed of two parts: (1) "Components" which are neither in excess nor deficiency for building up a structure corresponding to the molecular formula are selected from the designated "components"; this is provisionally called a "set" and all possible "sets" are assembled at this stage. (2) All possible structural formulas are built up on the basis of the number of bonds which the "components" in each "set" have.

The program is so arranged that the structural formulas thus computed, which differ only in graphical representation but are identical from the chemical point of view, for example, CH_3CH_2OH and $HOCH_2CH_3$, and

[structure: benzene ring with two CH_3 groups at ortho positions]

and

[structure: benzene ring with two CH_3 groups at ortho positions]

are considered as being the same in order to avoid unnecessary repetition. However, as chemical common sense beyond this is not contained in the KUMT program, structures which ignore bond angles or in which there is steric hindrance, that is, structures which violate the essence of organic chemistry, are also built up. This matter can be remedied to some extent by inserting a "bad list," as was done by Djerassi and co-workers,[8,11,147] in suitable places in KUMT to provide a sort of check point. This is one of the important improvements which should be taken up in the future.

When "components" designated by spectral information on ethyl levulinate are inserted in the above-mentioned KUMT program as input data, first the type and number of "components" which are neither in excess nor deficiency for forming a complete molecule are selected, and eight "sets" are formed. Next, one or two structural formulas are built up from each "set" so that a total of nine responses are sent out as output (Fig. 8).

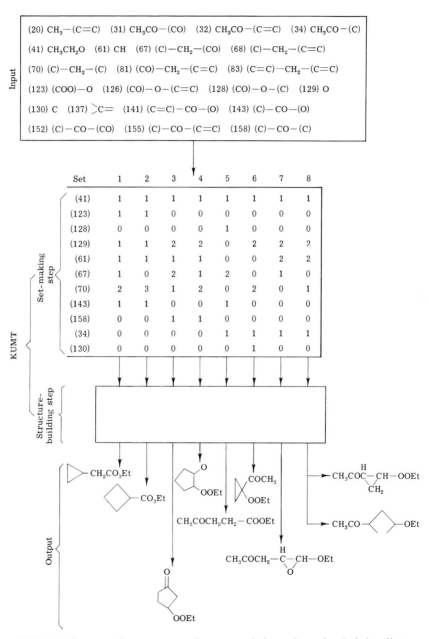

FIG. 8. Computation process of structural formulas of ethyl levulinate. Components, (129), (130), and (137) are added in all cases. Components, (123), (126), (128), and (141)–(158) are computed by the IR subroutine which was omitted in this text.

D. Results

The results given for several compounds tested to date are shown as follows:

Compound Tested	Structure(s) Computed
Allyl alcohol	$CH_2=CHCH_2OH$
Furoic acid	Three isomeric structures including
	![furoic acid structure] (furan ring with COOH)
2-Cyclopenten-1-one	(cyclopentenone structure)
Ethyl acrylate	$CH_2=CHCOOC_2H_5$
Pivalic acid	$(CH_3)_3CCOOH$
Diallyl ether	$CH_2=CHCH_2OCH_2CH=CH_2$
4,4-Dimethoxy-2-butanone	$(CH_3O)_2CHCH_2COCH_3$
2-Methylpentane	Three isomeric structures including
	$(CH_3)_2CH(CH_2)_2CH_3$
2,2-Dimethylbutane	$(CH_3)_3CCH_2CH_3$
2,3-Dimethylbutane	$(CH_3)_2CHCH(CH_3)_2$
Anisol	Four isomeric structures including
	(phenyl)–OCH_3
2,4-Heptadien-6-one	Four isomeric structures including
	$CH_3CH=CHCH=CHCOCH_3$
3-Methyl-2-cyclohexen-1-one	Seven isomeric structures including
	(cyclohexenone with CH_3)
2-Methoxycyclohexanone	Three isomeric structures including
	(cyclohexanone with OCH_3)
2-Heptanone	Two isomeric structures including
	$CH_3CO(CH_2)_4CH_3$
3-Heptanone	Two isomeric structures including
	$CH_3CH_2CO(CH_2)_3CH_3$
4-Heptanone	$CH_3(CH_2)_2CO(CH_2)_2CH_3$
2,4-Dimethylpentanone-2	$(CH_3)_2CHCOCH(CH_3)_2$

Compound Tested	Structure(s) Computed
1-Heptanol	Four isomeric structures including $CH_3(CH_2)_6OH$
2-Heptanol	Five isomeric structures including $CH_3(CH_2)_4CH(OH)CH_3$
4-Heptanol	Two isomeric structures including $CH_3(CH_2)_2CH(OH)(CH_2)_2CH_3$
3-Ethylpentanol-3	Two isomeric structures including $(C_2H_5)_3COH$
1,1-Dimethylpentanol-1	Four isomeric structures including $CH_3(CH_2)_3C(CH_3)_2OH$
2,2-Dimethylpentanol-1	Three isomeric structures including $CH_3(CH_2)_2C(CH_3)_2CH_2OH$
Acetophenone	Four isomeric structures including ⌬—COCH₃
4-Hydroxyacetophenone	Three isomeric structures including HO—⌬—COCH₃
Methyl salicylate	Twenty-nine isomeric structures including ⌬(OH)(COOCH₃)
1-Octyne	$CH_3(CH_2)_5C{\equiv}CH$
1-Octene	Eight isomeric structures including $CH_3(CH_2)_5CH{=}CH_2$
2-Ethylhexanal	Three isomeric structures including $CH_3(CH_2)_3CH(C_2H_5)CHO$
2-Octanone	Five isomeric structures including $CH_3CO(CH_2)_5CH_3$
5-Methylheptan-3-one	Five isomeric structures including $C_2H_5COCH_2CH(CH_3)C_2H_5$
Cyclooctanol	cyclooctane—OH
Ethyl diethoxyacetate	$(C_2H_5O)_2CHCOOC_2H_5$
2,5-Diethoxytetrahydrofuran	Five isomeric structures including C_2H_5O—(tetrahydrofuran-O)—OC_2H_5
Dibutyl ether	Two isomeric structures including $CH_3(CH_2)_3O(CH_2)_3CH_3$

Compound Tested	Structure(s) Computed
1-Octanol	Four isomeric structures including $CH_3(CH_2)_7OH$
2-Octanol	Seven isomeric structures including $CH_3(CH_2)_5CH(OH)CH_3$
2-Ethylhexanol-1	Five isomeric structures including $CH_3(CH_2)_3CH(C_2H_5)CH_2OH$
Mesitylene	Thirty-seven isomeric structures including
![mesitylene structure]
Nonanal	Four isomeric structures including $CH_3(CH_2)_7CHO$
2-Nonanone	Six isomeric structures including $CH_3CO(CH_2)_6CH_3$
2,6-Dimethylheptanone-4	Two isomeric structures including $(CH_3)_2CHCH_2COCH_2CH(CH_3)_2$
5-Nonanone	Six isomeric structures including $CH_3(CH_2)_3CO(CH_2)_3CH_3$
1-Nonanol	Twenty-one isomeric structures including $CH_3(CH_2)_8OH$
3-Nonanol	Twenty isomeric structures including $CH_3(CH_2)_5CH(OH)C_2H_5$
4-Nonanol	Forty-five isomeric structures including $CH_3(CH_2)_4CH(OH)(CH_2)_2CH_3$
5-Nonanol	Six isomeric structures including $CH_3(CH_2)_3CH(OH)(CH_2)_3CH_3$
Carvone	Two-hundred-twenty-one isomeric structures including
![carvone structure]
6-Heptynylmalonic acid	Eight isomeric structures including $(HOOC)_2CH(CH_2)_5C{\equiv}CH$
Decaline	Twenty-one isomeric structures including
![decaline structure]
Diamyl ether	Nine isomeric structures including $CH_3(CH_2)_4O(CH_2)_4CH_3$

Compound Tested	Structure(s) Computed
Diisoamyl ether	Four isomeric structures including $(CH_3)_2CH(CH_2)_2O(CH_2)_2CH(CH_3)_2$
Ethyl cinnamate	$C_6H_5CH{=}CHCOOC_2H_5$
Bornyl acetate	One-hundred-ninty-nine isomeric structures including ![bornyl acetate structure] OCOCH$_3$

References

[1] S. Abrahamsson, S. Stenhagen-Ställberg, and E. Stenhagen, *Biochem. J.* **2PB**, 92 (1964).
[2] B. Pettersson and R. Ryhage, *Ark. Kemi* **26**, 293 (1967).
[3] B. Pettersson and R. Ryhage, *Anal. Chem.* **39**, 790 (1967).
[4] M. C. Hamming and R. D. Grigsby, *12th Tetrasect. Okla. Sect. Amer. Chem. Soc.*, (1966).
[5] K. Biemann, C. Cone, and B. R. Webster, *J. Amer. Chem. Soc.* **88**, 2597 (1966). K. Biemann, C. Cone, B. R. Webster, and G. P. Arsenault, *J. Amer. Chem. Soc.* **88**, 5598 (1966).
[6] M. Senn, R. Venkataraghavan, and F. W. McLafferty, *J. Amer. Chem. Soc.* **88**, 5593 (1966).
[7] K. Heyns and H. F. Grützmacher, *Tetrahedron Lett.* p. 1761 (1963); *Justus Liebigs Ann. Chem.* **669**, 189 (1963). N. S. Wulfson, *Tetrahedron Lett.* p. 2805 (1965); p. 39 (1966).
[8] A. M. Duffield, A. V. Robertson, C. Djerassi, B. G. Buchanan, G. L. Sutherland, E. A. Feigenbaum, and J. Lederberg, *J. Amer. Chem. Soc.* **91**, 2977 (1969).
[9] B. Buchanan, G. Sutherland, and E. A. Feigenbaum, "Machine Intelligence," Vol. 4, p. 209. Edinburgh Univ. Press, Edinburgh, 1969.
[10] J. H. Beynon and A. E. Williams, "Mass and Abundance Tables for Use in Mass Spectrometry," p. 26. Elsevier, Amsterdam, 1963.
[11] J. Lederberg, G. L. Sutherland, B. G. Buchanan, E. A. Feigenbaum, A. V. Robertson, A. M. Duffield, and C. Djerassi, *J. Amer. Chem. Soc.* **91**, 2973 (1969).
[12] H. Budzikiewicz, C. Djerassi, and D. H. Williams, "Interpretation of Mass Spectra of Organic Compounds," p. 6. Holden-Day, San Francisco, California, 1964.
[13] K. Biemann, "Mass Spectrometry, Organic Chemical Application," p. 119. McGraw-Hill, New York, 1962.
[14] G. Schroll, A. M. Duffield, C. Djerassi, B. G. Buchanan, G. L. Sutherland, E. A. Feigenbaum, and J. Lederberg, *J. Amer. Chem. Soc.* **91**, 7440 (1969).
[15] K. Nakanishi, "Infrared Absorption Spectroscopy," Appendix I, p. 223. Holden-Day, San Francisco, California, 1962.

[16] S. Sasaki, Y. Kudo, S. Ochiai, Y. Ishida, and H. Abe, *Pittsburgh Conf. Anal. Chem. Appl. Spectrosc.*, **1970**, 148.
S. Sasaki, Y. Kudo, S. Ochiai, and H. Abe, *Mikrochim. Acta*, **1971**, 726.
[17] M. E. Munk, C. S. Sadano, R. L. McLean, and T. H. Haskel, *J. Amer. Chem. Soc.* **89**, 4158 (1967).
[18] S. Sasaki, H. Abe, T. Ouki, M. Sakamoto, and S. Ochiai, *Anal. Chem.* **40**, 2220 (1968).

APPENDIX TO CHAPTER 5

Newest Reports on NQR Applications in Organic and Metalloorganic Chemistry

M. G. VORONKOV AND V. P. FESHIN

Many new papers on the application of NQR have appeared since this chapter was sent to the publisher. It therefore seems very appropriate to complement the chapter with these new data.

For the last two years the attention of scientists studying the NQR method has been focused mainly on metalloorganic compounds and different homolytic and isostructural series of organic compounds. As in most previous papers, only qualitative concepts of the Townes–Dailey theory have been utilized. In some papers attempts have been made to estimate quantitatively the order of chemical bonds, their π-character, and the hybridization of the participating atoms. Some authors have also attempted to correlate the experimental NQR data with the results of quantum-chemical calculations of molecules. In this respect should be mentioned the attempt[1] to calculate the electronic structure of molecules of the $CH_{4-n}Cl_n$ series by the Pariser–Parr–Pople (PPP) and Wolfsberg–Helmholtz (WH) methods, and thus to determine the nuclear quadrupole-coupling constants eQq, dipole moments, and the spin-spin coupling constants J_{C-H} on the basis of the charge distribution in these molecules. The charges corresponding to $3p_z$ orbitals of the chlorine atom in the molecules of the $CH_{4-n}Cl_n$ series correlate with experimental values of eQq.

Calculations by the PPP method for molecules with σ-electrons show that good qualitative agreement with experimental data can be obtained if the hybrid orbitals and the degree of hybridization are properly chosen. However, making predictions on the basis of this method is handicapped by the uncertainty of the hybridization parameters. Calculations by the Wolfsberg–Helmholtz method utilized two approximations: (1) charge compensation on the central atom and (2) existence of charges on all atoms. The nuclear quadrupole-coupling constants and of spin–spin interactions obtained in these calculations are in better agreement with experimental data than those obtained by the simple WH method.[2] The bond electronegativity equalization method (BEEM) was used for verification of the NQR frequency interpretation of molecules of the RR'R"CCl series based on the Townes–Dailey theory. The nuclear quadrupole-coupling constants were determined from NQR data and gas microwave spectroscopy data. The s character of the σ-orbitals of the Cl atom differs slightly for the gas and the solid states. It depends on the type of atom bonded to chlorine and is independent (or more precisely it is dependent only to a small extent) of the rest of the molecule, having influence on the ionic character of the bond with the chlorine atom. The ^{35}Cl NQR spectra of RR'R''SiCl series molecules, as well as of molecules with the Cl atom bonded to P and S atoms, have been interpreted without taking acount of the $p\pi$–$d\pi$ conjugation.[3]

Also, the s- character of the σ-orbitals of the chlorine atom in some organic and inorganic compounds has been calculated from the nuclear quadrupole-coupling constants.[4]

Application of the results of calculations by the CNDO/2 method for molecules of the ROCOCl series (R = CH_3, CH_2F, CHF_2, and CF_3) to the Townes–Dailey equation showed[5] that the π-character of the C−Cl bond is insignificant (or nonexistent) and that it remains practically constant as the number of the fluorine atoms is changed. Consequently, the change of the NQR frequency of these compounds is explained in terms of the σ-inductive effect.

The insignificance of the π-character of the C—Cl bond in ROCOCl was also supported by the correlation of the ^{35}Cl NQR frequencies with the frequencies of the corresponding compounds of the RCH_2Cl series, which are entirely free of conjugation. Comparison of the calculated results with the experimental ^{35}Cl NQR frequencies of the ROCOCl molecules shows that a slight change in the population of the σ- and π-orbitals involves considerable changes in the NQR frequencies.

The ^{35}Cl NQR frequencies of the ROCOCl series have been compared with the spectra of RCOCl and RCl.[5]

A linear correlation has been observed between the ^{35}Cl NQR frequencies of 1,3-RC_6H_4COCl and 1,3-RC_6H_4Cl as well as of 1,4-RC_6H_4COCl and 1,4-RC_6H_4Cl. This correlation shows that the mechanism of influence on the chlorine atom in the COCl group of the R substituents is analogous to that in the aromatic ring. The sensitivity of the chlorine atom to the influence of substituents is higher in the aromatic ring than in the COCl group. The reason is evidently that changes in the electron density in the acid chlorides occur in the carbonyl group rather than in the C—Cl bond.[5a] The ^{35}Cl NQR frequencies of 1,3- and 1,4-RC_6H_4COCl also correlate with inductive constants, σ_I, of the substituents. It can be seen from this correlation that the inductive influence of the substituents is stronger in the meta than in the para position. The ^{35}Cl NQR spectra of substituted benzoic acid chlorides show that in these molecules the influence of substituents on the chlorine atom in the COCl group is transmitted mainly by the inductive mechanism. If such an effect occurs the change in the double-bond character of the C—Cl bond is either negligible or does not exert any noticeable influence on the q_{zz} component of the electric field-gradient along the C—Cl bond axis.[6]

The character of the bridging group X in molecules of the

$$\text{ClOC}-\!\!\left\langle\!\!\bigcirc\!\!\right\rangle\!\!-X-\!\!\left\langle\!\!\bigcirc\!\!\right\rangle\!\!-\text{COCl}$$

series exerts weak influence on the chlorine atom in the COCl group; e.g., the ^{35}Cl NQR frequencies of such compounds vary very slightly with changing X.[6]

A satisfactory linear correlation[7] has been observed between the ^{35}Cl NQR frequencies of the chloronitromethanes and the $\Sigma\sigma^*$ of the substituents. The deviation from linear dependence in the high-frequency region of the NQR frequency of chloronitrotoluene is explained in terms of hydrogen bonding. The dilution of $HC(NO_2)_2Cl$ with diglyme (1:1 by volume) results in a significant decrease in the NQR frequency. The value of ν^{77} of fluorochlorodinitromethane shows since a much larger deviation from the linear correlation toward the lower frequency range, since in this compound the intramolecular interaction is practically nonexistent. Dilution with cyclohexane (1:1 by weight) does not result in any change of ν^{77}. This deviation from the straight line is explained by the p,σ-conjugation in F—C—Cl where the fluorine atom acts as an electron donor to the chlorine atom. In order to prove this hypothesis the ^{35}Cl NQR spectra of chloro derivatives of the $RR'R''CCH_2Cl$ series were studied, in which the methylene group prevents conjugation between the fluorine and chlorine atoms. For this series of compounds, a good correlation has been observed between ^{35}Cl NQR frequencies and $\Sigma\sigma^*$. The NQR frequencies of $F(NO_2)_2CCH_2Cl$ correlate well with a straight-line dependence. Therefore, in this case the fluorine atom affects the chlorine atom by the inductive mechanism only.[7]

The ^{35}Cl NQR frequencies of chloroalkenes, $RR'C=R'R'''$ (R, R', R'', or R''' = Cl) vary according to the inductive effect of the substituents, but the contributions of the hybridization and the π-character of the C—Cl bond remain practically constant for all compounds in this series.[8] On this basis, the ionic character of the C—Cl bond in chloroalkenes was calculated and found to be only slightly different from the ionic character of the C—Cl bond of the corresponding chloroalkanes. The ^{35}Cl NQR data of some chloroalkenes have been compared with their microwave spectra.[8]

Assuming that substituent influence operates only by the inductive mechanism in the substituted benezenes and thiophenes, which supposition is not entirely true, the conclusion was drawn[9] that the inductive effect is better transmitted by the carbon atoms in the benzene ring than by the sulfur atom in the thiophene ring. It has also been shown that the ionic character of the C—Br bond in the benezene derivatives is higher than in the thiophene derivatives. The ^{79}Br NQR frequencies of the bromine atoms in the α-position in the thiophene ring correlate with the sum of the Hammet constants of substituents in the ring.[9]

A good correlation has been observed[10] between the ^{35}Cl NQR frequencies of the α-chlorine atom in α-chloroethylbenzene derivatives of the C_6H_5CClRR' series and the sum of the inductive Taft constants of R and R'. An identical correlation has been observed for the molecules of the ClCClRR' and HCClRR' series.[10] Two groups of compounds, $XCCl_2$-CCl_nH_{3-n} and $XCHClCCl_nH_{3-n}$, distinguish themselves in the XCClRR' series. In all three groups investigated (X = C_6H_5, Cl, and H) the transmittance of the substituent effect to the chlorine which is localized at the α-carbon atom is greater for the $XCHClCCl_nH_{3-n}$ than for the $XCCl_2$-CCl_nH_{3-n} type molecules. The data obtained indicate that accumulation of the electron-accepting substituents at the atom having a resonating nucleus leads to a decrease in its sensitivity to other substituents.[10]

The results of NQR investigations of benzyl chlorides have already been discussed. Recently, Dewar and Herr[11, 12] have reinvestigated the ^{35}Cl NQR spectra of these compounds. Their results as well as the formerly published ^{35}Cl NQR frequencies of substituted benzyl chlorides have been correlated[11] with the spectroscopic constants K_i.[12a] These authors explain the lower ^{35}Cl NQR frequency of benzyl as compared with methyl chloride, by the conjugation of the chlorine atom with the aromatic ring.[11, 12] However, as shown above, this sort of conjugation is just one of the reasons which might explain a decrease in the NQR frequency in these compounds.[12b]

Similarity of the values of ^{35}Cl NQR frequencies of 1,2-bis(chlorophenyl)-ethanes, $(ClC_6H_4CH_2—)_2$ indicates that intramolecular interaction between the chlorine atoms through the $—C_6H_4CH_2CH_2C_6H_4—$ system is transmitted to only an insignificant extent, if any. Therefore, it can be stated that the average ^{35}Cl NQR frequency of $ClC_6H_4(CH_2)_nX$ at $n \geq 2$ is equal to ~ 34.3 MHz.[13]

There is a linear correlation between the average values of the ^{35}Cl NQR frequencies of chloro-substituted ethylanthracenes and the absolute values of the dipole moments induced in the C—Cl bond. According to the authors,[14] this correlation confirms the significant influence of field effect on NQR frequency. However, the calculation of the C—Cl bond moments based on the assumption that electronic influence of the substituents is transmitted along the atomic chain, failed to enable prediction of the character of the frequency changes in the compounds in question. The possibility of the simultaneous action of the field effect and of the transmission of the influence along the atomic chain has been rejected by the above authors[14] for two reasons. First, they believe that the transmission of the influence through

Appendix: Newest Reports on NQR Applications 327

two or more linking atoms does not occur. Second, although in the *cis*- and *trans*-1,2-dichloro derivatives the participation of the transmission of the influence along the bonds should be identical, the relation between the NQR frequency and the calculated bond moments (or the induced moments) should not be linear if one accepts the idea that the contribution of the field effect is general for the entire series of linear relationships.

The influence of the substituent R on the ^{35}Cl NQR frequencies of trichloromethyl derivatives of the RCCl$_3$ series has been investigated.[15] An attempt has been made to establish the structure of *o*-, *m*-, and *p*-chlorobenzoyl chlorides from the ^{35}Cl NQR spectra, and from the dipole moments.[16] The NQR method was used[17] to confirm the structure of the products of free-radical telomerization of propylene with chloroform. In the spectra of the telomers NQR signals were found in the regions of 32, 36, and 38 MHz, corresponding to the ^{35}Cl resonance in the CH$_3$CHClCH$_2$CHCl$_2$ and CCl$_3$ groups, respectively. These results support the conclusion that these telomers have the following structures: CCl$_3$[CH$_2$CH(CH$_3$)]$_n$H ($n = 1$–3) and CHCl$_2$-[CH$_2$CH(CH$_3$)]$_n$Cl ($n = 1,2$).

^{35}Cl NQR spectra were used[18] to confirm the structures of products of photochemical chlorination of isomeric cresols and methylanisols by sulfuryl chloride.

The electronic structures of chloro-substituted cyclopropanes have been discussed[19] on the basis of the ^{35}Cl NQR spectra.

Assuming that the contributions of the separate chlorine atoms ^{35}Cl NQR frequencies are additive, the NQR frequencies have been assigned to various positions of chlorine in chloro-substituted norbornanes and norbornenes.[12] The differences in the ^{35}Cl NQR frequencies of *syn*- and *anti*-norbornenes can be explained[20] either by a change in hybridization of the C—Cl bond or by the formation of a weak hydrogen bond between the π-electrons of the double bond and the hydrogen atoms of the CH$_2$ bridge.

In the ^{35}Cl NQR spectrum of octachloropentafulvalene, which consists of four lines, the upper two lines correspond to the Cl atoms (two in each ring) involved in the intramolecular coordination.[21] This is in full agreement with X-ray structural data.[22]

The results of the examination of the NQR Zeeman spectra of ^{81}Br and ^{35}Cl of *p*-bromo- and *p*-chlorophenols[23] are in agreement with X-ray structural data for *p*-bromophenol and with previously reported data for NQR spectra of *p*-chlorophenol.[24] Four crystallographically unequivalent orientations, instead of the two reported earlier, were found for *p*-bromophenol.[25] The values of the assymetry parameters are as follows: $\eta_{Br} = 0.053 \pm 0.001$, $\eta_{Cl} = 0.071 \pm 0.001$ (34.12 MHz), and $\eta_{Cl} = 0.096 \pm 0.001$ (34.34 MHz).[23]

The value of $\eta_{Br} = 0.094 \pm 0.001$ was found from the NQR Zeeman spectrum of methyl-*p*-bromobenzoate. This indicates a considerable shift of the bromine atom's π-electrons toward the rest of the molecule, caused by the electron-accepting character of the carboxymethyl group.[26]

The basic directions of the electric field gradient as well as the η values and the angles between the bonds of the carbon atoms linking the chlorine atoms (ClCC = 118.1 ± 0.25°)[27] were determined from the Zeeman NQR spectra of 2,4-dichlorophenoxyacetic acid.

The symmetrical and asymmetrical positions of the atoms in phthalyl chloride molecules[28] were determined from ^{35}Cl NQR frequencies.

The NQR spectra of ^{35}Cl and ^{2}H of CDCl$_3$, CD$_2$Cl$_2$, CD$_2$ClCD$_2$Cl, CD$_3$CCl$_3$, and CD$_3$CCl$_2$CD$_3$ molecules have been obtained.[29] The value of η for deuterium connected to the sp^3-hybridized carbon atom is small. In contrast to ^{35}Cl, the NQR frequency of ^{2}H$_2$ increases with decreasing electronegativity of the substituents at the carbon atom. According to Bersohn[30] the observed tendency is entirely natural.[29]

Almost one-half of the papers published during recent years involving the NQR method were devoted to organometallic compounds. In addition to the use of chlorine and bromine as indicating resonating atoms, as is the case with most organic compounds, use has also been made of certain metals. Since many of them have nuclei spin greater than $\frac{3}{2}$ (^{55}Mn, ^{185}Re, ^{187}Re, ^{127}Sb, ^{123}Sb, ^{59}Co, ^{209}Bi, etc.), additional information concerning the electron density distribution in their organic derivatives could be derived from the η values. Nevertheless, the organic compounds involving these metals remain much less examined by the NQR method than their purely organic counterparts and, as a result, this examination is still far from systematic.

The parameters of asymmetry for chlorine atoms in polycrystalline samples of BCl$_3$ and C$_6$H$_5$BCl$_2$ have been measured by the Morino and Toyama method. The decrease in the NQR frequency of ^{35}Cl in the (C$_6$H$_5$)$_{3-n}$-BCl$_n$ series, when n is varied from 3 to 1, is explained by the increase in the π-bonding of boron and chlorine atoms and by the decrease in the electronegativity of the boron atom with the increasing number of phenyl groups.[31] The ^{35}Cl NQR spectra of some of the chlorine-containing boron organic compounds have been correlated with X-ray structural data.[32]

The e^2Qq_{zz} values of ^{127}I of o-, m-, and p-B-iodocarboranes as well as the ^{35}Cl NQR frequencies of the corresponding chlorine-containing compounds increase in the sequence from para to meta and ortho isomers. A linear relationship has been established between the values of e^2Qq_{zz} for ^{127}I and ν^{77} for ^{35}Cl in monohalocarboranes. This leads to the assumption that the asymmetry parameter in o-, m-, and p-B-monochlorocarboranes is just as small as it is in the B-iodocarboranes. The computation of the number of unbalanced electrons indicates that the electric field gradient along the axis of the B—I bond is slightly higher than along the axis of the B—Cl bond.[33]

Appendix: Newest Reports on NQR Applications   329

The NQR frequency of ^{75}As in *ortho*-carbarsborane (97.3 MHz) is much higher than in the meta (58.2 MHz) and para (61.36 MHz) isomers. This is reflected not only in the high value of the asymmetry parameter of the ortho isomers, but also in a much higher value of the nuclear quadrupole-coupling constant of the As atom.[34] Such a difference in the NQR frequencies of ortho, meta, and para isomers is in full agreement with their reactivities. The relatively large width of the *o*- and *m*-$B_{10}H_{10}$AsCH lines ($\Delta\nu^{77} \approx 2$ MHz) evidently indicates the statistically disordered distribution of molecules in the crystal. The NQR line width of the *p*-$B_{10}H_{10}$AsCH is smaller by one order of magnitude.[34]

The NQR spectra of ^{115}In and ^{69}Ga in organic compounds of indium and gallium, MR_3 (M = In, Ga, R = alkyl, aryl), indicate that these molecules have planar structures in the solid state.[35] The population of the p_z orbitals of the M atoms in $In(CH_3)_3$ molecules, where intramolecular interactions could be expected, and in $Ga(C_6H_5)_3$, where a possibility of $p\pi$–$p\pi$ interactions exists, is close to zero.[35]

The ^{27}Al and ^{35}Cl NQR data for molecules of the R_nAlX_{3-n} series (R = alkyl, aryl; X = halogen) have been utilized in structure determination, and led to the conclusion that many of these compounds have a dimeric bridged structure.[36]

The ^{35}Cl NQR frequencies of methylchlorosilanes correlate well with the frequencies of the stretching vibrations ν_{CH} as well as with the chemical shifts τ_{CH}.[37] As reported previously,[37] the straight line correlating the ^{35}Cl NQR frequencies of RR'R"SiCl series molecules with the sum of the Taft constants of the R substituents also applies to the NQR frequencies of the compounds with R = C_6H_5 and CH=CH_2 [38], whose π-electrons, according to contemporary concepts, may conjugate with the vacant 3*d* orbitals of the silicon atom. The frequencies of the organochlorosilanes with R = CH_2Cl (as shown previously by the present authors[38a] the chlorine atoms interact with the silicon atoms) also fall on the same line. As was the case previously[37a], this led to the conclusion that the interaction between the substituents connected to the Si atom is weak; this conclusion was confirmed by the small value of the ρ coefficient in Eq. (12) [38b, 37a*] in the case of organosilicon compounds of the RR'R"SiCH$_2$Cl series.[38] In organic compounds of the RR'R"MCl (M = C) series the sensitivity of ^{35}Cl frequency to an increase in the number of electronegative substituents at the central carbon atom decreases, while in the case of M = Si, Ge, and possibly Sn, a reverse relationship has been observed.[38] The decrease in the sensitivity of organic molecules indicates that a saturation effect takes place.[38, 38c-e] The increase in the sensitivity of the organometallic compounds can be explained in terms of an increase in the number of the electron acceptor substituents at the M

* See Chapter 5, page 178.

atom which are able to interact with Si, Ge, and Sn atoms, leading to a decrease in this interaction.[38] The transmittance of the electronic influence of substituents to the chlorine atom in molecules of the RR'R"CCl series decreases considerably if one of the substituents is silicoorganic in nature.[38]

The purity of compounds of the $R(C_6H_5)SnX_2$ type (X = halogen) has been established by means of ^{35}Cl NQR and PMR spectra as well as thin-layer chromatography. For example, the purity of ethylphenyldichlorostannate $C_2H_5(C_6H_5)SnCl_2$, was confirmed by the absence in its NQR spectrum (16.405 and 16.936 MHz) of the splitting frequency corresponding to diethyldichlorostannate (15.420 and 15.740 MHz) and to diphenyldichlorostannate (16.947 MHz).[39]

For compounds of the $SnCl_nR_{4-n}$ series (R = CH_3, C_2H_5, CH_2Cl, C_6H_5) the NQR frequency increases linearly with the increasing number of chlorine atoms. Larger splitting of the NQR frequencies in the spectra of $RSnCl_3$ indicates an occurrence in these compounds of intermolecular coordination. The ^{35}Cl NQR frequencies of tri-, di-, and monochlorostannates have been compared with the quadrupole splitting constants (ΔE_q) of the Mössbauer spectra of ^{119}Sn. The points corresponding to the stannic chloride and trichlorostannates deviate from the previously established correlation. This shows that intermolecular coordination either does not exist in these compounds or has an entirely different character than in compounds such as $RSnCl_3$ or R_2SnCl_2. The linear increase of ΔE_q (^{119}Sn) with decreasing ^{35}Cl NQR frequency means that the p-electron density of the chlorine atom increases with decreasing p-electron density of the Sn atom.[40]

Increasing the number of halogen atoms in $X_nSn[Co(CO)_4]_{4-n}$ molecules (X = Cl, Br, CH_3; n = 0, 1, 2, or 3) leads to an increase of the ^{59}Co NQR frequency; methyl groups exert a reverse effect.[41] An analysis of the NQR spectra of ^{59}Co, based on the assumption that the electric field gradient of ^{59}Co is determined mainly by the population of $3d$ orbitals of the metal, led to a conclusion that in cobalt the population of the $3d_{z^2}$ and d_{xz} or d_{yz} orbitals is lower than that of the $d_{x^2-y^2}$ and d_{xy} orbitals. The value eQq_{zz} of ^{59}Co correlates satisfactorily with the stretching frequencies of Co in IR spectra of hexane solutions of these compounds.[41]

The small or zero value of the asymmetry parameter in NQR spectra of ^{59}Co (at 25°C) of $Cl_3MCoL(Co)_3$ molecules (M = Si, Ge, Sn; L = C, phosphorus-containing ligand) is in agreement with the trans structure of these compounds.[42] An exchange of the CO group in $Cl_3MCo(CO)_4$ (M = Ge, Sn) for phosphorus-containing ligands does not result in any significant changes in the ^{59}Co eQq value. At the same time the corresponding changes in siliconorganic compounds (M = Si) are more significantly pronounced. The electronic effects in $Cl_2Sn[Co(CO)_4]_2$, $Cl(C_6H_5)Sn[Co(CO)_4]_2$, and $(C_6H_5)_2Sn[Co(CO)_4]_2$ have been investigated[43] by ^{35}Cl and ^{59}Co NQR.

Appendix: Newest Reports on NQR Applications   331

The NQR spectra of ^{59}Co of $Cl_2Ge[Co(CO)_4]_2$ and $Hg[Co(CO)_4]_2$ have been obtained.[41]

The temperature dependence of the NQR frequency of ^{59}Co in dicobalt-octacarbonyl has been investigated in the range from 77° to 303°K.[44] At 77°K, the two crystallographically independent positions of Co correspond to $e^2Qq/h = 90.18 \pm 0.15$ MHz, $\eta = 0.3149 \pm 0.001$ and $e^2Qq/h = 89.30 \pm 0.15$ MHz, $\eta = 0.4837 \pm 0.001$. The values of η at higher temperatures exhibit a greater spread. The population of the Co atom orbitals has been calculated.[44]

The ^{59}Co NQR frequencies of the salts of *trans*-dihalobis(ethylenediamine) cobalt $[Co(en)_2Hal_2]X$ vary considerably with changes in the nature of the anion.[45] For example, when passing from *trans*-$[Co(en)_2Cl_2]Cl$ to *trans*-$[Co(en)_2Cl_2]Cl \cdot HCl \cdot 2H_2O$, the ^{59}Co NQR frequency undergoes about a 15% increase, and when passing from *trans*-$[Co(en)_2Br_2]ClO_4$ to $[Co(en)_2Br_2]Br \cdot HBr \cdot 2H_2O$ this increase is about 25%. The difference between the experimental values of the ^{59}Co NQR frequencies of the *trans*-$[Co(en)_2Cl_2]Cl$ and *trans*-$[Co(en)_2Cl_2]Cl \cdot HCl \cdot 2H_2O$ is about the same as the difference between the values of the electric field gradients calculated on the basis of the models involving punctual charges and a contribution from the anti-screening factor. This led to the conclusion that the effect of the crystal lattice is responsible for the large NQR frequency shifts resulting from a change in the character of the anion.[45]

In molecules of the $(C_5H_5)_4TiBr_{4-n}$ series[46] an increase in the number of cyclopentadienyl groups leads to a sharp increase in the NQR frequency (about 20 MHz for each ring). Assuming that the cyclopentadienyl ring is a stronger π-donor than the bromine atom, then increasing n results in a decrease of the electron density accepted by d orbitals of the titanium atom from the p_x and p_y orbitals. An analogous relationship has been observed when the bromine atoms are replaced by ethoxy groups; however, in this case the interpretation of the NQR spectra is complicated by coordination interactions. The NQR frequencies indicate that the —OC_2H_5 groups exert a much weaker influence on the bromine atom than the C_5H_5 groups. The data obtained indicate that, in contrast to the halogenated derivatives of the elements of Group IVb, the basic influence on the NQR frequencies of Br in the Ti—Br bond is exerted not by polarization of the σ-bonds, but by the change in the population of the p_x and p_y orbitals of halogen which participate in π-bond formation. The observed changes in the NQR frequency of these molecules are in full agreement with the mass and the PMR spectrometry data of $(C_5H_5)TiCl_n(OC_2H_5)_{3-n}$ molecules.[47] The relatively large splitting of the NQR resonance lines of titanium ethoxy-bromides has been explained by coordination interaction and this was confirmed by X-ray structural investigation of the $Cl_2Ti(OC_2H_5)_2$,[48] which revealed its dimerization in the solid state through the ethoxy groups. Considerable differences in the frequencies of the ^{81}Br NQR spectrum of $Br_2(OC_2H_5)_2Ti$ (61.32 and 71.16 MHz) have been explained by the differences in the electronic distribution at the equatorial and axial bromine atoms. The NQR spectrum of $Br_3Ti(OC_2H_5)$, which consists of two groups of lines with average frequencies of 60.1 and 71.7 MHz,[46] has been explained in a similar fashion.

The NQR method has been applied [49] to the determination of the phenyl group orientation in phenylchlorophosphate molecules having the structure of the trigonal bipyramid. NQR frequencies of compounds with axial and equatorial chlorine atoms are considerably different. The NQR data show that replacing chlorine atoms by phenyl groups occurs in equatorial positions and that NQR frequencies of equatorial chlorine atoms undergo practically no change upon successive substitution. An increase in the number of the phenyl groups leads to a significant decrease in the NQR frequency of the axial chlorine atoms.[49]

The correlation between the NQR frequencies and σ-constants of molecules of the RSO_2Cl series[49a] has already been mentioned. A detailed examination of the NQR data[50] showed that it is impossible to derive a single correlation equation. Compounds containing substituents with unshared electron pairs (halogen, OR, NR_2) and substituents connected to the atom via the carbon atom (hydrocarbon radicals) have to be treated as different series. From the hydrocarbon series should be excluded substituents containing a halogen atom in α the position to the sulfur atom ($R \cdot = CH_2Cl, CCl_3$).

The very low value of ρ_c in Eq. (7)* for compounds of the $RSOCl_2$ series (independent of the type of R substituents) indicates that the influence of ρ_c on the chlorine atom is transmitted only by the inductive effect. It cannot be excluded that in the case of substituents with unshared electron pairs the $d\pi$–$p\pi$ interaction could inhibit transmission through the sulfur atom.[50]

At room temperature $\eta = 9\%$[51] for $Bi(C_6H_5)_3$ and $\eta = 58\%$[52] for $BiCl_3$. The quadrupole-coupling constant is more than twice as high for the first compound than for the second one. This variation in the value of η is due to structural differences. No substantial change in the ^{209}Bi NQR spectrum as compared to that of $Bi(C_6H_5)_3$ has been observed upon introduction of substituents into the aromatic ring. The invariability of the value of η indicates that the structure of the molecule remains intact.[53] Since the electric field gradient at the bismuth atom should as a rule be determined by the contribution of the unshared electron pairs,[54] consequently the variation in the degree of π-electron contribution is small.[53]

The values of e^2Qq of compounds of the R_3SbX_2 series (X = halogen) correlate with the sum of the Taft constants of the substituents. However,

* See Chapter 5, page 173.

Appendix: Newest Reports on NQR Applications 333

entirely different correlation equations apply in the case of alkyl (e^2Qq = 73.7 + 100,8$\Sigma\sigma$*) and electron acceptor substituents (e^2Qq = 298.6 + 73.9 $\Sigma\sigma$*). The high value of e^2Qq [13, 54a] in the latter case indicates[55] conjugation of the π-system of the equatorial substituents with the vacant orbitals of the central atom. This is reflected in the increased population of the axial sp^3d hybrid orbitals and as a consequence in the increased value of e^2Qq of the atom. Such an interaction is supported by the increased NQR frequencies of ^{79}Br and ^{81}Br in (cis-ClCH=CH)$_3$SbBr$_2$ relative to those in (n-C$_4$H$_9$)SbBr$_2$. When alkyl groups are equatorial substituents such conjugation does not occur.[55]

The NQR frequency of ^{123}Sb and consequently the value of $(e^2Qq/h)_{mol}$[56] are higher for (CD$_3$)$_3$SbX$_2$ than for (CH$_3$)$_3$SbX$_2$ (X = Cl, Br). This has been explained by the higher positive charge on the carbon atom in the H$_3$C—Sb bond as compared with the D$_3$C—Sb bond, which leads to an increased population in the p_x and p_y orbitals of antimony. The increase of the NQR frequency of ^{35}Cl and ^{79}Br in (CH$_3$)$_3$SbX$_2$ molecules as compared with the corresponding deutero derivatives has been explained in the same way.[56]

The significant increase of the ^{75}As NQR frequency of the R$_n$AsR'$_{3-n}$ series, in which the character of the electronic influence of both substituents is distinctly different, relative to compounds with identical R and R' groups is explained by the substantial increase of the angle between sp-hybridized orbitals of As, as it involves an increase of their p character.[57]

The NQR spectra of ^{35}Cl and ^{37}Cl of (C$_2$H$_5$)$_4$NAsCl$_6$[58] and the NQR spectrum of ^{75}As of (CH$_3$)$_4$NH$_2$AsO$_4$[59] have also been examined.

The Zeeman spectrum of dimethyltellurium dichloride[59a] indicates that two molecules with unequivalent chlorine atoms are located in a single unit cell, an arrangement which results in the formation of two components in the ^{35}Cl NQR spectrum. The Cl—Te—Cl bond angle has been determined for these molecules. The value (176 $\pm$ 3°) is close to that obtained from X-ray structural analysis (172 $\pm$ 0.3°).[60]

Introduction of a COR type substituent (R = CH$_3$, CF$_3$, C$_6$H$_5$) into the cyclopentadienyl ring of C$_5$H$_5$Mn(CO)$_3$ results in an increased η value, attributed to interaction between the substituents and the metal atom which results in distortion of the molecular geometry, or in the change of the cyclopentadienyl ring to a fulvene ring. An increase of the η value in the substituted cyclopentadienylrhenium tricarbonyls is higher than in the analogous manganese compounds.[60a, 60b, 61]

It can be deduced from the ^{55}Mn NQR spectra of Mn$_2$(CO)$_{10}$ that the value of η declines with decreasing temperature.[62]

NQR has become one of the most valuable methods of structure determination for charge-transfer complexes and coordination compounds.

During recent years a large number of new data concerning structure determination of such complexes and determination of electronic effects by the NQR method have been reported in the literature.

In most cases the NQR frequency of chloranil complexes with different organic donors is lower than that of the pure chloranil, which is in contrast to the expectation for complexes of π-acceptors.[63] In addition, no correlation between NQR frequencies and ionization potentials of the donors occurs for such complexes. This is explained by the fact that in weak complexes the contribution of crystal field effect to the NQR frequency is about the same as the contribution of charge transfer. Other contributions are also important.[63] The average values of the ^{35}Cl NQR frequencies of the chloranil complexes correlate with the characteristic C—Cl and C=O bond stretching frequencies.[63] The Zeeman spectrum of the monocrystalline complex of chloranil with 8-oxyquinoline has been obtained at room temperature, and the crystal structure the complex has been investigated.[63]

Since complexes of aniline and haloanilines with 2,4,6-trihalophenols of the same composition (1:1 or 1:2) give similar NQR and also IR spectra, it has been assumed[64] that their structures are identical. The spectra of complexes of different composition are different. The large difference in the NQR frequencies of ortho halogen atoms in the complex, as compared with the corresponding difference in the 2,4,6-halophenols, is explained in terms of either intramolecular hydrogen bonding, OH···X, or intermolecular hydrogen bonding,

$$-O-H\genfrac{}{}{0pt}{}{\cdots X \cdots}{\cdots N \cdots}H-O-$$

with participation of two molecules of substituted phenol. In the complex of 1:2 composition, two resonance frequencies of halogen atoms in the para position with respect to the OH group have been observed, one of them much higher than the frequency of the 2,4,6-trihalophenol and the other close to it. The increase of the first frequency has been explained by stronger intermolecular interaction of the halophenol with one of the aniline molecules. The large temperature coefficient of this high-frequency line indicates that $\pi-\pi$ interaction of trichlorophenol molecules with aniline is weakened at higher temperatures because of the higher amplitude of vibration of both aromatic rings. The proximity of the NQR frequencies of the para halogen atom of the second molecule of trihalophenol to those of the starting component may indicate a weak $\pi-\pi$ interaction between this trihalophenol molecule and the aniline, which is likely to be caused by free rotation with respect to the plane of the aniline molecule.[64]

The NQR frequencies of halogen derivatives of aniline, in complexes of 1:1 composition, are lower than those in haloanilines. NQR frequencies of

2,4,6-trihalophenols are changed similarly. This fact led to the conclusion that, in similar type complexes, a weak π–π interaction and a weak OH···N hydrogen bonding occurs.[64]

High-frequency shifts of the NQR signals of 3,5-bromopyridine as a result of complex formation with Br_2, IBr, and ICl have been explained in terms of an enhanced I-effect of the nitrogen atom which participates in coordination.[65] It is assumed that crystal effects in these complexes are negligible and that NQR frequencies depend mainly on the electron acceptor properties of Br_2, IBr, and ICl molecules (the magnitude of acceptor properties increases in the following sequence: Br_2 < IBr < ICl). An approximate distribution of charges in these complexes has been calculated on the basis of NQR data. The results show that the charge is transferred to the terminal atoms of the acceptor. The central halogen atom of the acceptor, which is close to the nitrogen atom, carries, apparently, a positive charge. This supposition is in agreement with the calculation of the charge distribution carried out by the LCAO-MO method utilizing the concept of three-center delocalized σ bonds.[65]

The NQR frequencies of chloroform complexes with nitrogen-containing donors (amines and heterocyclic compounds) are lower than that of chloroform. The magnitude of charge transfer has been calculated from the frequency shifts.[66]

The calculation for complexes of BCl_3 with amines and nitriles by the BEEM method shows that changes in the s-character of the σ-orbitals of the Cl atoms and in the ionic character of B—Cl bonds is small[67] in comparison with BCl_3, which is in agreement with the observed small changes of NQR frequencies of these compounds.[68] ^{35}Cl NQR frequencies of BCl_3 complexes with CH_3CN and $(CH_3)_3N$ or $(C_2H_5)_3N$ calculated by CNDO and SAVE-CNDO methods are in poor agreement with the experimental data.[68] BCl_3 forms weak complexes with aromatic and heterocyclic donors. as is shown by the small low- and high-frequency shifts in the NQR spectra (as compared with BCl_3).[69] The relative order of electron donor activity in these complexes is similar to that of the $GaCl_3$ complexes. Assuming an analogy between the BCl_3 molecule in the complex and the BCl_4^- anion, the charge distribution in BCl_3 was calculated on the basis of the Townes–Dailey theory. This calculation led to the absurd conclusion that the positive charge on the central boron atom increases from BCl_3 to BCl_4^-, which means that the Townes–Dailey theory is not applicable in this case.[69] BCl_3 does not form complexes with trichloroacetonitrile; the product of their interaction has the structure Cl_2B—N=C(Cl)R, since its NQR spectrum consists of four high-frequency lines corresponding to chlorine atoms in a C—Cl bond and two closely spaced low-frequency lines (23.08 and 23.16 MHz) corresponding to chlorine atoms in B—Cl bonds. The latter lines have different frequencies than the lines in the spectra of BCl_3 complexes with aromatic and heterocyclic donors (21-22 MHz). It is evident that the adduct of BCl_3 with monochloroacetonitrile may have several structures depending upon the conditions employed in the course of its preparation.[69]

The complexes of HgX_2 (X = Cl, Br, I) with N-oxides of pyridine and γ-picoline as well as the complexes of $HgBr_2$ with some organic oxygen-containing donors (tetrahydrofuran, acetophenone, benzophenone, phenoxanthin) and $HgCl_2 \cdot 2CH_3OH$ have been investigated by the NQR method.[70] Low-frequency shifts have been observed in the $HgCl_2$ complexes with all the above-mentioned donors (as compared with $HgCl_2$) but, in contrast, and rather unexpectedly, in complexes involving $HgBr_2$ and HgI_2 high-frequency shifts have been observed. According to the authors,[70] this indicates the donor properties of $HgBr_2$ and HgI_2. The structure of the $HgCl_2$ complexes has been determined on the basis of NQR spectra.[70]

The ν^{77} value of the ^{35}Cl NQR spectrum of the complex of $SnCl_4$ with methylal is more than 4 MHz lower than that of $SnCl_4$.[71] This indicates the formation of a complex of 1:2 composition with the six-coordinated stannic atom. The complex of 1:1 composition was not detected since only the intensive lines of free $SnCl_4$[71] appear in the ^{35}Cl NQR spectrum of the mixture of $SnCl_4$ with methylal.

It is believed that the trans configuration[71] is most probable for 1:2 complexes.

The ^{35}Cl NQR spectrum of $SnCl_4 \cdot 2CH_3OCH_2Cl$ reveals that the structure of this complex is octahedral and that the molecules of α-chloromethyl ether are in the cis position.[72] The high-frequency shift in ^{35}Cl NQR spectrum of CH_3OCH_2Cl in the complex (as compared with the starting compound) indicates that the oxygen atom, and not the chlorine atom, participates in the complex formation.[72] The character of the splitting of the spectrum lines of $SnCl_4 \cdot C_6H_5CH_2Cl$ (only the NQR spectrum of $SnCl_4$ was observed) indicates that, as in the preceding case, the structure of the 1:2 complex is octahedral and that donor molecules are in the cis position.[72]

Large low-frequency shifts of NQR signals of ^{75}As in the spectra of $AsCl_3$ complexes with aromatic donors (as compared with pure $AsCl_3$) indicate that the arsenic atom is the center of the complex formation and that it acts as an acceptor of the π-electrons from the aromatic donors.[73] The NQR frequency shifts correlate well with the ionization potentials of the donors.[73] The ^{75}As NQR spectra of $AsBr_3$ complexes with different donors have also been investigated. The transverse and the spin-lattice relaxation times have been measured. The low-frequency shifts of the NQR signals of the complexes correlate well with the electron donor properties of the aromatic donors, and the shifts of ^{75}As NQR frequencies are several times higher than the NQR frequency shifts of ^{121}Sb in the analogous complexes of SbBr$_3$.[74] The multiplicity of ^{35}Cl NQR spectra of the corresponding AsCl$_3$ and SbCl$_3$ complexes are not always equal. Therefore, not all the AsCl$_3$ and SbCl$_3$ complexes have similar structures. In many cases AsCl$_3$ complexes exhibit lower lattice symmetry.[75]

The Zeeman spectrum of 2SbBr$_3 \cdot$ C$_6$H$_5$ has been obtained.[76] The Br–Sb–Br angles in the complex-forming SbBr$_3$ molecule are equal to 92°24′ ± 11′, 97°7′ ± 14′, and 97°40′ ± 6′. The crystal has a monoclinic structure. The assymmetry parameters of the three bromine atoms in the complex are 169 ± 0.001, 0.063 ± 0.002, and 0.094 ± 0.002. The much higher value of η in the complex, as compared with pure SbBr$_3$, is caused by stronger intermolecular bonding in the complex. According to the quadrupole-coupling constants, the Sb—Br bond characteristics in SbBr$_3$ and in the complex 2SbBr$_3 \cdot$ C$_6$H$_6$ are essentially identical.[76] On the basis of analysis of the NQR spectra, the structures of the complexes BiCl$_3 \cdot$ D and SbCl$_3 \cdot$ D (D = ether, anisole, acetone, acetonitrile) were assumed to be similar.[77] Two values of η (14.2 and 34.0%) for BiCl$_3 \cdot$ O(C$_2$H$_5$)$_2$ indicate the existence of two rotational isomers of this complex having different orientation of the ethyl ether groups with respect to the chlorine atoms of BiCl$_3$. The BiCl$_3 \cdot$ C$_6$H$_5$OCH$_3$ complex appears to be the least stable of all the complexes investigated, as it decomposes upon storing to the starting components, and its spectrum shows the signals of BiCl$_3$ superimposed on the NQR frequencies of the complex. Three asymmetry parameters have been observed in the complex with acetonitrile, although there is, practically, only one constant of quadrupole bonding. This indicates that three molecules of the complex characterized by different small structure deformations are placed in each unit cell. The changes of the eQq_{zz} value are entirely independent of the donating properties of the donors in all complexes investigated[77] and this indicates that charge transfer exerts only a small influence on eQq_{zz}. The changes of eQq_{zz} in these complexes are related to the differences in the degree of d hybridization of the bismuth atom and these changes are determined, to some degree, by the geometric structure of the complex.[77]

The NQR method has been applied to structure determination of the CuCl$_2$ and CuBr$_2$ complexes with N-oxypyridine.[78] The NQR spectra ^{35}Cl of L$_2$MCl$_2$ complexes (M = Ni, Pd, Pt) have been obtained.[79]

References

[1] A. O. Litinskii, M. Z. Balyavichus, and A. B. Bolotin, *Teor. Eksp. Khim.* **5**, 316 (1969.)
[2] S. Sichel and M. A. Whitehead, *Theor. Chim. Acta.* **5**, 35 (1966).
[3] M. Kaplansky and M. A. Whitehead, *Mol. Phys.* **16**, 481 (1969).
[4] S. Chandra, *Trans. Faraday Soc.* **63**, 1569 (1967).

[5] R. M. Hart and M. A. Whitehead, *Can. J. Chem.* **49**, 2508 (1971).
[5a] A. A. Neimysheva, G. K. Semin, T. A. Babushkina, and I. L. Knunyants, *Dokl. Akad. Nauk SSSR* **173**, 585 (1967)
[6] E. V. Bryukhova, V. V. Korshak, V. A. Vasnev, and S. V. Vinogradova, *Izv. Akad. Nauk SSSR, Ser. Khim.* p. 599 (1972).
[7] L. V. Okhlobystin, T. A. Babushkina, V. M. Khutoretskii, and G. K. Semin, *Izv. Akad. Nauk SSSR, Ser. Khim.* p. 1899 (1970).
[8] D. E. Koltenbah and A. A. Silvidi, *J. Chem. Phys.* **52**, 1270 (1970)
[9] V. S. Grechishkin, S. I. Gushchin, and Yu. P. Dormidontov, *Teor. Eksp. Khim.* **7**, 706 (1971).
[10] M. G. Voronkov, E. P. Popova, I. P. Biryukov, and V. P. Feshin, *Teor. Eksp. Khim.* **8**, 6 (1972).
[11] M. J. S. Dewar and M. L. Herr, *Tetrahedron* **27**, 2377 (1971).
[12] M. J. S. Dewar, M. L. Herr, and A. P. Marchand, *Tetrahedron* **27**, 2371 (1971).
[12a] D. Biedenkapp and A. Weiss, *J. Chem. Phys.* **49**, 3933 (1968).
[12b] M. G. Voronkov, V. P. Feshin, and E. P. Popova, *Teor. Eksp. Khim.* **7**, 356 (1971).
[13] M. G. Voronkov, V. P. Feshin, and E. P. Popova, *Izv. Akad. Nauk SSSR, Ser. Khim.* p. 2700 (1972).
[14] B. A. Arbuzov, S. G. Vul'fson, I. A. Safin, I. P. Biryukov, and A. N. Vereshchagin, *Izv. Akad. Nauk SSSR, Ser. Khim.* p. 1243 (1970).
[15] E. G. Brame, *Anal. Chem.* **1**, 35 (1971).
[16] V. G. Vasil'ev, V. N. Vasil'eva, Yu. K. Maksyutin, and A. F. Volkov, *Zh. Org. Khim.* **6**, 1864 (1970).
[17] B. A. Englin, V. A. Valovoi, L. G. Zelenskaya, T. A. Babushkina, G. K. Semin, V. B. Bondarev, B. N. Osipov, and R. Kh. Freidlina, *Izv. Akad. Nauk SSSR, Ser. Khim.* p. 2700 (1970).
[18] V. P. Feshin, E. P. Popova, and M. G. Voronkov, *Izv. Akad. Nauk SSSR, Ser. Khim.* p. 187 (1972).
[19] V. S. Grechishkin, I. V. Murin, V. P. Sivkov, and M. Z. Yusupov, *Zh. Fiz. Khim.* **45**, 2891 (1971).
[20] H. Chihara, N. Nakamura, and T. Irie, *Bull. Chem. Soc. Jap.* **42**, 3034 (1969).
[21] D. Gill, M. Hayck, and R. M. J. Loewenstein, *J. Chem. Phys.* **51**, 2756 (1969).
[22] P. J. Weatley, *J. Chem. Soc., London*, p. 4936 (1961).
[23] P. Bucci, P. Cecchi, and A. Colligiani, *J. Chem. Phys.* **50**, 530 (1969).
[24] G. E. Peterson and P. M. Bridenbaugh, *J. Chem. Phys.* **46**, 2644 (1967).
[25] K. V. S. R. Rao and C. R. K. Murty, *Curr. Sci.* **34**, 660 (1965).
[26] R. Angelone, P. Cecchi, and A. Colligiani, *J. Chem. Phys.* **53**, 4096 (1970).
[27] G. E. Peterson, N. Steed, and P. M. Bridenbaugh, *J. Chem. Phys.* **47**, 2262 (1967).
[28] S. Kinaslowski and J. Peitrzak, *Bull. Acad. Pol. Sci., Ser. Sci. Chim.* **16**, 155 (1968).
[29] J. L. Ragle and K. L. Sherk, *J. Chem. Phys.* **50**, 3557 (1969).
[30] R. Bersohn, *J. Chem. Phys.* **32**, 85 (1960).
[31] J. A. S. Smith and D. A. Tong, *J. Chem. Soc., A* p. 173 (1971).
[32] J. A. S. Smith and D. A. Tong, *J. Chem. Soc., A* p. 178 (1971).
[33] E. V. Bryukhova, Yu. V. Gol'tyapin, V. I. Stanko, and G. K. Semin, *Zh. Strukt. Khim.* **12**, 1095 (1971).
[34] E. V. Bryukhova, V. P. Kyskin, and L. I. Zakharkin, *Izv. Akad. Nauk SSSR, Ser. Khim.* p. 532 (1972).

Appendix: Newest Reports on NQR Applications 339

[35] V. I. Svergun, L. M. Bednova, O. Yu. Okhlobystin, and G. K. Semin, *Izv. Akad. Nauk, Ser. Khim.* p. 1449 (1970).

[36] M. J. S. Dewar, D. B. Patterson, and W. I. Simpson, *J. Amer. Chem. Soc.* **93**, 1030 (1971).

[37] H. Kriegsmann, G. Engelhardt, R. Radeglia, and H. Geisler, *Z. Phys. Chem.* **240**, 294 (1969).

[37a] I. P. Biryukov, M. G. Voronkov, and I. A. Safin, *Dokl. Akad. Nauk SSSR.* **165**, 857 (1965).

[38] M. G. Voronkov, V. P. Feshin, V. O. Reikhsfeld, and L. S. Romanenko, *Zh. Obshch. Khim.* **43**, 7 (1973).

[38a] M. G. Voronkov, V. P. Feshin, V F. Mironov, S. A. Mikhaiilyants, and T. K. Gar, *Zh. Obshch. Khim.* **1**, 2211 (1971).

[38b] I. P. Biryukov, M. G. Voronkov, and I. A. Safin, *Izv. Akad. Nauk Latv. SSR, Ser. Khim.* p. 706 (1965).

[38c] R. W. Taft, *in* "Steric Effects in Organic Chemistry" (M. S. Newman, ed.), p. 536. Wiley, New York, 1956; *in* "Prostranstvennye effekty v organicheskoi khimii" (A. N. Nesmeyanov, ed.), p. 562. IL, Moskva, 1960.

[38d] Yu. A. Zhdanov and V. I. Minkin, "Korrelyatsionnyi analiz v organicheskoi khimii." Izd. Rostovskogo Universiteta, Rostov na Donu, 1966.

[38e] V. A. Pal'm, "Osnovy kolichestvennoi teorii organicheskikh reaktsii." Khimiya, Leningrad, 1967.

[39] I. P. Gol'dshtein, E. N. Guryanova, L. S. Mel'nichenko, N. N. Zemlyanskii, T. I. Perepelkova, Yu. K. Maksyutin, and K. A. Kocheshkov, *Dokl. Akad. Nauk SSSR* **201**, 105 (1971).

[40] Yu. K. Maksyutin, V. V. Khrapov, L. S. Mel'nichenko, G. K. Semin, N. N. Zemlyanskii, and K. A. Kocheshkov, *Izv. Akad. Nauk SSSR, Ser. Khim.* p. 602 (1972).

[41] D. D. Spenser, J. D. Kirsch, and T. L. Brown, *Inorg. Chem.* **9**, 235 (1970).

[42] K. Ogino and T. L. Brown, *Inorg. Chem.* **10**, 517 (1971).

[43] J. D. Graybeal, S. D. Ing, and M. W. Hsu, *Inorg. Chem.* **9**, 678 (1970).

[44] E. S. Mooberry, H. W. Spiess, B. B. Garrett, and R. K. Sheline, *J. Chem. Phys.* **51**, 1970 (1969).

[45] T. B. Brill and Z. Z. Hugus, *J. Phys. Chem.* **74**, 3022 (1970).

[46] G. K. Semin, O. V. Nogina, V. A. Dubovitskii, T. A. Babushkina, and A. N. Nesmeyanov, *Dokl. Akad. Nauk SSSR*, **194**, 101 (1970).

[47] A. N. Nesmeyanov, E. I. Fedin *et al.*, *Tetrahedron, Suppl.* **8**, Part 2, 389 (1969).

[48] W. Haase and H. Hoppe, *Acta Crystallogr., Sect. B* **24**, 281 (1968).

[49] V. I. Svergun, V. G. Rozinov, E. F. Grechkin, V. G. Timokhin, Yu. K. Maksyutin, and G. K. Semin, *Izv. Akad. Nauk SSSR, Ser. Khim.* p. 1918 (1970).

[49a] G. K. Semin, A. A. Neimysheva, and T. A. Babushkina, *Izv. Akad. Nauk SSSR, Ser. Khim.* p. 486 (1970).

[50] G. K. Semin, A. A. Neimysheva, and T. A. Babushkina, *Izv. Akad. Nauk SSSR, Ser. Khim.* p. 486 (1970).

[51] H. G. Robinsohn, H. G. Dehmelt, and W. Gordy, *Phys. Rev.* **89**, 1305 (1953).

[52] H. G. Robinsohn, *Phys. Rev.* **100**, 1731 (1955).

[53] L. N. Petrov, I. A. Kyuntsel', and V. S. Grechishkin, *Vestn. Leningrad. Univ., Fiz., Khim.* **1**, 167 (1969).

[54] T. P. Das and E. L. Hahn, *Solid State Phys., Suppl.* **1**, 152 (1958).

[54a] T. P. Das and E. L. Hahn, "Nuclear Quadrupole Resonance Spectroscopy." Academic Press, New York, 1958.

[55] V. I. Svergun, A. E. Borisov, N. V. Novikova, T. A. Babushkina, E. V. Bryukhova, and G. K. Semin, *Izv. Akad. Nauk SSSR, Ser. Khim.* p. 484 (1970).
[56] T. B. Brill and Z. Z. Hugus, *J. Chem. Phys.* **53**, 1291 (1971).
[57] V. I. Svergun, T. A. Babushkina, G. N. Shvedova, L. V. Kudryavtsev, and G. K. Semin, *Izv. Akad. Nauk SSSR, Ser. Khim.* p. 482 (1970).
[58] J. V. DiLorenzo and R. F. Schneider, *J. Phys. Chem.* **72**, 761 (1968).
[59] A. P. Zhukov, L. S. Golovchenko, and G. K. Semin, *Izv. Akad. Nauk SSSR, Ser. Khim.* p. 1399 (1968).
[59a] W. J. Orvill-Thomas, *Quart. Rev., Chem. Soc.* **11**, 162 (1957); *Usp. Khim.* **27**, 731 (1958).
[60] C. D. Christofferson, R. A. Sparks, and J. D. McCullough, *Acta Crystallogr.* **11**, 782 (1958).
[60a] G. K. Semin, "Nekotorye primeneniya yadernogo kvadrupol'nogo rezonansa v khimii." Avtoreferat doktorskoi diss., Moskva, 1970.
[60b] E. V. Bryukhova, "Issledovanie elektronnykh effektov i koordinatsionnogo vzaimodeistviya v elementoorganicheskikh soedineniyakh metodom YaKR." Avtoreferat kand. diss., Moskva, 1969.
[61] A. N. Nesmeyanov, G. K. Semin, E. V. Bryukhova, T. A. Babushkina, K. N. Anisimov, N. E. Kolobova, and Yu. V. Makarov, *Tetrahedron Lett.* p. 3987 (1968).
[62] S. L. Segel, *J. Chem. Phys.* **51**, 848 (1969).
[63] H. Chihara and N. Nakamura, *Bull. Chem. Soc. Jap.* **44**, 2676 (1971).
[64] A. V. Korshunov, T. A. Babushkina, V. F. Shabanov, V. E. Volkov, and G. K. Semin, *Opt. Spektrosk.* **30**, 887 (1971).
[65] G. A. Bowmaker and S. Hacobian, *Aust. J. Chem.* **22**, 2047 (1969).
[66] J. P. Lucas and L. Guibe, *Mol. Phys.* **19**, 85 (1970).
[67] M. Kaplansky and M. A. Whitehead, *Mol. Phys.* **16**, 461 (1969).
[68] M. Kaplansky and M. A. Whitehead, *Can. J. Chem.* **48**, 697 (1970).
[69] S. Ardjomand and E. A. C. Lucken, *Helv. Chim. Acta* **54**, 176 (1971).
[70] T. B. Brill and Z. Z. Hugus, *J. Inorg. Nucl. Chem.* **33**, 371 (1971).
[71] I. S. Morozova, G. M. Tarasova, V. V. Ivanov, E. V. Bryukhova, and N. S. Enikolopyan, *Dokl. Akad. Nauk SSSR* **199**, 654 (1971).
[72] Yu. K. Maksyutin, V. P. Makridin, E. N. Guryanova, and G. K. Semin, *Izv. Akad. Nauk SSSR, Ser. Khim.* p. 1634 (1970).
[73] V. S. Grechishkin and M. Z. Yusupov, *Opt. Spectrosk.* **29**, 804 (1970).
[74] A. D. Gordeev, V. S. Grechishkin, and M. Z. Yusupov, *Zh. Strukt. Khim.* **12**, 725 (1971).
[75] V. P. Anferov, V. S. Verzilov, V. S. Grechishkin, and M. Z. Yusupov, *Zh. Strukt. Khim.* **12**, 924 (1971).
[76] T. Okuda, H. Terao, O. Ege, and H. Negita, *Bull. Chem. Soc. Jap.* **43**, 2398 (1970).
[77] Yu. K. Maksyutin, E. N. Gur'yanova, and G. K. Semin, *Izv. Akad. Nauk SSSR, Ser. Khim.* p. 1632 (1970).
[78] J. J. R. Frausto da Silva, L. F. Vilas Boas, and R. Wootton, *J. Inorg. Nucl. Chem.* **33**, 2029 (1971).
[79] S. Kondo, E. Kaiuchi, and T. Shimizu, *Bull. Chem. Soc. Jap.* **42**, 2050 (1969).

Author Index

Numbers in parentheses are reference numbers, and are included to assist in locating references when the authors' names are not mentioned in the text. Numbers in italics refer to the page on which the full reference is listed.

A

Abe, H., 295, 304(16, 18), 305(16), *321*
Abe, M. M., 55, *73*
Abe, Y., 54(188), 55(188, 189, 191), *73*
Abragam, A., 92(29), *97,* 100(7), 102(18), 108(23), *145, 146*
Abrahamsson, S., 285(1), *320*
Adman, E., 59(202), *73*
Adrian, F. J., 115(31), 141(40), *146*
Afanas'ev, V. A., 170(30), *225*
Agron, P. A., 274(171), *284*
Ainbinder, M. E., 202(176), 217(235), *230, 232*
Ainsworth, J., 11(42), 13, 18(42), 51, *69*
Akishin, P. A., 47(161), 48(161), 67(242, 243, 245, 246), *72, 74*
Aleksandrov, A. Yu., 248(11, 18), 253(28, 30, 35, 38), 256(61), 258(71, 73), 261(35), *279, 280, 281*
Aliev, L. A., 256(65), *281*
Allen, H. C., 183(106), 200(106), 216(106, 226), *228, 232*
Allen, R., 133(73), *149*
Allerhand, A., 94(34), 95(34), 96(34), *97*
Allred, A. L., 183(112), *228*
Almenningen, A., 18(64, 65), 24, 29(92), 36, 37(134), 41(144), 46(144), 59, 60(212), *69, 70, 71, 72, 73, 74,* 262(99), *282*
Ambe, F., 258(75), *281*
Anders, L. R., 151(2, 3), 155(2, 3), *166*
Andersen, B., 60(212), 61, *74*
Andersen, P., 61, 62, 63, *74*
Anderson, A. S., 96(40), *97*
Ando, K. J., 240(4), *278*
Andreassen, A. L., 35(124), 49(169), *71, 72*
Andrews, L. J., 218(247), *232*
Anet, F. A. L., 79(11), 81, 93(11), *96*
Anferov, V. P., 336(75), 337(75), *340*
Angelone, R., 328(26), *338*
Anisimov, K. N., 333(61), *340*

Anokhina, I. K., 191(129), *229*
Antheunis, D., 138(85), *149*
Arbuzov, B. A., 326(14), *338*
Ardjomand, S., 335(69), 336(69), *340*
Arkhipova, T. M., 178(78), 181(78), 205(78), 208(78), *227*
Arsenault, G. P., 289(5), *320*
Aston, J. G., 60, *74*
Aten, C. F., 40, 41, *72*
Atkins, P. W., 144(94, 94a), *150*
Aue, D. H., 166(38), *167*
Ausloos, P., 164(31), *167*
Avery, E. C., 144(94a), *150*

B

Babeshkin, A. M., 258(74, 79), *281*
Babich, E. D., 178(78), 181(78), 205(78), 208(78), *227*
Babushkina, T. A., 175(39), 178(64, 71, 72, 83), 179(72, 94, 95), 180(72), 181(64, 97), 183(71, 105, 110), 185(72, 97, 121), 186(97), 188(72), 189(105), 200(166, 167, 168, 169), 203(183), 209(210, 212), 210(64, 94, 212, 213, 216), 211(64, 105, 216, 217), 212(95, 216, 219, 220), 213(220, 223), 214(216, 220), 215(220), 216(227), 218(227, 239, 240), 219(121, 227), 222(227, 272, 273), 223(272, 273, 274), 224(276, 277, 278, 282, 283, 284), *226, 227, 228, 230, 231, 232, 233,* 324(5a), 325(7), 327(17), 331(46), 332(46, 49a, 50), 333(55, 57, 61), 334(64), 335(64), *338, 339, 340*
Baenziger, M. C., 32(121), *71*
Baisa, D. F., 224(283), *233*
Bak, B., 29, 30, 47, *71, 72*
Baker, J., 191(134), *229*
Baldeschwieler, J. D., 151(1, 2, 3, 10, 12), 152(1), 154(4), 155(1, 2, 3, 10, 12), 163(10), 166, *166, 167*
Baldin, V. I., 136(80a), *149*

Baldwin, J. E., 129(59, 63, 66, 67), 136 (59, 63, 66, 67), *148, 149*
Balhausen, C. F., 260(88), *281*
Balyavichus, M. Z., 323(1), *337*
Bancroft, G. M., 265(120), *282*
Baranovskii, V. I., 263(105), *282*
Barbas, J. T., 140(37b), 143(37b), *147*
Bargon, J., 100(11), 101(12), 111(11), 124(12), 136(79), *145, 147, 149*
Barnes, R. G., 170(14), 199(154, 155, 157, 158, 159), 200(154, 155), 201(157, 158), 202(154, 157, 158), 204(159), 205(154, 159, 194), 206(194), 207(154, 159, 194), *225, 229, 230*
Bartell, L. S., 3, 4(31), 9, 15(43), 16(43), 17, 20, 23(86, 87), 28(86, 87, 107), 35, 40(62, 107), 41(107), 46(86, 87), 54(184), 55, 56, 57, 58, 61, 65, 66(239), 67, *69, 70, 71, 73, 74*
Bartlett, P. D., 115(26), *146*
Basch, H., 22, *70*
Bastiansen, O., 3, 4(30), 18, 23, 24, 29(92), 36(128), 37(128, 134), 41(144), 46(144), 59(206), *69, 70, 71, 72, 73*
Bauer, S. H., 2(13), 3, 6(36), 20, 24(96), 25(102), 26(102), 27(102, 104), 29(96), 30(96), 31, 35(124), 37, 38, 42(116), 43(151), 46(157), 48(164, 165), 49(169), 59(93), 63, *68, 69, 70, 71, 72, 74*
Baurova, Yu. V., 25(97), *70*
Beamer, W., 61(222), *74*
Bearden, A. J., 253(36), 256(36), *280*
Beauchamp, J. L., 151(2, 3, 4, 17, 18), 155(2, 3, 4, 17), 156(4), 160(4), 166, *166, 167*
Becconsall, J. K., 82(17), *96*
Becker, E., 273(145), *283*
Becker, E. D., 143(93), *150*
Bednova, L. M., 329(35), *339*
Bekker, A., 258(74), *281*
Belakhasky, M., 255(46), *280*
Belov, V. F., 256(63), *281*
Belov, Yu. M., 263(118), *282*
Benedekand, G. B., 224(280), *233*
Bennet, R. A., 218(252), *232*
Bennett, L. H., 100(8), *145*

Bennett, W. E., 263(109), *282*
Berkman, S., 273(152), *283*
Bernstein, H. J., 52, *73*
Bersohn, R., 176, 202(43), *226*, 328, *338*
Berta, A., 265(124, 125), *282*
Beynon, J. H., 294(10), *320*
Biedenkapp, D., 199(156), 200(156), 201(156), 202(156), 218(253), 219(253, 254, 271), 220(254), 221(254), *229, 232, 233,* 326(12a), *338*
Biemann, K., 289, 292, 296(13), 301(13), *320*
Bilevich, K. A., 253(29), *279*
Binsch, G., 95(37), *97*
Biros, F. J., 198(140), 199(140), *229*
Biryukov, I. P., 170(29), 174(29), 175(29), 178(57, 58, 59, 63, 65, 66, 67, 68, 69, 73, 74, 75, 76, 77, 78, 79, 80, 81, 82), 179(58, 63, 65, 66, 67, 75, 77, 81), 180(58, 59, 63, 65, 66, 68, 73, 74, 80), 181(58, 59, 63, 78, 81), 182(58, 67), 183(59, 65, 66, 68, 74, 76, 79, 80), 185(68, 81), 186(68, 77), 188(77), 191(57, 58, 59), 193(29), 195(29), 200(63), 201(63), 200(29), 205(29, 78, 197), 206(29, 63, 197), 208(78), 213(29), 214(67), *225, 226, 227,* 326(10, 14), 329(37a, 38b), 333(10), *338, 339*
Blair, L. K., 155(23, 24, 25), 156(23), 162(23, 24, 25), *167*
Blank, B., 131(70), *149*
Bloch, F., 101(16), 102(16), *145,* 277(190), *284*
Bloembergen, N., 101(17), 106(17), *145*
Blomstrom, D. C., 260(85), *281*
Blount, J. F., 274(165), *283*
Bobovich, Ya. S., 209(206), *231*
Bodenseh, H.-K., 51(175), *73*
Boembergen, N., 224(280), *233*
Boguslavskii, A. A., 179(96), *228*
Bohn, R. K., 63, *74*
Bolotin, A. B., 323(1), *337*
Bolton, J. R., 108(22), 114(22), *145*
Bondarev, V. B., 224(276), *233,* 327(17), *338*
Bondoris, G., 170(13), *225*
Bonham, R. A., 3(25), 9(25), 20, 23, 28, 35, 36, 37, 46(86, 87), 47, 48, 49,

52, 53(180, 181), 54(183, 184), 69, 70, 71, 72, 73
Borisov, A. E., 183(110), 228, 333(55), 340
Borshagovskii, B. V., 253(39, 40), 280
Bothner-By, A. A., 118(38b), 128(38b), 147
Boudart, M., 255(47, 48, 52), 280
Bourne, A. J. R., 79(11), 81, 93(11), 96
Bowers, M. T., 151(16), 155(16), 166(38), 166, 167
Bowmaker, G. A., 335(65), 340
Braig, W., 127(53), 148
Brame, E. G., 170(24), 198(24), 225, 327(15), 338
Bratcev, V. A., 190(128), 209(128), 229
Brattsev, V. A., 185(120), 190(120), 199(120), 201(120), 202(120), 205(120, 198), 208(120, 198), 209(120), 228, 231
Brauman, J. I., 82(21), 96, 155(23, 24, 25), 156(23), 162(23, 24, 25), 165(33), 167
Bray, P. J., 170(11), 176(11), 179(11), 185(11), 186(11), 194(11), 199(11, 154, 155, 157, 158, 159, 161), 200(11, 154, 155, 161, 170), 201(157, 158), 202(11, 154, 157, 158, 161, 179), 203(11), 204(11, 159), 205(11, 154, 159, 161, 194, 195), 206(11, 194), 207(154, 159, 161, 194, 195), 208(194), 225, 229, 230
Bregadze, V. I., 184(115), 212(219), 215(115), 228, 231, 253(35), 258(73), 261(35), 280, 281
Briegleb, G., 217(231), 232
Brill, R., 273(149), 283
Brill, T. B., 331(45), 333(56), 336(70), 339, 340
Britton, D., 257(67), 281
Brocklehurst, B., 115(33), 146
Brockway, L. O., 3, 4(31), 9(39), 23, 61(220), 68, 69, 70, 74
Brown, D. A., 273(154), 283
Brown, E., 170(28), 225
Brown, J. E., 129(59, 63, 66), 136(59, 63, 66), 148
Brown, T. L., 330(41, 42), 331(41), 339
Brune, H. A., 66, 67(241), 74

Bryukhova, E. V., 174(37), 175(37), 177(54), 178(37, 54, 83), 180(37), 181(37), 182(37), 183(37, 110), 184(115), 188(37), 190(37), 202(182), 205(199), 209(199), 212(37, 220), 213(37, 220, 222, 223), 214(37, 220), 215(37, 115, 220), 216(37), 219(222, 255), 224(275, 278), 226, 227, 228, 230, 231, 232, 233, 324(6), 325(6), 328(33), 329(34), 333(55, 60b, 61), 336(71), 338, 340
Bryukhanov, V. A., 248(12), 279
Bucci, P., 200(162), 205(192), 207(192), 229, 230, 327(23), 338
Bridenbaugh, P. M., 327(24), 328(27), 338
Buchachenko, A. L., 106(21b), 136(80, 80a,b,c,d), 139(80b), 142(90), 143(90), 145, 149
Buchanan, B. G., 293(8, 9), 294(11), 295, 296, 298, 299(14), 300, 301, 303, 315(14), 320
Buchanan, D. N. E., 258(78), 281
Buchanan, I. C., 144(94), 150
Buck, F. A. M., 37, 71
Budzikiewicz, H., 163(29), 164(29), 167, 294(12), 320
Bukshpan, S., 253(32, 33), 279
Bullock, R. J., 259(83a), 281
Burkhart, R. D., 115(26), 146
Burnham, B. F., 263(108), 282
Butcher, S. S., 42(47), 43(47), 72
Buttrill, S. E., Jr., 151(4, 9, 19), 154(4), 155(4, 9, 19), 156(4), 160(4), 163(29), 164(29), 166, 167
Bylina, G. C., 136(80b), 149

C

Cable, J. W., 273(147), 283
Cahay, R., 218(244, 245, 246), 219(245), 232
Calder, I. C., 82(14, 15, 16), 96
Carboni, R. A., 32(118), 34(118), 71
Carr, H. Y., 92(28), 97
Carrington, A., 128(55), 148
Carroll, B. L., 66(239), 67, 74
Carver, T. R., 100(5, 9), 145
Casabella, P. A., 199(158), 201(158), 202(158), 229

Casserio, M. C., 83(22), *96*
Cecchi, P., 205(192), 207(192), *230,* 327(23), 328(26), *338*
Cefola, M., 260(90), *281*
Chandra, S., 324(4), *337*
Chang, C. H., 24, 29, 30, 37, 38, 46, 49 (169), *70, 71*
Chapovskii, Yu. A., 190(128), 205(198), 208(198), 209(128), *229, 231*
Charlton, J. L., *147*
Chemiak, E. A., 44(154), 45(154), *72*
Chene, M., 255(46), *280*
Chevalier, R., 255(46), *280*
Chiang, J. F., 24, 25(102), 26(102), 27(102), 29, 31, 37(110), 43(151), 59(93), *70, 71, 72*
Chiba, T., 198(141), 206(202), 207(202), *229, 231*
Chihara, H., 327(20), 334(63), *338, 340*
Chivers, T., 268(140), *283*
Christensen, D., 47(160), *72*
Christofferson, C. D., 333(60), *340*
Christophorou, L. G., 156(26), *167*
Christov, T., 255(46), *280*
Clark, H. C., 257(66), *281*
Clark, M. G., 268(135), *283*
Closs, G. L., 101(14, 15a), 113(25a), 118(14, 15a, 34c), 119(14, 15a), 121 (14, 15a), 122(14, 15a), 123(41), 124(14, 15a), 126(43b), 127(41, 44), 135(15a, 25a, 77, 78), 136(77), 137(25a), 138(77), 141, 143, *145, 147, 148*
Closs, L. E., 113(25a), 135(25a), 137(25a), *145*
Clow, R. P., 151(15), 155(15), *166*
Cocivera, M., 138(86), 143(86), *149*
Coetzer, J., 46, 47, *72*
Coffin, K. P., 2(14), *68*
Coffman, D. D., 127(46), *148*
Colligiani, A., 205(192), 207(192), *230,* 327(23), 328(26), *338*
Collins, R. L., 248(21), 255(41), 260(87), 272(21), *279, 280, 281*
Compton, R. N., 156(26, 27), *167*
Cone, C., 289(5), 292, *320*
Conger, R. L., 92(30), *97*
Coogan, H. M., 240(4), *278*
Cooper, R. A., 117(37), 127(45, 48), 129(56), 131(68a), 133(71), 139(87), 143(37), *147, 148, 149*
Corbet, H. C., 5(32), *69*
Cordell, R. W., 129(66), 136(66), *148*
Cordey-Hayes, M., 268(138), *283*
Corio, P. L., 118(38a), *147*
Cornil, P., 218(244, 245, 246), 219(245), *232*
Cornwell, C. D., 177(48), 179(48), *226*
Costain, C. C., 17, 20, 44(154), 45(154), 50(171, 172), *69, 72*
Coulson, C. A., 204(188), *230*
Cox, H. L., Jr., 3(25), 9(25), 61(217), *69, 74*
Cox, R. H., 140(37b), 143(37b), *147*
Curran, C., 263(112), 268(136), *282, 283*
Cutforth, H. G., 273(153), *283*
Cuy, E. J., 273(151), *283*
Cyvin, S. J., 18(68), 60(70, 211), *70, 74*
Czerlinski, G. H., 76(9), 95(9), *96*

D

Dahl, J. P., 260(88), *281*
Dahl, L. F., 273(144, 155, 156), 274(144, 165), *283*
Dahm, D. J., 274(167, 168), *284*
Dailey, B. P., 170(2), 176(44), 177(2, 50, 55), 176, 183(55), *224, 226,* 278(195), *284*
Dakkouri, M., 23(88), 51(176), *70, 73*
Dallinga, G., 5(32), 22, 23, 24, 25(99, 101, 103), 26, 27(101), 31, 32(120), 34, 42(150), 43, 54(186), 55(186), *69, 70, 71, 72, 73*
Daniels, J. M., 100(6), 103(6), 108(6), *145*
Danilkin, V. T., 178(63), 179(63), 180(63), 181(63), 200(63), 201(63), 206(63), *226*
D'Antonio, P., 37(131a), 38(131a, 136), 39(136), 52(179), 57, *71, 73*
Darnall, K. R., 136(82), *149*
Das, T. P., 170(6), 176(6), *225,* 332(54), 333(54a), *339*
Dave, L. D., 262(98), *282*
Davies, A. G., 265(123), *282*
Davis, J. P., 95(39), *97*
Davis, M. I., 32(119), 34, *71*
Davison, A., 276(182), *284*

Author Index 345

Dean, C., 224(279), *233*
DeBoer, E., 106(21a), *145*, 277(184), *284*
Debye, N. W. G., 268(134, 173), 274 (178), 275(179), *283, 284*
Debye, P., 2, 7, *68*
Degard, C., 10, *69*
Dehmelt, H. G., 216(229), 219(229), *232*, 332(51), *339*
Deich, A. Ya., 184(113), *228*
Delgass, W. N., 255(47, 48, 52), *280*
Delyagin, N. N., 248(11, 12), 258(71), *279, 281*
DeMaeyer, L., 76(5, 6), 95(5, 6), *96*
Demelt, H., 170(1), *224*
deNeui, R. J., 17(55), *69*
den Hollander, J. A., 138(85), *149*
Denisovich, L. I., 253(40), *280*
Depireux, J., 198(144), *229*
Derendyaev, B. G., 90(27), *97*
Derissen, J. L., 29, 36, 37, 49, *71*
De Vries, J. L. K. F., 277(184), *284*
Dewar, M. J. S., 191(135), 204(135), 205(193, 196), 206(193, 196), *229, 230, 231*, 326(11, 12), 327(12), 329(36), *338, 339*
Dharmawardena, K. G., 265(120), *282*
Diekman, J., 151(10, 14), 155(10, 14), 163(10), 164(14), *166*
DiLorenzo, J. V., 219(260), 222(260), *233*, 333(58), *340*
Djerassi, C., 151(10, 14), 155(10, 14), 163(10, 29), 164(14, 29), *166, 167*, 293, 294(11, 12), 295, 296, 298, 299 (14), 300, 301, 303, 315, *320*
Dobyns, Sr. V., 47, *72*
Dodonov, V. A., 136(80c), *149*
Donaldson, J. D., 261(95), 263(95), 266 (127), *282*
Dorfman, Y. A. G., 253(28), *279*
Dorko, E. A., 46, *72*
Dormidontov, Yu. P., 325(9), 326(9), *338*
Doubleday, C. E., 135(78), 141(78), *149*
Douglass, D. C., 217(237), *232*
Dows, D. A., 24, *70*
Doyle, J. R., 32(121), *71*
Drago, R. S., 170(19), *225*
Dreizler, H., 57(197), *73*
Dresdner, R. D., 61, *74*

Drew, E. H., 131(68), *149*
Dubovitskii, V. A., 331(46), 332(46), *339*
Duchesne, J., 198(144), 201(173, 175), 218(244, 245, 246), 219(245), *229, 230, 232*
Duffield, A. M., 293(8), 294(11), 295, 296, 298, 299(14), 300, 301, 303, 315(14), *320*
Dunbar, R. D., 151(2, 8), 155(2, 8), *166*
Duncan, J. F., 250(23), *279*
Dunitz, J. D., 28, 37(109), *71*
Dwek, R. A., 100(10a), 105(10a), 107 (10a), *145*
Dzevitskii, B. E., 263(105), *282*

E

Eadon, G., 151(14), 155(14), 164(14), *166*
Eaton, D. R., 106(21), *145*
Ebsworth, E. A. V., 177(56), 183(56), *226*
Edelstein, N., 276(182), *284*
Eframov, E. N., 258(74), *281*
Ege, O., 337(76), *340*
Eggers, D. F., Jr., 28(108), 37(108), *71*
Egorov, Yu. P., 209(207), *231*
Eicher, H., 274(164), *283*
Eigen, M., 76, 95(5, 6, 7, 8), *96*
Einstein, F. W. B., 275(176, 177), *284*
Elleman, D. D., 151(16), 155(16), *166*
Ellenbogen, P., 133(73), *149*
Ellison, R. D., 274(171), *284*
Engel, P. S., 115(26), *146*
Engelhardt, G., 329(37), *339*
Englin, B. A., 224(276, 277), *233*, 327 (17), *338*
Enikolopyan, N. S., 336(71), *340*
Epstein, L. M., 248(22), 250(22), 256 (59), *279, 280*
Erickson, N. E., 258(77), 260(90, 91), 274(166), *281, 282, 284*
Erickson, W. F., 129(67), 136(67), *149*
Ermolaeva, M. V., 210(215), 211(215), *231*
Ernst, R., 143(93), *150*
Esteva, D., 202(179), *230*
Ettinger, R., 75(4), *96*
Evans, G. T., 127(44), *148*

Author Index

Evdokimov, V. V., 66(237), *74*
Eyler, J. R., 166(38), *167*

F

Faingor, V. A., 184(115), 213(221), 215(115), *228, 231*
Fainzil'berg, A. A., 179(90, 91), 181(91, 97), 183(90, 91), *227*
Fairhall, A. W., 274(166), *284*
Faller, J. W., 84(23), 86(24), 88(25), 95(38), 96(40), *96, 97*
Farenholtz, S. R., 136(81), *149*
Farmery, K., 274(170), *284*
Farrar, T. C., 143(93), *150*
Fedin, E. I., 170(9, 10, 18), 176(9, 18), 199(10), 201(10), 217(238), *225, 232,* 333(47), *339*
Feeney, J., 82(18), *96*
Feigenbaum, E. A., 293(8, 9), 294(11), 295, 296, 298, 299(14), 300, 301, 303, 315(14), *320*
Feldman, H. G., 28(109), 37(109), *71*
Fenger, J., 258(76), *281*
Fenton, D. E., 263(106, 107), *282*
Fernholt, L., 41(144), 46(144), *72*
Ferrer-Correia, A. J., 166(36), *167*
Feshin, V. P., 175(38), 178(84), 181(38, 98, 99, 101, 102), 182(84), 183(84), 185(38, 98, 99, 101, 102), 186(98, 99, 101, 102), 187(101, 125, 126), 188(84, 125), 189(127), 191(126, 127), 192(38), 193(38), 194(38, 101, 125), 195(38, 101), 204(127), 205(126), 209(126), 212(84), 223, *226, 227, 228, 229, 231,* 326(10, 12b, 13), 327(18), 329(38, 38a), 330(38), 333(10, 13), *338, 339*
Fessenden, R. W., 108(24), 114(24), 142(89), *145, 149*
Field, F. H., 154(22), *167*
Finbak, C., 2(6), 4(6), *68*
Fischer, E. O., 262(97), *282*
Fischer, H., 100(11), 101(12, 15), 111(11), 115(31), 117(35), 118(15, 34b, 35), 119(15), 121(15), 122(15), 124(12, 15), 127(52), 128(35), 131, 136(37a, 79), 142, 144(94a), *145, 146, 147, 148, 149*
Fish, R. H., 83(22), *96*

Fitzsimmons, B. W., 258(77), 260(92), 268(132, 133), *281, 282, 283*
Fluck, E., 250(24), 274(159, 160, 161, 162, 163), 275(175), *279, 283, 284*
Ford, B. F. E., 267(128), *283*
Forsen, S., 75, 80, 82, 85, 90, 95(2), *96*
Fort, R. C., 191, 194(132), *229*
Franklin, J. L., 154(22), *167*
Frausto da Silva, J. J. R., 337(78), *340*
Freidlina, R. Kh., 223(274), 224(276, 277), *233,* 327(17), *338*
Freudlich, H., 273(151), *283*
Freyland, W., 18, *70*
Friebolin, H., 96(42), *97*
Fritsch, F. N., 23(90), *70*
Frovlova, A. A., 217(236), *232*
Fukui, K., 206(204), *231*
Fukuyama, T., 17(59), 22(59), 40(59), 42(59), 44(156), 45(156), 46(59), 49(59, 168), 50(59, 156, 168), *69, 72*
Fung, B. M., 82, 93(32), *96, 97*
Furlani, C., 276(183), *284*
Futrell, J. H., 151(15), 154(22), 155(15), *166, 167*

G

Galishevskii, Yu. A., 218(242), *232*
Gaoni, Y., 82(16), *96*
Gar, T. K., 178(84), 182(84), 183(84), 188(84), 212(84), 223, *227*
Garratt, P. J., 82(14, 15, 16), *96*
Garrett, B. B., 331(44), *339*
Garst, J. F., 140(37b), 143(37b), *147*
Garten, R. L., 255(47, 48), *280*
Gegenheimer, R., 51(175), *73*
Geisler, H., 329(37), *339*
George, C. F., 37(131a), 38(131a, 136), 39(136), 52, 57(195), 60(209), *71, 73*
Gerhart, F., 123(41a), 129(65), 133(41a), 136(65), *148*
Gibb, T. C., 260(93), 261(94), 268(142), 272(142), *282, 283*
Gill, D., 327(21), *338*
Giller, S. A., 181(98), 185(98), 186(98), *228*
Gilson, D. F. R., 218(249), *232*
Glarum, S., 115(31), *146*

Author Index 347

Glarum, S. H., 115(30), 140(30), 144 (94a), *146, 150*
Glauber, R., 3, *68, 69*
Glentworth, P., 259(83a), *281*
Goel, R. G., 267(128), *283*
Goldanskii, V. I., 177(54), 178(54), *226,* 240(2), 248(12, 13, 14, 15, 16), 253 (27, 29, 30, 31, 34, 35, 37, 38, 39, 40), 255(50), 256(62, 63, 64), 258 (72, 73), 261(35, 96), 263(116, 118), 275(175), *278, 279, 280, 281, 282, 284*
Gol'dshtein, I. P., 330(39), *339*
Goldstein, C., 253(33), *279*
Goldstein, J. H., 177(51, 53), *226*
Golovchenko, L. S., 184(115), 215(115), *228,* 333(59), *340*
Gol'tyapin, Yu. V., 224(275), *233,* 328 (33), *338*
Goncharova, I. N., 181(98, 99), 185(98, 99), 186(98, 99), *228*
Gonser, U., 263(117), *282*
Goode, G. C., 166(36), *167*
Goodman, B. A., 247(7), 260(89, 93), *279, 281, 282*
Gordeev, A. D., 218(242), 219(267), *232, 233,* 337(74), *340*
Gordy, W., 170(5), 176, 178(5), *225, 226,* 332(51), *339*
Gorodinsky, G. M., 248(13), 253(37), 258(72), *279, 280, 281*
Grant, R. W., 263(117), *282*
Gray, G. A., 151(7), 155(7), 162, 163 (7), 166, *166, 167*
Gray, H. B., 22(83), *70*
Graybeal, J. D., 177(48, 49), 179(48), *226,* 265(124, 126), *282,* 330(43), *339*
Greatrey, R., 274(170), *284*
Grechishkin, V. S., 170(8, 17), 202(176), 203(184, 185, 186, 187), 204(184), 205(191), 217(234), 218(234, 242, 250), 219(234, 250, 261, 262, 263, 264, 265, 266, 267, 270), 221(250, 262, 264, 265, 266), 224(281, 285), *225, 230, 232, 233,* 325(9), 326(9), 327(19), 332(53), 336(73, 75), 337 (74, 75), *338, 339, 340*
Grechkin, E. F., 332(49), *339*
Green, J., 265(126), *282*
Green, W. H., 57, *73*
Greenberg, B., 59(202), *73*
Greenwood, N. N., 247(7), 260(89, 93), 268(141, 142, 143), 272(142, 143), 274(170), *279, 281, 282, 283, 284*
Grigsby, R. D., 286(4), *320*
Grinberg, P. L., 25(97), *70*
Grubert, H., 262(97), *282*
Grützmacher, H. F., 289(7), *320*
Gruverman, I. J., 240(3), *278*
Gubin, S. P., 253(40), *280*
Guggenheim, E. A., 94, *97*
Guibe, L., 335(66), *340*
Guillory, J. P., 55, 56(194), 57, 58(194), *73*
Gurd, R. C., 144(94, 94a), *150*
Gur'yanova, E. N., 216(230), 218(239, 240, 241), 219(241, 255, 259), 220 (241, 259), 221(259), 222(241), *232, 233,* 330(39), 336(72), 337 (77), *339, 340*
Gushchin, S. I., 219(270), *233,* 325(9), 326(9), *338*
Gutlich, P., 258(77), *281*
Gutowsky, H. S., 94(34), 95(34), 96 (34), *97,* 108(25), *145,* 179(88), *227*

H

Haaland, A., 262(99), *282*
Haas, B., 177(52), *226*
Haase, J., 6(34), 18(69), 23(88), 51 (174, 176), *69, 70, 72, 73*
Haase, W., 331(48), *339*
Hackler, R. E., 129(67), 136(67), *149*
Hacobian, S., 335(65), *340*
Hager, F. D., 127(46), *148*
Hahn, E. L., 332(54), 333(54a), *339*
Haikin, L. C., 66(237), *74*
Halfon, M., 127(45), *148*
Hamilton, W. C., 20, 32(118), 34, *70, 71,* 265(122), 276(181), *282, 284*
Hamming, M. C., 286(4), *320*
Hansen, K. W., 65, *74*
Hansen-Nygaard, L., 47(160), *72*
Hargittai, I., 64, *74*
Harmony, M. D., 42(48), *72*
Harriman, J. E., 115(29), 140(29), *146*
Harris, R. K., 96(41), *97*

Harrison, P. G., 262(100, 101, 102, 103), 277(186, 187), 278(196), *282, 284*
Harshbarger, W. R., 48(164), 49, *72*
Hart, P. A., 95(39), *97*
Hart, R. M., 324(5), *338*
Harvey, A. B., 37, 57, *71, 73*
Harvey, R. B., 2(13), *68*
Haskel, T. H., 304(17), *321*
Hassel, O., 2(6, 8), 4(6, 30), 59(203, 204, 205), *68, 73*
Hatton, J., 179(93), 199(93), *227*
Haugen, W., 40, 46(139), 49(139), 50(139), *72*
Hauptman, H., 2(18), *68*
Hausser, K. H., 100(10), 105(10, 20), 108(10), *145*
Hayashi, H., 115(30), 140(30), *146*
Hayck, M., 327(21), *338*
Hazony, Y., 255(53), *280*
Hedberg, K., 16, 20, 23(90), 40(141), 41(141), *69, 70, 72*
Hedberg, L., 40(141), 41(141), *72*
Heilbronner, E., 191(133), *229*
Heinrich, A., 82(18), *96*
Hencher, J. L., 46(157), *72*
Henderson, A. T., 61(226), *74*
Henis, J. M. S., 151(5), 155(5), 165, *166, 167*
Herber, R. H., 240(2), 248(10), 256(54), 259(80, 81), 263(110, 111), 265(119), 268(119), 274(157, 158), 277(189, 191), *278, 279, 280, 281, 282, 283, 284*
Herr, M. L., 326(11, 12), 327(12), *338*
Herschbach, D. R., 17(53, 54), *69*
Herzberg, G., 21, *70*
Heyns, K., 289(7), *320*
Hieber, W., 273(145), *283*
Higginbotham, H. K., 17, 23, 40(62), 61, *69, 74*
Hilderbrandt, R. L., 20, 35, *70, 71*
Hirota, E., 16, 18(45), 54, 60(45, 210), *69, 73*
Hirota, N., 115(30), 140(30), *146*
Hirschfelder, J. O., 115(29), 140(29), *146*
Ho, B. Y. K., 263(114), *282*
Hobson, M. C., Jr., 255(42), *280*
Hoffman, R. A., 75, 80, 82, 90, 95(2), *96*

Hoffmann, A. K., 61(226), *74*
Höfle, G., 129(63, 66), 136(63, 66), *148*
Hohenemser, C., 248(8), *279*
Hollaender, J., 136(84), *149*
Holland, R. J., 83(22), *96*
Hollowell, C. D., 28(107), 40(107), 41(107), *71*
Holm, R. H., 276(182), *284*
Holtz, D., 151(17), 155(17), *166*
Hoober, D., 2(12), *68*
Hooper, H. O., 170(11), 176(11), 179(11), 185(11), 186(11), 194(11), 199(11, 159), 200(11), 202(11), 203(11), 204(11, 159), 205(11, 159, 195), 206(11), 207(159, 195), 218(243, 252), *225, 229, 230, 232*
Hoppe, H., 331(48), *339*
Hori, Y., 50(173), *72*
Hoskins, B. F., 276(180), *284*
Hsu, M. W., 330(43), *339*
Hudson, A., 108(22), 114(22), *145*
Hüther, H., 67(241), *74*
Hugus, Z. Z., 331(45), 333(56), 336(70), *339, 340*
Hume, D. N., 92(31), 94(31), *97*
Huntress, W. T., Jr., 151(12, 18), 155(12), *166*

I

Ianakieva, M., 255(46), *280*
Iijima, T., 9, 13, 17(60), 18(68), 32(60), 33(60), 35(60), 61(223), *69, 74*
Incorvia, M. J., 84(23), 86(24), 88(25), *96, 97*
Ing, S. D., 330(43), *339*
Inoue, M., 151(13), 155(13), *166*
Irie, T., 327(20), *338*
Ishida, Y., 304(16), *321*
Ito, A., 250(25, 26), *279*
Itoh, K., 115(30), 140(30), *146*, 198(143), 201(172), 202(172), *229, 230*
Ivanov, I. D., 263(118), *282*
Ivanov, V. V., 336(71), *340*
Iwamura, H., 129(60), 130(60), 136(60), *148*
Iwamura, M., 129(60), 130(60), 136(60, 84), *148, 149*

Iwasaki, M., 16, 20, 69, 70
Izmest'ev, I. V., 205(191), 230

J

Jacob, E. J., 37, 71
Jacobson, R. A., 274(167, 168), 284
Jaeschke, A., 96(42), 97
Jeffries, C. D., 100(6), 103(6), 108(6), 145
Jemison, R. W., 129(61), 136(61), 148
Jenkins, H. O., 61(220), 74
Jennings, K. R., 151(11), 155(11), 166, 166, 167
Jensen, H. H., 63, 74
Johansson, A., 255(51), 280
Johnsen, U., 100(11), 111(11), 145
Johnson, C. S., Jr., 95(35), 97, 100(2), 115(33), 144, 146
Johnson, R. F., 273(148), 283
Jones, J., 94(34), 95(34), 96(34), 97
Jones, L. T., 218(248), 232

K

Kabachnik, M. I., 178(60, 61, 62), 181 (61, 62), 182(60, 61, 62), 183(60, 61), 191(60, 61, 62), 200(60, 171), 201(60, 171), 206(60), 209(212), 210(212), 226, 230, 231
Kabakoff, D. S., 82(13), 96
Kadina, M. A., 178(72), 179(72), 180 (72), 185(72), 188(72), 227
Kaiuchi, E., 337(79), 340
Kalinkin, M. I., 183(105), 189(105), 210 (216), 211(105, 216), 212(216), 214(216), 228, 231
Kalinin, V. N., 187(126), 191(126), 205 (126), 209(126, 209), 229, 231
Kalvius, M., 274(164), 283
Kanazawa, Y., 82(21), 96
Kaplan, F., 151(6), 155(6), 166
Kaplansky, M., 324(3), 335(67, 68), 337, 340
Kaptein, R., 101(13), 114(13), 118(13), 119(13), 121(13), 122(13), 124(13, 42), 126(43a), 131, 136(69), 138, 140(69), 145, 148, 149
Karachevtsev, G. V., 154(22), 166
Karasev, A. N., 255(44, 45), 280
Karle, I. L., 1(1), 2(3, 4, 5, 9, 12, 15), 4(4, 5), 9(3, 4), 10(4, 15), 11(4, 9), 12, 18(4), 19, 35(125), 36(125), 37(9), 41(71), 46(71), 60(213, 214), 68, 71, 74
Karle, J., 1(1), 2(3, 4, 5, 9, 12, 15, 16, 17, 18, 19), 4(4, 5), 9(3, 4), 10(4, 15), 11(4, 9, 42), 12, 13, 18(4, 18, 42), 37(9), 38(136), 39(136), 51, 52(179), 57(195), 59(208), 60(208, 209, 213), 68, 69, 71, 73, 74
Karyagin, S. V., 248(13, 19), 253(37), 258(72), 279, 281
Kasai, P. H., 28, 37(108), 71
Kashima, M., 219(268, 269), 233
Kashiwagi, H., 208(205), 231
Kaska, W., 127(54), 148
Katada, K., 48(165), 72
Kato, C., 17, 32, 33, 35, 69
Kazimir, E. O., 260(90), 281
Kazitsyna, L. A., 224(278), 233
Keefer, R. M., 218(247), 232
Keeley, D. F., 273(148), 283
Keidel, F. A., 2(13), 68
Kerler, W., 250(24), 274(159, 160, 161, 162, 163), 279, 283
Kessenikh, A. V., 136(80b,c,d), 139 (80b), 149, 179(91), 181(91, 97), 183(91), 227, 228
Khrapov, V. V., 177(54), 178(54), 226, 248(13, 14, 15, 16), 253(27, 29, 31, 34, 35, 37), 256(62), 258(72, 73), 261(35, 96), 279, 280, 281, 282, 330(40), 339
Khrlakyan, S. P., 185(121), 219(121), 228
Khutoretskii, V. M., 181(97), 228, 325 (7), 338
Kienle, P., 274(164), 283
Kilner, M., 274(170), 284
Kimura, K., 6(36), 41(145), 46(145), 48(165), 54, 55, 69, 72, 73
Kimura, M., 6(35), 9, 13, 17(60), 32 (60), 33(60), 35(60), 69
Kinaslowski, S., 328(28), 338
King, J., Jr., 151(16), 155(16), 166
King, R. B., 274(157), 283
Kingston, W. R., 259(80, 81), 274(158), 277(191), 281, 283, 284
Kinner, L., 127(54), 148
Kirsch, J. D., 330(41), 331(41), 339
Kiser, R. W., 152(21), 166

Kitaigorodskii, A. I., 200(163, 165), 217 (236, 238), *230*
Kivelson, D., 115(30), 140(30), *146*
Klein, D. J., 115(26), *146*
Klimova, A. I., 205(199), 209(199), *231*
Knight, E., Jr., 260(85), *281*
Knudsen, R. E., 60(209), *73*
Knunyants, I. L., 178(64, 70, 85), 181(64, 70, 100), 185(121), 210(64, 70, 85, 100, 215), 211(64, 70, 100, 215), 219(121), *226, 227, 228, 231,* 324(5a), *338*
Kobrina, L. S., 200(164, 166, 167, 168, 169), 201(164), 202(177), *230*
Kocheshkov, K. A., 330(39, 40), *339*
Kochetkova, N. S., 253(39), *280*
Kocken, J. W. M., 49(164a), *72*
Koenig, T., 135, *149*
Kohl, D. A., 23(87), 28(87), 54(184), 70, 73
Kojima, S., 185(122, 123), *228*
Kokolas, J. J., 277(188, 190), *284*
Kokot, E., 277(185), *284*
Kolesnikov, S. P., 216(227), 218(227), 219(227), 222(227), *232*
Kolobova, N. E., 333(61), *340*
Kolos, W., 115(29), 140(29), *146*
Koltenbah, D. E., 325(8), *338*
Komissarova, V. A., 248(17), *279*
Konaka, S., 16, 17(60), 32(60), 33(60), 35(60), *69*
Kondo, S., 337(79), *340*
Koptyug, V. A., 90(27), *97*
Korshak, V. V., 202(182), *230,* 324(6), 325(6), *338*
Korshunov, A. V., 334(64), 335(64), *340*
Korte, W. D., 127(54), *148*
Korte, W. O., 127(54), *148*
Korytko, L. A., 248(12, 13), 253(30, 37, 38), 258(72), *279, 280, 281*
Kostyanovskii, R. G., 179(95), 212(95), *228,* 253(31), *279*
Kotkhar, V., 278(194), *284*
Kozima, K., 198(147), *229*
Kraitchman, J., 17, *69*
Kramer, D. N., 277(188, 190), *284*
Kravchenko, E. A., 219, 220, 221, *233*
Kravtsov, D. N., 184(115), 213(221), 215(115), *228, 231*

Kravtsova, I. D., 209(207), *231*
Kregzde, J., 263(117), *282*
Kriegsmann, H., 329(37), *339*
Kriemler, P., 151(19), 155(19), *166*
Krizhansky, L. M., 248(13), 253(37), 258(72), *279, 280, 281*
Krüger, H., 170(1), *224*
Kruse, W., 76(6), 95(6), *96*
Kubo, M., 41(145), 46(145), 54, 55, 72, 73, 170(22), 202(181), 208 (205), 213(22), *225, 230, 231*
Kuchitsu, K., 2(20, 21), 6(35), 16(44), 17(55, 57), 18(68), 20(75), 22, 25 (100), 26, 27(100), 28(107), 32 (100), 34(100), 40(107), 41(107), 42(59), 44(156), 45(156), 46(59), 49(59, 166, 167, 168), 50(59, 156, 166, 167, 168, 173), 53, 54(188), 55(188, 190), 60(20, 21), *68, 69, 70, 71, 72, 73*
Kudo, Y., 295, 304(16), 305(16), *321*
Kudryavtsev, L. V., 333(57), *340*
Kuhn, P., 250(24), 274(162), *279, 283*
Kulikov, L. A., 258(79), *281*
Kurita, Y., 202(181), *230*
Kurkovskaya, L. A., 179(91), 181(91), 183(91), *227*
Kushida, T., 224(280), *233*
Kusuda, K., 199(150), *229*
Kwart, H., 191, *229*
Kyono, S., 201(172), 202(172), *230*
Kysakov, M. M., 25(98), *70*
Kyskin, V. P., 329(34), *338*
Kyuntsel', I. A., 217(234), 218(234, 250), 219(234, 250, 261, 262, 263, 264, 265, 266), 221(250, 262, 264, 265, 266), *232, 233,* 332(53), *339*

L

Lampe, F. W., 154(22), *167*
Lamson, D. W., 127(51), *148*
Landau, R. L., 127(49), *148*
Lane, A. G., 136(84), *149*
Lane, B. C., 278(196), *284*
Lang, G., 260(86), *281*
Large, N. R., 259(83a), *281*
Laurie, V. W., 17(53, 54), 37(133), *69, 71*
Lawler, R. G., 100(11a), 101(12a), 113 (11a), 114(12a), 115(34a), 117

(34a, 37), 118(34a), 119(34a), 122 (34a), 124(12a, 34a), 127(11a, 48), 129(56, 57), 131(68a), 133(71, 73), 135(74), 136(11a), 139(87), 143 (37), *145, 147, 148, 149*
Led, J. J., 29(112), 30(112), *71*
Lederberg, J., 293(8), 294(11), 295, 296, 298, 299(14), 300, 301, 303, 315(14), *320*
Lehn, J. M., 170(23, 27), *225*
Lehnig, M., 136(37a), 142(88), *147, 149*
Leibfritz, D., 136(84), *149*
Leites, L. A., 209(207, 209), *231*
Lependina, O. L., 253(28), *279*
Lepley, A. R., 126(43), 127(47, 49, 50), 129(58, 64), 136(58, 64, 83), *148, 149*
Levin, R., 277(188, 190), *284*
Levin, V. S., 183(105), 189(105), 211 (105), *228*
Levy, H. A., 274(171), *284*
Leyte, J. C., 38(135), 39(135), *71*
Lide, D. R., Jr., 28, 37, *71*
Lindauer, W., 274(169), *284*
Lippmaa, E., 142, 143, *149*
Litinskii, A. O., 323(1), *337*
Livingston, R. L., 37, 61(221), *71, 74,* 144(94a), *150,* 176(45, 46), 179 (45, 46, 86, 87, 92), 185(45, 87, 92), 191(45, 87), 193(92), *226, 227*
Lobanov, D. I., 178(60, 61, 62), 181(61, 62), 182(60, 61, 62), 183(60, 61), 191(60, 61, 62), 200(60, 171), 201 (60, 171), 206(60), 209(212), 210 (212), *226, 230, 231*
Loewenstein, R. M. J., 327(21), *338*
Loken, H. Y., 117(37), 129(56, 57), 135 (74), 143(37), *147, 148, 149*
Long, T. V., II, 277(186, 187), *284*
Longuet-Higgins, H. C., 82(15), *96,* 206 (200), *231*
Lonsdale, K., 23, *70*
Lopatina, G. P., 223(274), *233*
Loshadkin, N. A., 178(85), 210(85), *227*
Low, M. J. D., 255(43), *280*
Lowrey, A. H., 37(131a), 38(131a), 52 (179), 57(195), *71, 73*
Lucas, J. P., 335(66), *340*
Lucken, E. A. C., 170(15, 16, 31), 176

Author Index 351

(15, 31, 47), 177(47), 179(15, 31), 181(15, 31, 47), 183(108), 185(31), 188(31), 190(47), 195(31), 198(31, 148), 199(15, 31), 200(31), 205 (16, 193, 196), 206(15, 16, 31, 193, 196), 207(31), 209(108), 210 (108), 222, *225, 226, 228, 229, 230, 231,* 267(130), *283,* 335(69), 336 (69), *340*
Luckhurst, G. R., 108(22), 114(22), 115 (30), 140(30), *145, 146*
Ludwig, G. W., 199(160), 201(160), 202 (160), *229*
Ludwig, U., 129(63), 136(63), *148*
Lukevits, E. Ya., 178(81, 82), 179(81), 181(81), 185(81), *227*
Luss, G., 42(48), *72*
Lutskii, A. E., 209(208), *231*

M

Maass, G., 76(6), 95(6), *96*
Mabey, W. R., 135(76), *149*
McConnell, H. M., 75(3), 78(10), *96*
McCoy, W. H., 54(183), *73*
McCullough, J. D., 333(60), *340*
MacDonald, A. C., 32, *71*
Machmer, P., 183(107), *228*
McHugh, J. P., 28, *70*
McIver, R. T., Jr., 151(20), 166(20, 38), *166, 167*
McLachlan, A. D., 128(55), *148*
McLafferty, F. W., 163(29, 30), 164 (29), *167,* 289, 290, 291, *320*
McLauchlan, K. A., 144(94, 94a), *150*
McLean, R. L., 304(17), *321*
MacLeod, J. K., 151(10), 155(10), 163 (10), *166*
McMillen, D. F., 82(21), *96*
Maeda, K., 2(21), 60(21), *68, 73*
Maestro, M., 200(162), *229*
Mair, H. J., 29, *71*
Mairinger, F., 170(20, 25), 220(25), *225*
Makarov, E. F., 248(12, 13, 14, 15, 16), 253(37, 39), 256(62, 63, 64), 258 (72), 263(116), *279, 280, 281, 282*
Makarov, Yu. V., 333(61), *340*
Maki, A. H., 276(182), *284*
Makridin, V. P., 336(72), *340*
Maksyutin, Yu. K., 216(230), 218(239, 240, 241), 219(241, 255, 259), 220

(241, 259), 221(259), 222(241), 232, 233, 327(16), 330(39, 40), 332(49), 336(72), 337(77), 338, 339, 340
Malizev, V. A., 253(38), 280
Maltsev, V. A., 253(30), 279
Mamaeva, G. I., 67(246), 74
Mamatyuk, V. I., 90(27), 97
Mann, D. E., 28, 60, 71, 74
Mannschreck, A., 96(42), 97
Marchand, A. P., 326(12), 327(12), 338
Mardanyan, S. S., 263(116, 118), 282
Margulis, T. N., 59(202), 73
Marsh, H. S., 253(36), 256(36), 280
Marshall, J. H., 115(30), 140(30), 144(94a), 146, 150
Martin, J. C., 131(68), 149
Martin, R. L., 276(180), 277(185), 284
Marvel, C. S., 127(46), 148
Mastryukov, V. S., 25(97), 67(242, 243, 244), 70, 74
Matchanov, G. I., 263(118), 282
Matsen, F. A., 115(26), 146
Mays, J. M., 177(55), 183(55), 226
Mazeline, C., 198(148), 199, 229
Meal, H. C., 199(152), 204(190), 229, 230
Meiboom, S., 24, 70
Meinzer, R. A., 94(34), 95(34), 96(34), 97
Mel'nichenko, L. S., 330(39, 40), 339
Melton, C. E., 154(22), 167
Memory, J. D., 118(38), 147
Mennicke, J., 51(175), 73
Merrifield, R. E., 115(33), 146
Merrill, J. C., 115(26), 145
Metzler, R. B., 136(84), 149
Meyer, L. H., 179(88), 227
Mijlhoff, F. C., 38(135), 39(135), 71
Mikami, M., 55(189, 191), 73
Mikhaiilyants, S. A., 178(84), 182(84), 183(84), 188(84), 212(84), 223, 227, 329(38a), 339
Mikhailov, B. M., 191(136), 229
Milledge, H. J., 265(123), 282
Miller, L. I., 191, 229
Milleur, M. B., 115(29), 140(29), 146
Mills, O. S., 273(150), 283
Minkin, V. I., 173(33), 182(33), 225, 329(38d), 339

Mironov, V. F., 178(69, 80, 84), 180(80), 182(84), 183(80, 84), 188(84), 209(207), 212(84), 223, 227, 231, 329(38a), 339
Mironova, L. I., 181(98, 99), 185(98, 99), 186(98, 99), 228
Mitrofanov, K. P., 248(11), 253(28), 256(61), 258(71), 279, 280, 281
Miyagawa, I., 198(141, 146), 229
Mizushima, S., 52, 73
Mochalov, S. S., 58(198), 73
Momany, F. A., 47, 48, 49, 53(181), 54, 72
Monfils, A., 198(144), 201(173, 175), 229, 230
Montgomery, L. K., 46, 47, 72
Mooberry, E. S., 331(44), 339
Moore, P. W., 18(63), 69
Morice, J., 255(49), 280
Morino, Y., 2(21), 16(44), 17(52, 59), 18, 20, 22(59), 40(59), 42(59), 44(156), 45(156), 46(59), 49(59, 166, 167, 168), 50(59, 156, 166, 167, 168, 173), 52(177), 53, 60(210), 61(223), 68, 69, 70, 72, 73, 74, 198(141, 143, 145), 201(172), 202(172), 206(202), 207(202), 229, 230, 231
Morozova, I. S., 336(71), 340
Morris, D. G., 129(61, 62), 136(61, 62), 148
Morris, J. I., 140(37b), 143(37b), 147
Morrison, R. C., 140(37b), 143(37b), 147
Morton, J. R., 50(171), 72
Moshkovskii, Yu. Sh., 263(115, 116, 118), 282
Moskowitz, S., 199(159), 204(159), 205(159), 207(159), 229
Motzfeldt, T., 36(128), 37(128), 71, 262(99), 282
Muecke, T. W., 32(119), 34, 71
Müller, K., 126(43b), 148
Muensch, H., 96(42), 97
Muir, A. H., 240(4), 278
Munk, M. E., 304(17), 321
Munthe-Kaas, T., 18(64), 69
Murata, Y., 6(35), 20(75), 61(223), 69, 70, 74
Murin, I. V., 327(19), 338

Murphy, R. B., 191(131), 229
Murrell, J. N., 115(29), 140(29), 146
Murty, C. R. K., 327(25), 338
Myers, R. J., 28(108), 37(108), 71

N

Nagakura, S., 115(30), 140(30), 146
Naik, V., 268(136), 283
Nakamura, D., 170(21, 22), 208(205), 213(22), 225, 231
Nakamura, J., 18(63), 69
Nakamura, N., 327(20), 334(63), 338, 340
Nakanishi, K., 299, 320
Nakayama, J., 129(60), 148
Namanworth, E., 82(13), 96
Nametkin, N. S., 25(98), 70, 178(78), 181(78), 205(78), 208(78), 227
Nathan, W., 191(134), 229
Nazarenko, I. I., 64, 74
Nefedov, O. M., 216(227), 218(227), 219(227), 222(227), 232
Negita, H., 206(203, 204), 207(203), 231, 337(76), 340
Negita, M., 219(268, 269), 233
Neimysheva, A. A., 178(64, 70, 71, 85), 181(64, 70, 100), 183(71), 209(70), 210(64, 70, 85, 100, 215), 211(64, 70, 100, 215), 226, 227, 228, 231, 324(5a), 332(49a, 50), 338, 339
Nelsen, S. F., 136(84), 149
Nelson, L. Y., 37, 71
Nesmeyanov, A. N., 184(115), 213(221), 215(115), 228, 231, 253(39), 258(74, 79), 280, 281, 331(46, 47), 332(46), 333(61), 339, 340
Neuert, H., 156(28), 167
Neumann, W. P., 136(84), 149
Neuwirth, W., 250(24), 274(159, 160, 161, 162, 163), 279, 283
Nicholas, D., 3(26), 9(26), 69
Nichols, A. L., 259(83a), 281
Nicholson, D. G., 266(127), 282
Niederer, P., 136(84), 149
Nikonova, L. A., 214(225), 232
Nishida, T., 129(60), 130(60), 136(60), 148
Noggle, J. H., 95(39), 96(43), 97
Nogina, O. V., 331(46), 332(46), 339

Noltes, J. G., 262(101), 282
North, A. M., 117(36), 147
Novikova, N. V., 183(110), 228, 333(55), 340
Noyes, R. M., 117(36), 141(40), 147
Nygaard, L., 29(112), 30(112), 71

O

Oberhammer, H., 31, 42(116), 43, 64, 65, 66, 67, 71, 74
O'Brien, R. J., 257(66), 281
O'Brien, S., 82(17), 96
Ochiai, S., 295, 304(16, 18), 305(16), 321
Odar, S., 258(77), 281
Ogāwa, S., 185(122, 123), 228
Ogino, K., 330(42), 339
Oka, T., 16(44), 17(51, 52), 69
Okawara, R., 256(57), 280
Okhlobystin, O. Yu., 184(115), 212(219), 213(223), 215(115), 228, 231, 248(18), 253(27, 35), 256(61), 258(73), 261(35), 279, 280, 281, 329(35), 339
Okhlobystina, L. V., 179(91), 181(91), 97), 183(91), 227, 228, 325(7), 338
O'Konski, C. T., 170(12), 176(12), 179(12), 181(12), 199(12), 200(12), 201(12), 202(12), 218(249), 225, 232
Okuda, T., 219(268, 269), 233, 337(76), 340
Oltmans, F., 5(32), 69
O'Malley, R. M., 151(11), 155(11), 166(36), 166, 167
Onishchenko, T. A., 224(277), 233
Ono, K., 250(26), 279
Oosterhoff, L. J., 101(13), 114(13), 118(13), 119(13), 121(13), 122(13), 124(13, 42), 138(85), 145, 148, 149
Oppenheim, V. D., 25(98), 70
Orlova, E. K., 209(211), 231
O'Rourke, M., 263(112), 282
Orvill-Thomas, W. J., 170(4), 178(4), 224, 333(59a), 340
Osipenko, A. N., 202(176), 217(235), 230, 232
Osipov, B. N., 224(276), 233, 327(17), 338

Author Index

Ostermann, G., 123(41a), 129(60, 63), 130(60), 133(41a), 136(60, 63), *148*
Otto, K., 250(25), *279*
Ouki, T., 304(18), *321*
Overhauser, A. W., 100, 108, *145*

P

Pace, E. L., 60, *74*
Pakhomov, V. I., 212(220), 213(220), 214(220), 215(220), *231*
Pal'm, V. A., 173(34), 178(85), 187(34), 191(34), 210(85), *225, 227, 228,* 329(38e), *339*
Parish, R. V., 268(131, 137, 139), *283*
Parks, A. T., 56(194), 58(194), *73*
Parravano, G., 255(52), *280*
Patsch, M., 129(63), 136(63), *148*
Patterson, A., Jr., 75(4), *96*
Patterson, D. B., 329(36), *339*
Paul, H., 144(94a), *150*
Pauli, G. H., 53, *73*
Pauling, L., 23, *70,* 206(201), *231*
Paulson, D. R., 135(78), 141(78), 143(91), *149*
Paushkin, Ya. M., 256(63, 64, 65), *281*
Pavlov, B. N., 178(67), 179(67), 182(67), 214(67), *227*
Peacock, R. D., 268(138), *283*
Pehk, T., 142(90), 143(90), *149*
Peitrzak, J., 328(28), *338*
Pence, D. T., 37(133), *71*
Pendin, A. A., 256(65), *281*
Penfold, B. R., 275(176, 177), *284*
Perekalin, V. V., 209(206), *231*
Perepelkova, T. I., 330(39), *339*
Perfilev, Y. D., 258(79), *281*
Perkins, P. G., 268(143), 272(143), *283*
Pervova, E. Ya., 185(121), 219(121), *228*
Peterson, G. E., 327(24), 328(27), *338*
Pettersson, B., 285(2), 286(3), *320*
Petrov, L. N., 332(53), *339*
Phillips, W. D., 106(21), *145,* 260(85), *281*
Pierce, L., 47, *72*
Pike, W. T., 163(30), *167*
Pitts, J. N., Jr., 136(82), *149*
Plachinda, A. S., 255(50), *280*
Plachy, W., 115(30), 140(30), *146*
Planje, M. C., 54, 55(186), *73*
Plate, N. A., 253(30, 38), *279, 280*
Platt, R. H., 268(131, 137, 139), *283*
Plotnikova, M. V., 253(28), *279*
Plyler, E. K., 60, *74*
Poeth, T. P., 277(186, 187), *284*
Polak, L. S., 248(11), 253(28), 255(44, 45), 256(61), 258(71), *279, 280, 281*
Polis, Ya. Yu., 187(125), 188(125), 189(125), 194(125), *229*
Poller, R., 256(58), *280*
Poller, R. C., 256(56), *280*
Ponnamperuma, C., 151(12), 155(12), *166*
Popova, E. P., 175(38), 181(38, 102), 185(38, 102), 186(102), 189(127), 191(127), 192(38), 193(38), 194(38), 195(38), 204(127), *226, 228, 229,* 326(10, 12b, 13), 327(18), 333(10, 13), *338*
Porter, R. F., 24(96), 29(96), 30(96), *70*
Post, B., 59(202), *73*
Pound, R. V., 101(17), 106(17), *145,* 224(279), *233*
Pozdnyakova, M. V., 184(114), *228*
Presnjakova, B. M., 47(161), 48(161), *72*
Prins, R., 22(84), *70*
Pritchard, H. O., 133(72), *149*
Proctor, W. G., 102(18), *145*
Prokofiev, A. K., 179(95), 212(95), *228,* 253(31), *279*
Purcell, E. M., 92(28), *97,* 101(17), 106(17), *145*
Purdela, D., 210(214), *231*
Puxley, D. C., 265(123), *282*

Q

Quiles, J. P., 255(46), *280*
Quitman, W., 202(178), *230*

R

Radeglia, R., 329(37), *339*
Ragle, J. L., 328(29), *338*
Rakshys, J. W., Jr., 143(92), *149*
Rambidi, N. G., 6(33), *69*

Rao, K. V. S. R., 327(25), *338*
Rastrup-Andersen, J., 29(112), 30(112), 47(160), *71*
Razuvaev, G. A., 136(80c), *149*
Read, M., 218(244, 245, 246), 219(245), *232*
Rebbert, R. E., 164(31), *167*
Rees, L. V. C., 255(49), *280*
Reeves, L. W., 39, *71*, 95(36), *97*
Reichle, W. T., 265(119), 268(119), *282*
Reikhsfeld, V. O., 329(38), 330(38), *339*
Reinhardt, P. W., 156(27), *167*
Remko, J. R., 144(94a), *150*
Rich, N., 24, *70*
Richards, G. F., 32(121), *71*
Richards, R. E., 100(10a,b), 105(10a,b), 107(10a), *145*
Ridley, D. R., 267(128), *283*
Rieker, A., 136(84), *149*
Risberg, E., 4(30), *69*
Rithie, C. D., *228*
Robas, V. I., 178(72), 179(72), 180(72), 185(72, 119, 120), 188(72), 190(120), 199(120), 201(120), 202(120, 177), 203(183), 205(120), 208(120), 209(120), 212(218), 214(218), 216(218), 217(232, 233), 222(272, 273), 223(272, 273), 224(278, 282, 284), *227, 228, 230, 231, 232, 233*
Roberts, J. D., 32(118), 34(118), *71*
Roberts, R. D., 140(37b), 143(37b), *147*
Robertson, A. V., 293(8), 294(11), 295, 296, 298, *320*
Robins, J., 191(131), *229*
Robinsohn, H. G., 332(51, 52), *339*
Rochev, V. Ya., 253(27, 29, 31, 34), 261(96), *279, 282*
Rochow, E. G., 183(112), *228*
Rogers, M., 219(258), *233*
Rokhlina, E. M., 213(221), *231*
Roley, D. L. B., 256(56), *280*
Roll, D. B., 198(140), 199(140), *229*
Rollin, V. Y., 179(93), 199(93), *227*
Romanenko, L. S., 329(38), 330(38), *339*
Romers, C., 38, 39, *71*
Roos, B., 41(143), *72*
Roper, R., 277(185), *284*
Ros, P., 22(82), 24, *70*

Rose, M. E., 100(6), 103(6), 108(6), *145*
Rosenberg, E., 275(178), *284*
Roth, E. A., 28(107), 40(107), 41(107), *71*
Roth, H. D., 138(86), 143(86), *149*
Rouault, M., 3(26), 9(26), *69*
Rozinov, V. G., 332(49), *339*
Ruby, S. L., 255(53), *280*
Ruddick, J. N. R., 256(56), 268(141), *280, 283*
Rudolph, H. D., 57(197), *73*
Rüchardt, C., 136(84), *149*
Rundle, R. E., 273(155, 156), *283*
Russell, G. A., 127(51), *148*
Ryan, K. R., 154(22), *167*
Ryan, J. A., 219(258), *233*
Ryansnyi, G. K., 248(18), *279*
Rybinskaya, M. I., 253(39), *280*
Ryhage, R., 285(2), 286(3), *320*
Rykov, S. V., 136(80, 80a,b,c,d), 139(80b), 142(90), 143(90), *149*

S

Sadano, C. S., 304(17), *321*
Sadova, N. I., 40, 58(198, 199, 201), *72, 73*
Safin, I. A., 170(29), 174(29), 175(29), 178(65, 66, 68, 69, 73, 74, 77, 80, 81, 82), 179(65, 66, 77, 81), 180(65, 66, 68, 74, 80), 181(81), 183(65, 66, 68, 74, 80), 185(68, 81), 186(68, 77), 188(77), 193(29), 195(29), 202(29), 205(29, 197), 206(29, 197), 213(29), 224(284), *225, 226, 227, 231, 233,* 326(14), 329(37a, 38b), *338, 339*
Saito, N., 258(75), 259(82), *281*
Saito, S., 198(147), 204(189), 205(189), *229, 230*
Sakamoto, M., 304(18), *321*
Sakurai, I., 216(228), *232*
Sams, J. R., 267(128), 268(140), *283*
Sano, H., 248(9, 10), 256(55), 257(68, 69, 70), 258(75), 259(82, 83), *279, 280, 281*
Sano, M., 259(83), *281*
Sasaki, S., 295, 304(18), 305, *321*
Satoy, S., 206(203, 204), 207(203), *231*
Sauer, J., 127(53), *148*

Saunders, M., 89(26), *97*
Savchuk, V. I., 178(70), 181(70), 209(70), 210(70, 215), 211(70, 215), *227, 231*
Schachtschneider, J. H., 22, *70*
Schaefer, J. P., 39, *71*
Schäfer, L., 63, *74*
Schawlow, A. L., 17(48), 69, 170(3), 176(3), 183(111), 202(111), *224, 228*
Schirmer, R. E., 95(39), 96(43), *97*
Schlemper, E. O., 257(67), 265(122), *281, 282*
Schleyer, P. R., 191, 194(132), *229*
Schmid, H. G., 96(42), *97*
Schneider, R. F., 219(260), 222(260), *233*, 333(58), *340*
Schöllkopf, U., 129(60, 63), 130(60), 136(60, 63), *148*
Schomaker, V., 3, 10, 28(109), 32(118), 34, 37(109), 68, 69, 70, 71
Schossig, J., 129(60), 130(60), 136(60), *148*
Schroll, G., 299(14), 300, 301, 303, 315 (14), *320*
Schubert, W. M., 191, *229*
Schuler, R. H., 108(24), 114(24), 142(89), *145*
Schwendemann, R. H., 9(39), *69*
Scott, R. M., 129(67), 136(67), *149*
Scrocco, E., 200(162), *229*
Seeley, N. J., 268(132, 133), *283*
Segel, S. L., 170(14), 199(158, 159), 201 (158), 202(158), 204(159), 205 (159, 194), 206(194), 207(159, 194), 208(194), *225, 229, 230*, 333 (62), *340*
Seidel, C. W., 240(5), *278*
Seip, H. M., 20, *70*
Selwood, P. W., 92(30), *97*, 273(153), *283*
Semin, G. K., 170(9, 10, 18), 174(35, 36), 175(35, 36), 176(9, 18), 177 (54), 178(54, 60, 61, 62, 64, 71, 72, 83, 85), 179(72, 89, 90, 91, 94, 95, 96), 180(35, 72, 89), 181(61, 62, 64, 91, 97), 182(60, 61, 62), 183 (36, 60, 61, 71, 90, 91, 105, 110), 184(115), 185(72, 119, 120, 121), 186(89, 116), 188(72), 189(105), 190(36, 120), 191(60, 61, 62), 199 (10, 120, 153), 200(60, 153, 163), 164, 165, 166, 167, 168, 169, 171), 201(10, 60, 64, 85, 94, 120, 164, 171), 202(120, 177, 182), 203(183), 205 (120, 199), 206(60), 208(120), 209 (120, 199, 210, 212), 210(212, 213), 211(64, 105, 217), 212(36, 95, 218, 219, 220), 213(36, 220, 221, 222, 223), 214(36, 218, 220, 225, 241), 215(36, 115, 220), 216(36, 218, 227, 230), 217(36, 232, 233), 218 (36, 227, 239, 240, 241), 219(36, 121, 222, 227, 241, 255, 259), 220 (36, 259), 221(36, 259), 222(36, 116, 227, 241, 272, 273), 223(272, 273, 274), 224(275, 276, 277, 278, 282, 284), *225, 226, 227, 228, 229, 230, 231, 232, 233*, 324(5a), 325 (7), 327(17), 328(33), 329(35), 330(40), 331(46), 332(46, 49, 49a, 50), 333(55, 57, 59, 60a, 61), 334 (64), 335(64), 336(72), 337(77), *338, 339, 340*
Senn, M., 289(6), 290, 291, *320*
Sergeev, V. P., 263(105), *282*
Shabanov, V. F., 334(64), 335(64), *340*
Shapet'ko, N. N., 179(91), 181(91), 183 (91), *227*
Shashkov, A. S., 223(274), *233*
Sheline, R. K., 273(146, 147), 274(169), *283, 284*, 331(44), *339*
Sheppard, N., 96(41), *97*
Sherk, K. L., 328(29), *338*
Shimanouchi, T., 2(20), 54, 55(188, 189, 190, 191), 60(20), 68, 73
Shimauchi, A., 185(122, 123), *228*
Shimizu, H., 115(33), *146*
Shimizu, T., 337(79), *340*
Shimorava, T., 198(141), *229*
Shimozawa, T., 206(202), 207(202), *229, 231*
Shirley, D. A., 278(193), *284*
Shishkin, V. A., 219(270), *233*
Shlikhter, E. B., 255(44, 45), *280*
Shokina, V. V., 185(121), 219(121), *228*
Shpinel, V. S., 248(11, 12, 17, 18), 253 (28), 255(44, 45), 256(61), 258 (71), 278(194), *279, 280, 281, 284*

Author Index 357

Shteingarts, V. D., 217(232, 233), *232*
Shtern, D. Ya., 178(67), 179(67), 182(67), 214(67), *227*
Shvedova, G. N., 333(57), *340*
Sichel, S., 323(2), *337*
Siekierska, K. E., 258(76), *281*
Silvidi, A. A., 325(8), *338*
Simanouti, T., 52(177), *73*
Simpson, A. F., 144(94, 94a), *150*
Simpson, W. I., 329(36), *339*
Sivkov, V. P., 327(19), *338*
Skancke, P. N., 3, 41(143), *69, 72*
Slichter, C. P., 100(5, 9), *145*
Smallcombe, S. H., 83(22), *96*
Smaller, B., 144(94a), *150*
Smith, A. W., 268(132, 133), *283*
Smith, J. A. S., 328(31, 32), *338*
Smith, P. J., 265(123), *282*
Smith, R. M., 198(149), 199(149), 205(149), *229*
Smith, W. J., 170(7), *225*
Smyth, K. C., 165(33), *167*
Snyakin, A. P., 187(126), 191(126), 205(126), 209(126), *229*
Snyder, L. C., 24, *70*
Soboleva, T. A., 223(274), *233*
Soda, G., 198(145), *229*
Soifer, G. B., 170(17), 203(184, 185, 186), 203(187), 204(184), 205(191), 224(187, 285), *225, 230, 233*
Sokolinskaya, T. A., 256(63, 64, 65), *281*
Sokolov, N. D., *228*
Sokolov, S. D., 209(210), *231*
Solomon, I., 104(19), *145*
Sommer, L. H., 127(54), *148*
Sondheimer, F., 82(14, 15, 16), *96*
Sonnino, T., 253(32, 33), *279*
Sorensen, G. O., 29(112), 30(112), *71*
Sorokin, A. A., 248(17), *279*
Sparks, R. A., 333(60), *340*
Spenser, P. D., 330(41), 331(41), *339*
Spiess, H. W., 274(169), *284*, 331(44), *339*
Spiridonov, V. P., 6(33), *69*
Spratley, R., 276(181), *284*
Srivastava, T. S., 219(256), *232*
Stanko, V. I., 185(120), 190(120, 128), 199(120), 201(120), 202(120), 205(120, 198, 199), 208(120, 198), 209(120, 128, 199), 224(275), *228, 229, 231, 233,* 328(33), *338*
Steed, N., 328(27), *338*
Stehlik, D., 100(10), 105(10, 20), 108(10), *145*
Steingross, W., 51(174), *72*
Stenhagen, E., 285(1), *320*
Stenhagen-Ställberg, S., 285(1), *320*
Sternheimer, R. M., 267(129), *283*
Stevenson, D. P., 154(22), *167*
Stockdale, J. A. D., 156(27), *167*
Stöckler, H. A., 248(9, 10), 256(54, 55), 257(68, 69, 70), 263(110, 111), 265(119), 268(119), *279, 280, 281, 282*
Stoicheff, B. P., 21, 50(172), *70, 72*
Stolevik, R., 20(78), *70*
Stone, A. J., 265(120), *282*
Strand, T. G., 20(78), 61(217, 218), *70, 74*
Straub, D. K., 256(59), *280*
Strel'tsova, I. N., 201(174), *230*
Strieter, F. G., 218(251), *232*
Struchkov, Yu. T., 201(174), *230*
Stukan, R. A., 248(16), 253(39, 40), 256(62, 63, 64), 263(118), 275(172, 174), *279, 280, 281, 282, 284*
Su, L. S., 36, 37, *71*
Suenaga, M., 250(26), *279*
Suhara, M., 202(180), *230*
Sullivan, P. D., 108(22), 114(22), *145*
Suprun, A. P., 223(274), *233*
Sutherland, G. L., 293(8, 9), 294(11), 295, 296, 298, 299(14), 300, 301, 303, 315(14), *320*
Sutler, D., 57, *73*
Suzdalev, I. P., 248(12, 13), 253(37), 255(50), 258(72), *279, 280, 281*
Svergun, V. I., 178(72, 83), 179(72), 180(72), 183(110), 185(72), 188(72), 210(213), 216(227), 218(227), 219(227), 222(227), *227, 228, 231, 232,* 329(35), 332(49), 333(55, 57), *339, 340*
Swick, D. A., 35(125), 36(125), 60(213, 214), *71, 74*
Swiger, E. D., 177(49), *226,* 265(125), *282*
Syrkin, Ya. K., 191(137), *229*
Sysoeva, N. A., 106(21b), *145*

T

Taft, R. W., 173(32), 191(32), 225, 329 (38c), 339
Tai, J. C., 108(25), 145
Takahashi, A., 2(21), 60(21), 68, 73
Tal'rose, V. L., 154(22), 166
Tanimoto, M., 49, 50(166, 173), 72
Tarasova, G. M., 336(71), 340
Tatsuzaki, I., 198(142), 229
Tavard, C., 3(26), 9(26), 69
Taylor, B., 256(56), 280
Taylor, D., 100(10a), 105(10a), 107(10a), 145
Taylor, J. W., 131(68), 149
Teixeira-Dias, J. J. C., 115(29), 140(29), 146
Temkin, A. Ya., 253(28), 279
Temnikova, T. I., 191(138), 229
Temperley, A. A., 277(188), 284
Templeton, D. H., 218(251), 232
Terao, H., 337(76), 340
Thompson, D. D., 75(3), 96
Thomsen, M. E., 88(25), 95(38), 97
Timasheva, T. P., 58(200), 73
Timokhin, V. G., 232(49), 339
Titova, N. S., 205(199), 209(199), 231
Todd, J. E., 195(139), 198(139), 229
Tolman, R. C., 119(39), 147
Tominaga, T., 258(75), 259(82), 281
Tomkinson, D. M., 133(72), 149
Toneman, L. H., 5(32), 23(91), 24(91), 25(99, 101, 103), 26, 27(101), 31, 32(120), 34, 42(150), 43, 54(186), 55(186), 69, 70, 71, 72, 73
Tong, D. A., 219(257), 232, 328(31, 32), 338
Tonomura, T., 202(180), 230
Torrey, H. C., 100(8), 145
Townes, C. H., 17(48), 69, 170(3, 44), 176, 224, 226, 278(195), 284
Toyama, M., 198(143, 145), 201(172), 202(172), 206(202), 207(202), 229, 230, 231
Traetteberg, M., 18(65, 66), 37(134), 40, 41, 42(142, 146, 149), 43(152), 44(155), 45(155), 46(139), 49(139), 50(139, 155), 60(212), 69, 70, 71, 72, 74
Travis, J. C., 248(21), 272(21), 279

Trifunac, A. D., 101(15a), 118(15a), 119(15a), 121(15a), 122(15a), 123(41), 124(15a), 127(41, 44), 135(15a, 77), 136(77), 138(77), 143(91), 145, 148, 149
Trooster, J. M., 277(184), 284
Troitskaya, V. S., 209(211), 231
Trotter, J., 32, 71, 257(66), 281
Trozzolo, A. M., 134, 136(81), 149, 260(84), 281
Trukhtanov, V. A., 248(16), 256(62, 63), 279, 280, 281
Tsukada, K., 185(122, 123), 228
Tsvetkov, E. N., 178(60, 61, 62), 181(61, 62), 182(60, 61, 62), 183(60, 61), 191(60, 61, 62), 200(60, 171), 201(60, 171), 206(60), 209(211, 212), 210(212), 226, 230, 231
Twerdochlib, M., 115(29), 140(29), 146

U

Ukaji, T., 9, 52, 53(180), 69, 73
Umanskii, S. Ya., 228
Upadysheva, A. V., 224(278), 233

V

Valovoi, V. A., 224(276, 277), 233, 327(17), 338
van der Draai, R. K., 23(91), 24(91), 70
Van Geet, A. L., 92(31), 94(31), 97
Van Vleck, J. H., 115(27), 146
van Weelden, R. H., 49(164a), 72
van Willigen, H., 106(21a), 145
Vasil'ev, V. G., 327(16), 338
Vasil'eva, V. N., 327(16), 338
Vasil'yeva, T. T., 223(274), 233
Vasnev, V. A., 202(182), 230, 324(6), 325(6), 338
Vaughan, G., 61(221), 74
Vdovin, V. M., 25(97), 70, 178(78), 181(78), 205(78), 208(78), 227
Vedenina, N. Ya., 263(116), 282
Velichko, F. K., 212(220), 213(220, 222), 214(220, 225), 215(220), 219(222), 231, 232
Venkataraghavan, R., 289(6), 290, 291, 320
Vereshchagin, A. N., 228, 326(14), 338
Verzilov, V. S., 336(75), 337(75), 340

Author Index 359

Viervoll, H., 2(7, 8), *68*
VilasBoas, L. F., 337(78), *340*
Vilkov, L. V., 6(33), 25, 40, 47, 48, 58 (198, 199, 200, 201), 64, 66, 67, *69, 70, 72, 73, 74*
Vinogradova, S. V., 202(182), 209(209), *230, 231,* 324(6), 325(6), *338*
Vinokurov, V. G., 209(211), *231*
Vishnyakova, T. P., 256(63, 64, 65), *281*
Viste, A., 22(83), *70*
Vlasov, V. M., 211(217), *231*
Vledder, H. J., 38(135), 39(135), *71*
Vogt, D., 156(28), *167*
Volkov, A. F., 327(16), *338*
Volkov, V. E., 334(64), 335(65), *340*
von Trepka, V. L., 156(28), *167*
Voronkov, M. G., 170(29), 174(29), 175 (29, 38), 178(57, 58, 59, 63, 65, 66, 67, 68, 69, 73, 74, 75, 76, 77, 78, 79, 80, 81, 82, 84), 179(58, 63, 65, 66, 67, 75, 77, 81), 180(58, 59, 63, 65, 66, 68, 73, 74, 80), 181(38, 58, 59, 63, 78, 81, 98, 99, 101, 102), 182(58, 67, 84), 183(59, 65, 66, 68, 74, 76, 79, 80, 84), 184(113, 114), 185(38, 68, 81, 98, 99, 101, 102), 186(68, 77, 98, 99, 101, 102), 187 (101, 125, 126), 188(77, 84, 125), 189(125, 127), 191(57, 58, 59, 126, 127), 192(38), 193(29, 38), 194(38, 101, 125), 195(29, 38, 101), 200 (63), 201(63), 202(29), 204(127), 205(29, 78, 126, 197), 206(29, 63, 197), 208(78), 209(126), 212(84), 213(29), 214(67, 224), 223, *225, 226, 227, 228, 229,* 326(10, 12b, 13), 327(18), 329(37a, 38, 38a,b), 330(38), 333(10, 13), *338, 339*
Vucelic, M., 268(138), *283*
Vul'fson, S. G., *228,* 326(14), *338*

W

Wada, M., 256(57), *280*
Wall, D. H., 268(143), 272(143), *283*
Walling, C., 136(83), *149*
Walloe, L., 59(206), *73*
Ward, H. R., 100(11a), 113(11a), 115 (34), 117(37), 127(11a, 48), 129 (56, 57), 131(68a), 133(71, 73), 135 (74), 136(11a), 139(87), 143(37), *145, 147, 148, 149*
Watanabe, I., 52(177), *73*
Waterman, H., 277(185), *284*
Weatherly, T. L., 202(178), *230*
Webb, G. A., 106(21c), *145*
Webb, H. M., 166(38), *167*
Weber, K. E., 195(139), 198(139), *229*
Webster, B. R., 289(5), 292, *320*
Wegmann, L., 6(34), *69*
Weiher, J. F., 260(85), *281*
Weiss, A., 199(156), 200(156), 201 (156), 202(156), 218(253), 219 (253, 254, 271), 220(254), 221 (254), *229, 232, 233,* 326(12a), *338*
Weissman, S. I., 115(30), 140(30), *146*
Wendling, E., 170(26), *225*
Werner, R., 136(84), *149*
Wertheim, G. K., 240(6), 258(78), 259 (80, 81), 274(157, 158), 277(191), *279, 281, 283, 284*
West, R., 198(149), 199(149, 150), 205 (149), *229*
Westenberg, A. A., 177(53), *226*
Wexler, S., 151(13), 155(13), *166*
Wheatley, P. J., 39, *71,* 327(22), *338*
White, A. H., 276(180), 277(185), *284*
Whitehead, M. A., 183(108), 195(139), 198(139), 209(108), 210(108), *228, 229,* 323(2), 324(3, 5), 335(67, 68), *337, 338, 340*
Whitten, D. G., 263(113), *282*
Wiberg, K. B., 28(108), 37(108), *71*
Wickman, H. H., 260(84), *281*
Wigley, P. W. R., 250(23), *279*
Wignall, J. W. G., 277(192), *284*
Wilcox, C. F., Jr., 25(102), 26(102), 27 (102, 104), *70*
Wilkins, R. G., 76(7), 95(7), *96*
Wilkinson, G., 262(98), *282*
Willeford, B. R., 277(186, 187), *284*
Williams, A. E., 294(10), *320*
Williams, D. H., 163(29), 164(29), *167,* 294(12), *320*
Wilson, E. B., Jr., 31, 59, *71, 73,* 199 (151), 200(151), *229*
Wilson, J. B., 177(53), *226*
Wishnok, J. S., 127(45), *148*
Wolff, H. P., 67(241), *74*
Wolniewicz, L., 115(29), 140(29), *146*

Wolowski, R., 82(15), 96
Woodgate, S. S., 166, 167
Woods, W. G., 32(118), 34(118), 71
Wootton, R., 337(78), 340
Wulfson, N. S., 289(7), 320

Y

Yakobson, G. G., 200(163, 164, 166, 167, 168, 169), 201(164), 202(177), 211(217), 217(232, 233), 230, 231
Yamaguchi, S., 52(177), 73
Yokozeki, A., 25(100), 26, 27(100), 32(100), 34(100), 70
Yonemitsu, T., 202(180), 230
Yonezawa, T., 206(204), 231
Yoshida, M., 129(60), 130(60), 136(60), 148
Young, J. A., 61, 74
Young, J. E., 28(107), 40(107), 41(107), 71
Young, J. F., 263(104), 282
Yurieva, L. P., 275(172), 284
Yusupov, M. Z., 327(19), 336(73, 75), 337(74, 75), 338, 340

Z

Zagorevskii, V. A., 209(211), 231
Zahn, V., 274(164), 283
Zakharkin, L. I., 187(126), 190(128), 191(126), 205(126, 198), 208(198), 209(126, 128, 209), 229, 231, 253(35), 258(73), 261(35), 280, 281, 329(34), 338
Zalukaev, L. P., 187(124), 191(124), 228
Zavarzin, G. A., 263(116), 282
Zeil, W., 6(34), 18(69), 23, 51(174, 175, 176), 64, 65, 69, 70, 72, 73, 74, 177(52), 226
Zeldes, H., 144(94a), 150, 179(92), 185(92), 193(92), 227
Zelenskaya, L. G., 224(277), 233, 327(17), 338
Zemlyanskii, N. N., 330(39, 40), 339
Zhagata, L. A., 184(114), 214(224), 228, 231
Zhdanov, Yu. A., 173(33), 182(33), 225, 329(38d), 339
Zhukov, A. P., 213(221), 231, 333(59), 340
Zilberg, I. Yu., 58(201), 73
Zimmermann, B., 250(24), 274(162), 279, 283
Zocher, H., 273(152), 283
Zuckerman, J. J., 235(1), 253(36), 256(36, 60), 262(100, 101, 102, 103), 263(106, 107, 108, 114), 268(134), 272(1), 275(178), 277(186, 187), 278(196), 278, 280, 282, 283, 284
Zueva, G. Ya., 178(72), 179(72), 180(72), 185(72), 188(72), 227

Subject Index*

A

Acetaldehyde, ED, 17
  error limits of bond lengths, 33
Acetic acid, structural parameters, 36
Acetic anhydride, structure, 38, 39
Acetone, ED, 17
  bond distances, 35
  error limits of bond lengths, 33
Acetophenone, AAS, 318
Acetylacetone,
  keto–enol tautomerism, 85
  SSL, 86
  structure, 38, 39
tris(Acetylacetonato)iron(III), MöS, 265, 277
Acidities from proton transfer, ICR, 161
Acrolein, ED, 17
  bond distances, 45
$AlBr_3$ ethers, NQR, 220
$AlBr_3$ pyridines, NQR, 220
$AlBr_3$ sulfides, NQR, 220
Alicyclic compounds, NQR
Alkanes, AAS, 286
Alkoxides, ICR, 156
Allyl alcohol, AAS, 317
Aluminum compounds, NQR, 329
Amino acids, MS, 291
Anisol, AAS, 317
Annulenes, conformational equilibria, 82
Anthracene, DNP, 138
Antimony compounds, NQR, 332, 333, 337
Aromatic systems, NQR
Arsenic tribromide complexes, NQR, 336
Arsenic trichloride complexes, NQR, 336
Arylsulfenyl chlorides, NQR, 211
Azobacteria, MöS, 263
Azocompounds, decompositions, 136

B

Benzene antimony tribromide complex, NQR, 337
Benzenes, chlorosubstituted, NQR, 203

Benzoyl chlorides, substituted, NQR, 324
Benzoyl peroxide, DNP, 112, 132
  decomposition, 139
Benzylchloride tin tetrachloride complex, NQR, 336
Benzyloxy radical, DNP, 131
Bicyclo[2,2,1]heptane, ED, 25, 26
Bicyclo[3,1,1]heptane, ED, 25, 27
Bicyclo[2,2,1]hepta-2,5-diene, error limits of bond lengths, 34
  structure, 32
Bicyclo[2,1,1]hexane, ED, 25, 26
Bicyclo[1,1,1]pentane, ED, 25
Bidentate organosulfur ligands, MöS, 276
Bisallylpalladium, conformational equilibria, 82
1,2-Bis(chlorophenyl)ethanes, NQR, 326
Bismuth compounds, NQR, 332
Bloch equations, 78
Blood, MöS, 264
Bond lengths, from ED, 20
Bornyl acetate, AAS, 320
Boron compounds, NQR, 186, 208
Boron trichloride complexes, NQR, 335
Bromine–bromobenzene complex, NQR, 219
N-bromoacetamide, NQR, 208
Bromobenzenes, NQR, 201
(2-Bromoethyl)benzene, DNP, 110
N-Bromo-ϵ-caprolactam, NQR, 208
Bromo-iodobenzene derivatives, NQR, 199
p-Bromophenol, NQR, 327
3,5-Bromopyridine, NQR, 335
N-Bromosuccinimide, NQR, 208
2-Bromothiophene, ED, 48, 49
N-Bromophthalimide, NQR, 208
1-Bromo-2-pyrrolidinone, NQR, 208
1,3-Butadiene, ED, 17, 49
  structure, 40
n-Butane, rotational isomers, 54
2-Butanone, AAS, 298
Butatriene, bond lengths, 37

* Letters after a compound indicate the technique applied, viz., MS = mass spectrometry, RD = optical dispersion, NMR = nuclear magnetic resonance, NQR = nuclear quadrupole resonance, ICR = ion cyclotron resonance, ED = electron diffraction, MöS = Mössbauer spectroscopy, AAS = Automated Analysis System, DNP = dynamic nuclear polarization, SSL = spin saturation labeling.

Subject Index

tert-Butyl alcohol, ICR, 159
tert-Butyl chloride, NQR, 192
n-Butyl chloride, rotational isomers, 52, 54
sec-Butyl chloride, NAR, 192
  rotational isomers, 52, 54
t-Butyl lithium, DNP, 110

C

o-Carbarsborane, NQR, 329
Carbene intermediates, DNP, 138
Carbon disulfide, ED, 17
Carboranes, NQR, 205
Carvone, AAS, 319
Chloranline complexes, NQR, 334
Chloroalkenes, NQR, 325
Chloroaniline complex with methylene, NQR, 217
Chlorobenzenes, NQR, 201
2-Chlorobenzothiazole, NQR, 205
Chlorobenzoylchlorides, NQR, 327
α-Chlorobenzyl radical, 129
1-Chloro-4-bromobenzene, NQR, 197
cis-1,3-Chlorobromocyclobutane, 59
trans-1,3-Chlorobromocyclobutane, 59
Chlorocarboranes, NQR, 328
Chloro(dimethylamino)sulfone, structure, 64
1-Chloro-2,6-dinitrobenzene, NQR, 197
α-Chloroethylbenzene derivatives, NQR, 326
Chloroform, DNP, 135
Chloroform complexes with ethers, NQR, 219
1-Chloro-4-fluorobenzene, NQR, 197
o-, m-, and p-Chlorobenzyl chloride, NQR, 205
α-Chlorocyclohexanone, NQR, 198
Chloroderivates, NQR, 205
2-Chloroethyl chlorosulfinate, NQR, 205
2-Chloroethyl p-toluenesulfonate, NQR, 206
1-Chloro-4-iodobenzene, NQR, 197
Chloromaleic anhydride, NQR, 205
Chloromethanes, deuterated, NQR
2-Chloronaphthalene, NQR, 206
1-Chloro-4-nitrobenzene, NQR, 197
Chloronitromethanes, NQR, 325
Chloronitrotoluene, NQR, 325
p-Chlorophenol, NQR, 327
1-Chloropiperidine, NQR, 208

α-Chloropropylbenzene, DNP, 129
2-Chloropyridine, NQR, 206
3- and 4-Chloropyridines, NQR, 206
2-Chloroquinoline, NQR, 206
6- and 7-Chloroquinolines, NQR, 206
N-Chlorosuccinimide, NQR, 207, 208
2-Chlorothiophene, ED, 48, 49
Chlorothiophenes, NQR, 206
α-Chlorotoluene, DNP, 127
o-Chlorotoluene, NQR, 197
p-Chlorotoluene, NQR, 197
α-Chlorotaluene, NQR, 204
Cobalt(III) acetylacetonate, MöS, 259
Cobalt carbonyl complexes, NQR, 330
Cobalticinium tetraphenylborate, MöS, 259
Complex of dimethyltin dichloride and 2,2′, 2″-terpyridyl, MöS, 275
Complexes, molecular, NQR, 216
Conformers, ED, 51
Coordination in metalloorganic Compounds, NQR
Copper compounds, NQR, 337
Copper halide complexes with pyridine N-oxide, NQR, 337
Cross-relaxation, 103
π-Crotylpalladium, chloride dimer, PMR, 87
cis-π-Crotyl(2-picoline)palladium, PMR, 87
trans-π-Crotyl(2-picoline)palladium, PMR, 87
π-Crotylpalladium complexes, cis-trans isomerization, 86
Cumene, conformation, 58
1,3-Cycloalkadienes, conjugated, bond parameters, 43
Cyclobutadiene iron tricarbonyl structure, ED, 67
Cyclobutene, ED, 30
1,3-Cyclohexadiene, bond distances, 43
1,4-Cyclohexadiene, ED, 31
Cyclohexane, spin saturation transfer, 81
Cyclohexene, ED, 31
Cyclooctanol, AAS, 318
Cyclopentadiene derivatives, NQR, 214
  tin derivatives, MöS, 262
π-Cyclopentadienyldicarbonylmolybdenum-π-methallyl, PMR, 84
  isomeric configurations, 84

Subject Index 363

Cyclopentadienylrhenium tricarbonyl, NQR, 333
Cyclopentadienyltricarbonyl compounds, NQR, 214
2-Cyclopenten-1-one, AAS, 317
Cyclopropane, ED, 23, 27
 carboxylic acid chloride, rotational isomers, 58
Cyclopropene, bond lengths, 37
 structure, 28
Cyclopropyl carboxaldehyde, rotational isomers, 56
Cyclopropyl methyl ketone, rotational isomers, 58
Cytochrome system of these bacteria, MöS, 262

D

Decalin, AAS, 319
Dehydro-[1,6]annulenes, conformational equilibria, 82
Deuterated, peroxides, decompositions, 134
Deuterium oxide, ICR, 157
Deuterochloroform, NQR, 328
Deuteroethane, ED, 17
Deuteromethane, ED, 17
Diamyl ether, AAS, 319
bis(N,N-Dialkyldithiocarbamato)iron, MöS, 259
Dialkyltin oxides, MöS, 256
Diallyl ether, AAS, 317
Diborane, configuration, 66
o-Dibromobenzene, ED, 61
cis-1,3-Dibromocyclobutane, 59
di-n-Butyl(dipicolinato)tin(IV), MöS, 268
Dibutyl ether, AAS, 318
Dibutylin, MöS, 254
Dibutyltin dimaleate, MöS, 258
3,5-Dichloroaniline, NQR, 197
2,4-Dichlorobenzaldehyde, NQR, 197
3,4-Dichlorobenzaldehyde, NQR, 197
α,α-Dichlorobenzene, DNP, 132
cis-3,4-Dichlorocyclobutene-1, ED, 29
2,2-Dichlorocyclohexanone, NQR, 198
1,2-Dichloroethane, ED, 11, 12, 13
cis-1,2-Dichloroethylene, NQR, 198, 205
1,2-Dichloro-3,3-difluorocyclopropene, NQR, 199
Dichloromaleic anhydrides, NQR, 205

Dichloromethyl radical, DNP, 142
2,3-Dichloronaphthoquinone, NQR, 205
2,4-Dichlorophenol, NQR, 197
2,4-Dichlorophenoxyacetic acid, NQR, 328
1,1-Dichlorosilacyclobutane, ED, 24
α,α-Dichlorotoluene, DNP, 129
1,6-Dichlorotoluene, NQR, 197
Dicobaltoctacarbonyl, NQR, 331
Dicyclopentadienyl tin, MöS, 262
2,5-Diethoxytetrahydrofuran, AAS, 318
Diethyldichlorostannate, NQR, 330
bis(N,N-Diethyldithiocarbamato)iron, MöS, 276, 277
Diffusive encounters, 135
1,1-Difluoroethylene, ED, 10, 12
 bond lengths, 37
$trans$-Dihalobis(ethylenediamine)cobalt, NQR, 331
1,3-Dihalocyclobutane conformation, 59
Diisoamyl ether, AAS, 320
4,4-Dimethoxy-2-butanone, AAS, 317
Dimethylacetylene, ED, 49
$tris$-(Dimethylamino)phosphene, ED, 65
 structure, 66
N-Dimethylaniline, conformation, 58
2,3-Dimethylbutadiene, bond distances 41
2,2-Dimethylbutane, AAS, 317
2,3-Dimethylbutane, AAS, 317
Dimethylether, rotational isomers, 55
Dimethylcyclopropylcarbinyl cation, barrier to rotation, 82
Dimethylfulvene, bond distances, 46
cis,cis-3,4,-Dimethyl-2,4-hexadiene, bond distances, 44
2,4-Dimethyl-3-hexanone, AAS, 298
2,6-Dimethyl-4-heptanone, AAS, 319
1,1-Dimethyl-1-pentanol, AAS, 318
2,2-Dimethyl-1-pentanol, AAS, 318
2,4-Dimethyl-2-pentanone, AAS, 317
2,2-Dimethylpropane, ED, 23
Dimethylsulfodiimine, ED, 64
 structure, 65
Dimethyltellurium dichloride, NQR, 333
Dimethyltin dichloride, MöS, 266
Dimethyltin difluoride, MöS, 265
Dimethyltrifluorophosphine, structures, 65
Diphenylchloromethane, NQR, 204
1,2-Diphenyl-3,3-dichlorocyclopropene, NQR, 199

Subject Index

Diphenyl(dipicolinato)tin(IV), MöS, 268
Dipole–dipole interaction, 106
Di-*tert*-butylnitroxide free radical, 61, 62
Divinyl(dipicolinato)tin(IV), MöS, 268

E

Electron polarization, 144
bis(1,2-Ethanedithiolato)tin, MSS, 256
Ethanol, ICR, 158
Ethanol-2,2,2d, ICR, 157
Ethoxytitanium bromides, NQR, 214
Ethyl acrylate, AAS, 317
Ethers, aliphatic, AAS, 299
Ethylbenzene, DNP, 110
Ethyl chloride, NQR, 192
Ethyl cinnamate, AAS, 320
Ethyl diethoxyacetate, AAS, 318
tris(Ethyleneimino)phosphene structure, 66
2-Ethylhexanal, AAS, 318
2-Ethyl-1-hexanol, AAS, 319
Ethyl iodide, DNP, 132
Ethyllithium, DNP, 127, 129
3-Ethyl-3-pentanol, AAS, 318
Ethylphenychlorostannate, NQR, 330
Ethyl radical, 129, 131

F

Ferredoxin, MöS, 260
Ferricinium ferrichloride, MöS, 273, 274
Ferrocene, MöS, 256, 261, 274
Fluorene, proton exchange, 82
Fluorenyllithium, proton exchange, 82
Formaldehyde, ED, 17
 error limits of bond lengths, 33
Formic acid, structural parameters, 36
Free radicals, ED, 61
Furan derivatives, NQR, 205
Furoic acid, AAS, 317

G

Gallium, NQR, 220
Gallium compounds, NQR, 329
Geminate encounters, 127
 DNP, 130
Germanium compounds, NQR, 182
Glyoxal, ED, 17
Gold cyanide complexes, MöS, 254
Greenwood's Rule, MöS, 272

H

Haloanilines, NQR, 334
Halo compounds, NQR, 175
Hemoglobin, MöS, 260
2,4-Heptadien-6-one, AAS, 317
Heptamethylbenzenonium ion, PMR spectrum, 89
1-Heptanol, AAS, 318
2-Heptanol, AAS, 318
4-Heptanol, AAS, 318
2-Heptanone, AAS, 317
3-Heptanone, AAS, 298, 317
Heterocyclic compounds, NQR, 205
Hexabromobenzene, ED, 61
1,2,3,4,5,6-Hexachlorocyclohexane, NQR, 198
Hexachloroethane, potential barrier, 60
Hexafluoroacetone, bond distances, 35
Hexafluoropropene, structure, 38
3-Hexanone, AAS, 298
1,3,5-*cis*-Hexatriene, bond distances, 42
1,3,5-*trans*-Hexatriene, bond distances, 41
Hydrocarbons, AAS, 287
B. *Hydrogenomonas Z-I*, MöS, 263
Hydroxide show, MöS, 257
4-Hydroxyacetophenone, AAS, 318
2-Hydroxy-3-chloroisoxazoline-4, NQR, 209
Hyperfine field, 117
Hyperfine splittings, 141
6-Heptynylmalonic acid, AAS, 319

I

Indenyl free radical, ED, 63
Indium compounds, NQR, 329
Intersystem crossing, 115
Iodoalkanes, free radical reactions, 133
Iodocarboranes, NQR, 328
N-Iodosuccinimides, NQR, 208
Ion–molecule reactions, ICR, 155
Ion structures, ICR, 163
Iron compounds, MöS, 259
Isobutylchloride, rotational isomers, 53
Isobutylene, error limits of bond lengths, 35
Isobutylene, DNP, 110
Isomer Shift, MöS, 242
Isopentyl chloride, NQR, 192
Isoprene, bond distances, 40
Isopropyl alcohol, ICR, 159

Subject Index   365

Isopropyl carboxaldehyde, rotational isomers, 57
Isopropyl chloride, NQR, 192

K

$K_6Sn_2(C_2O_4)_7 \cdot 4H_2O$, MöS, 259
Ketones, aliphatic, AAS, 293
 rearrangement, 296

M

bis(Maleonitrildeithiolate)iron, MöS, 276
Manganese, NQR, 214
Manganese compounds, NQR, 333
McLafferty ions, 165
 rearrangement, 163, 165, 296, 297
Mercury complex with pyridine N-oxide, NQR, 336
Mesitylene, AAS, 319
Methane, ED, 17
1,6-Methano-1,3,5,7,9-cyclodecamentaene, structure, ED, 47
Methanol, ICR, 157, 158
2-Methoxycarbonyl-2-(2-oxo-1-cyclohexyl)-4-chloro-2H1-benzopyran, NQR, 209
2-Methoxycyclohexanone, AAS, 317
2-Methyl-1-butene
 bond distances, 35
 rotational isomers, 55
Methyl chloride, NQR, 192
Methylcyclobutanol, 164
3-Methyl-2-cyclohexen-1-one, AAS, 317
3-Methyl-3-heptanone, AAS, 298
5-Methyl-3-heptanone, AAS, 318
2-Methyl-3-hexanone, AAS, 298
Methyl iodide, DNP, 109
tris(Methylmethylphosphonato)iron(III), MöS, 277
4-Methyloctane, MS, 288
2-Methyloctan-3-one, AAS, 298
N-Methylpyrrole, bond distances, 47, 48
Methyl salicylate, AAS, 318
Methyltetrafluorophosphine, structure, 65
$Mn_2Fe(CO)_{14}$, MöS, 274
Molecular complexes, NQR

N

$Na_2[Fe(CO)_4]$ MöS, 274
Naphthalene chlorosubstituted, NQR, 203
NMR data processing, AAS, 308

Nonanal, AAS, 319
1-Nonanol, AAS, 319
3-Nonanol, AAS, 319
4-Nonanol, AAS, 319
5-Nonanol, AAS, 319
2-Nonanone, AAS, 319
3-Nonanone, AAS, 298
4-Nonanone, AAS, 298
5-Nonanone, AAS, 319
4-Nonanone-1,1,1-d, 164
Norbornane, ED, 28
Nuclear Overhauser effect, 96, 100, 103

O

Octachlorocycloheptatriene, NQR, 199
1,7-Octadiyne, DNP, 135
1-Octanol, AAS, 319
2-Octanol, AAS, 319
2-Octanone, AAS, 318
3-Octanone, AAS, 298
4-Octanone, AAS, 298
1-Octene, AAS, 318
1-Octyne, AAS, 318
Oligopeptides, AAS, 289
Organolithium compounds, DNP, 126
Organometallic compounds, NQR, 329, 330
Organotin compounds, MöS, 253, 256
Overhauser effect; see Nuclear Overhauser effect
Overhauser model, 111, 113
Oxyhemoglobin, MöS, 260

P

3-Pentanone, AAS, 298
Pentachloro-N-methylaniline, NQR, 197
Peptides, MS, 290, 292
Perfluorocyclobutene, ED, 30
Perfluorodimethylacctylene, ED, 49
Peroxides, deuterated, decompositions, 134, 136
Picryl chloride benzenes, NQR, 218
Pivalic acid, AAS, 317
Phase transitions in organic crystals, NQR, 224
Phenylcyclopropane, conformation, 58
o-Phenylenedioxytin(II), MöS, 256
Phenyl radical, DNP, 131
Phosphinous chlorides, NQR, 209
Phosphorus compounds, NQR, 209

Subject Index

Phthalyl chloride, NQR, 328
Polychloral, NQR, 223
Polyferrocene, MöS, 256
Polymerization product of 3,3,3-trichloro-1-propene, NQR, 223
Polymeric compounds, NQR, 222
Polytrifluorochloroethylene, NQR, 223
Polyvinvyldene chloride, NQR, 223
Potassium ferrocyanide, MöS, 250
Propionic acid, structural parameters, 36
Propionoxy radical, DNP, 131
Propionyl peroxide, DNP, 132
n-Propylbenzene, DNP, 128, 129
n-Propyl chloride, rotational isomers, 54
tris(i-Propylmethylphosphonato)iron(III), MöS, 277
Proton transfer, ICR, 161
Prussian blue, MöS, 250
Pyridine derivatives, NQR, 205, 207
Pyrimidines, NQR, 205, 207
Pyridine N-oxide complexes, NQR, 336

Q

Quadrupole splitting, solvent effect, MöS, 251

R

Radical pair model, 114
  precursors, DNP, 137
Radical pairs, caged, 140
  lifetimes, 141
Rate constants, ICR, 155
  via spin labeling, 81
Reaction mechanisms, ICR, 155
Reaction pathways, ICR, 162
Red cells, MöS, 264
Rhenium, NQR, 214
Rhenium compounds, NQR, 333
Relaxation times, 91

S

SbBr$_3$, NQR, 221
SbCl$_3$, NQR, 221
Scalar interaction, time-dependent, DNP, 107
Silacyclobutane, ED, 25
Silicon compounds, NQR, 182, 208
Spiropentane, ED, 24
Stannous halide complexes
  2-Morpholine, MöS, 266
  Piperazine, MöS, 266
  2-Piperidine, MöS, 266
Steric effects and ED, 60
Styrene, DNP, 110
2-Substituted 1- chloromethyl-o-carboranes, NQR, 209
Sulfur heterocycles, NQR, 205
Sulfur compounds, NQR, 210

T

Tellurium compounds, NQR, 333
2,3,4,6-Tetrachloroaniline, NQR, 197
1,2,4,5-tetrachlorobenzene, ED, 61
Tetrachlorocyclopropene, NQR, 198
  structure, 29
1,2,3,4-Tetrachloro-3,5-dinitrobenzene, NQR, 197
2,4,5,6-Tetrachloro-m-xylene, NQR, 197
2,3,5,6-Tetrachloro-p-xylene, NQR, 197
1,1,3,3-Tetrachlorosilacyclobutane, ED, 24
Tetrafluoroethylene, bond lengths, 37
Tetrakis(trifluoromethyl)hydrazine, steric crowding, 61
Tetramethyldiborane, structure, ED, 67
Tetramethylethylene, DNP, 134
Tetraphenyltin, MöS, 258
1,2,5-Thiadiazole, bond distances, 47, 48
Thiophene, bond distances, 47
  NQR, 205
Tin compounds, NQR, 182, 330
Tin porphyrin derivatives, MöS, 263
Titanium compounds, NQR, 332
bis(Toluene-3,4-dithiolate)iron, MöS, 277
bis(Toluene-3,4-dithiolato)tin, MöS, 256
Tribromotitaniumethoxide, NQR, 332
Trichloroacetonitrile complexes, NQR, 335
Trichloroacetyl peroxide, decomposition, DNP, 134, 135
2,3,4-Trichloroaniline, NQR, 197
1,2,4-Trichloro-3,5-dinitrobenzene, NQR, 197
1,3,5-Trichloro-2,6-dinitrobenzene, NQR, 197
1,1,1-Trichloroethane DNP, 109
1,2,3-Trichloro-5-nitrobenzene, NQR, 197
1,2,4-Trichloro-5-nitrobenzene, NQR, 197
Trichlorostannates, NQR, 330
3,3,3-Trifluoro-2-(trifluoromethyl)-1-butene, bond distances, 35

2,4,6-Trihalophenols, NQR, 334
Triiron dodecacarbonyl, MöS, 272
Trimethyltin cyanide, MöS, 256
Trimethyltin fluoride, MöS, 256
Triphenylbismuth, NQR, 332
1,1,2-Triphenylethanol, DNP, 137
Triphenyltin Lithium. MöS, 252
Triphenylmethyl free radical, 63
Triphenyltin hydroxide, MöS, 257
Turnbull's blue, MöS, 250

V

Venous rat red cells, MöS, 264

Vinylacetylene, ED, 49, 50
Vinyl chloride, NQR, 205
Vinyl radical, DNP, 108

W

Water, ICR, 158

X

m-Xylene, AAS, 315

Z

Zeeman spectra, 327